Lecture Notes in Artificial Intelligence   5494

Edited by R. Goebel, J. Siekmann, and W. Wahlster

Subseries of Lecture Notes in Computer Science

T0242511

Peter Bruza   Donald Sofge
William Lawless   Keith van Rijsbergen
Matthias Klusch (Eds.)

Quantum Interaction

Third International Symposium, QI 2009
Saarbrücken, Germany, March 25-27, 2009
Proceedings

 Springer

Volume Editors

Peter Bruza
Queensland University of Technology, Faculty of Science and Technology
GPO Box 2434, Brisbane 4001, Australia
E-mail: p.bruza@qut.edu.au

Donald Sofge
Navy Center for Applied Research in Artificial Intelligence
Naval Research Laboratory
4555 Overlook Avenue S.W., Washington, DC 20375, USA
E-mail: donald.sofge@nrl.navy.mil

William Lawless
Paine College, 1235 15th Street, Augusta, GA 30901-3182, USA
E-mail: lawlessw@mail.paine.edu

Keith van Rijsbergen
University of Glasgow, Department of Computing Science
17 Lilybank Gardens, Glasgow, G12 8QQ, UK
E-mail: keith@dcs.gla.ac.uk

Matthias Klusch
German Research Center for Artificial Intelligence
Stuhlsatzenhausweg 3, 66123 Saarbrücken, Germany
E-mail: klusch@dfki.de

Library of Congress Control Number: Applied for

CR Subject Classification (1998): I.2, F.1, F.2.1-2, F.4.1, I.6, H.3

LNCS Sublibrary: SL 7 – Artificial Intelligence

ISSN   0302-9743
ISBN-10   3-642-00833-X Springer Berlin Heidelberg New York
ISBN-13   978-3-642-00833-7 Springer Berlin Heidelberg New York

Typesetting: Camera-ready by author, data conversion by Scientific Publishing Services, Chennai, India
Printed on acid-free paper   SPIN: 12643547   06/3180   5 4 3 2 1 0

Preface

These are the proceedings of the Third International Symposium on Quantum Interaction (QI-2009), held at the German Research Centre for Artificial Intelligence (DFKI), Saarbrücken during March 25-27, 2009.

Quantum theory (QT) is being applied to domains such as artificial intelligence, human language, cognition, information retrieval, biology, political science, economics, organizations, and social interaction. After highly successful meetings at Stanford (QI-2007) and Oxford (QI-2008), QI-2009 brought together researchers interested in advancing and applying the methods and structures of QT to these and other domains outside of quantum physics:

- Advancement of theory and experimentation for applying quantum theory to non-quantum domains
- Use of quantum algorithms to address, or to more efficiently solve, problems in non-quantum domains (including contrasts between classical vs. quantum methods)
- Practical applications to quantum domains, such as implementation of AI, or information retrieval (IR) techniques

The proceedings include 21 long papers and 3 position papers. Each paper was thoroughly reviewed by at least two members of the international Program Committee. The proceedings highlight the cross-disciplinary nature of quantum interaction with papers covering topics such as computation, cognition, decision theory, information retrieval, information systems, social interaction, computational linguistics and finance. In addition, we were honored to receive a keynote presentation by Dagmar Bruss (Institute for Theoretical Physics, University of Düsseldorf). We gratefully acknowledge the support of Earl Research and Kirsty Kitto for helping prepare these proceedings.

January 2009

Peter Bruza
Donald Sofge
William Lawless
Keith van Rijsbergen
Matthias Klusch

Organization

Organizing Committee

Peter Bruza	Queensland University of Technology, Australia
Donald Sofge	Naval Research Laboratory, USA
William Lawless	Paine College, USA
Keith van Rijsbergen	University of Glasgow, UK
Matthias Klusch	DFKI, Germany

Program Committee

Sven Aerts	Brussels Free University, Belgium
Diederik Aerts	Brussels Free University, Belgium
Salvador Venegas-Andraca	Tecnológico de Monterrey, Mexico
Dagmar Bruss	University of Düsseldorf, Germany
Peter Bruza	QUT, Australia
Jerome Busemeyer	Indiana University, USA
Stephen Clark	Oxford University, UK
Bob Coecke	Oxford University, UK
Charles Fox	Oxford University, UK
Riccardo Franco	ISIF, Italy
Liane Gabora	University of British Columbia, Canada
Emmanuel Haven	University of Leicester, UK
Kirsty Kitto	QUT, Australia
Andre Khrennikov	Växjö University, Sweden
Matthias Klusch	DFKI, Germany
Pierfrancesco La Mura	Leipzig School of Management, Germany
Marco Lanzagorta	ITT Corporation, USA)
William Lawless	Paine College, USA
Michael Leyton	Rutgers University, USA
Massimo Melucci	University of Padua, Italy
Dusko Pavlovic	Oxford University, UK
Keith van Rijsbergen	University of Glasgow, UK)
Donald Sofge	Naval Research Laboratory, USA)
Giuseppe Vitiello	Universitá di Salerno, Italy
Dominic Widdows	Google, USA
John Woods	University of British Columbia, Canada

Table of Contents

Quantum Mechanics and Decision Theory

Quantum Mechanics and Computation

Quantum Mechanics and Social Interaction

Quantum Mechanics and Semantic Space

Quantum Mechanics and Information Retrieval

Quantum Mechanics and Economics

Introduction to Quantum Probability for Social and Behavioral Scientists

Jerome R. Busemeyer

Indiana University

1 Introduction to Quantum Probability for Social and Behavioral Scientists

There are two related purposes of this tutorial. One is to generate interest in a new and fascinating approach to understanding behavioral measures based on quantum probability principles. The second is to introduce and provide a tutorial of the basic ideas in a manner that is interesting and easy for social and behavioral scientists to understand.

It is important to point out from the beginning that in this tutorial, quantum probability theory is viewed simply as an alternative mathematical approach for generating probability models. Quantum probability may be viewed as a generalization of classic probability. No assumptions about the biological substrates are made. Instead this is an exploration into new conceptual tools for constructing social and behavioral science theories. Why should one even consider this idea? The answer is simply this (cf., Khrennikov, 2007). Humans as well as groups and societies are extremely complex systems that have a tremendously large number of unobservable states, and we are severely limited in our ability to measure all of these states. Also human and social systems are highly sensitive to context and they are easily disturbed and disrupted by our measurements. Finally, the measurements that we obtain from the human and social systems are very noisy and filled with uncertainty. It turns out that classical logic, classic probability, and classic information processing force highly restrictive assumptions on the representation of these complex systems. Quantum information processing theory provides principles that are more general and powerful for representing and analyzing complex systems of this type.

Although the field is still in a nascent stage, applications of quantum probability theory have already begun to appear in areas including information retrieval, language, concepts, decision making, economics, and game theory (see Bruza, Lawless, van Rijsbergen, and Sofge, 2007; Bruza, Lawless, van Rijsbergen, and Sofge, 2008; also see the Special Issue on Quantum Cognition and Decision to appear in Journal of Mathematical Psychology in 2008).

The tutorial is organized as follows. First we describe a hypothetical yet typical type of behavioral experiment to provide a concrete setting for introducing the basic concepts. Second, we introduce the basic principles of quantum logic and quantum probability theory. Third we discuss basic quantum concepts

P. Bruza et al. (Eds.): QI 2009, LNAI 5494, pp. 1–2, 2009.

including compatible and incompatible measurements, superposition, measurement and collapse of state vectors. Classic probability and quantum probability will be compared, side by side, so that we can see exactly where these theories agree and disagree. All of the material can be downloaded from my web site http://mypage.iu.edu/ jbusemey/quantum/Quantum%20Cognition%20Notes. htm.

Revealing Quantum Entanglement via Locally Noneffective Operations

Dagmar Bruß[1], Sevag Gharibian[2], and Hermann Kampermann[1]

[1] Institut für Theoretische Physik III,
Heinrich-Heine-Universität Düsseldorf,
Düsseldorf, Germany
{bruss,kampermann}@thphy.uni-duesseldorf.de
[2] School of Computer Science and Institute for Quantum Computing,
University of Waterloo,
Waterloo, Canada
sggharib@cs.uwaterloo.ca

Abstract. Quantum entanglement is at the heart of quantum information processing. Various methods for the detection of entanglement have been developed. Here, we will explain an approach that uses locally noneffective unitary operations which, however, do cause a change of the global density matrix - an indication for the existence of correlations. We investigate whether this method can distinguish between classical and quantum correlations.

Revealing Quantum Entanglement via Locally Noneffective Operations

Composite quantum systems can exhibit the property of being *entangled*, i.e. the subsystems are correlated "stronger" than any classical correlation allows. Inspite of being intensively studied over the last two decades [1], entanglement is not yet fully understood. In particular, no complete constructive method to answer the following question is known: "given a quantum state ϱ, is it entangled or separable?" For mixed bipartite quantum states (i.e. quantum systems composed of two subsystems), entanglement is defined as follows. Let ϱ be a state acting on the Hilbert space $\mathcal{H}^M \otimes \mathcal{H}^N$, where M and N denote the respective dimensions of the subspaces, and $N \leq M$ without loss of generality. If ϱ can be written in the form

$$\varrho = \sum_{k=1}^{n} p_k |a^k\rangle\langle a^k| \otimes |b^k\rangle\langle b^k|, \tag{1}$$

for $p_k \in \mathbb{R}^+$, $\sum_{k=1}^{n} p_k = 1$, $n \geq 1$, and normalised vectors $|a^k\rangle \in \mathcal{H}^M$, $|b^k\rangle \in \mathcal{H}^N$, then ϱ is *separable* [2], otherwise ϱ is *entangled*. In this talk, we will focus on a particular theme for bipartite states: we will study the *global* action of certain *locally noneffective* operations, and try to use them as a tool to detect entanglement. Locally noneffective unitary operations have been introduced by

P. Bruza et al. (Eds.): QI 2009, LNAI 5494, pp. 3–5, 2009.

Fu [3], under the name of "local cyclic operations". Given ϱ, acting on $\mathcal{H}^M \otimes \mathcal{H}^N$, a locally noneffective (or cyclic) unitary operation U^B is defined by the condition

$$U^B \varrho^B U^{B\dagger} = \varrho^B , \tag{2}$$

where $\varrho^B = \mathrm{Tr}_A(\varrho)$ is the reduced density matrix of the second subsystem. The above condition is equivalent to $[\varrho^B, U^B] = 0$. Although U^B has no effect on the reduced density matrix ϱ^B, the action of $(\mathbf{1} \otimes U_B)$ may change the global density matrix ϱ. We denote the final global density matrix as $\varrho_f = (\mathbf{1} \otimes U^B)\varrho(\mathbf{1} \otimes U^{B\dagger})$. The distance $\mathrm{d}(\varrho, U^B)$ between the original global state and the transformed one is given by

$$\mathrm{d}(\varrho, U^B) := \frac{1}{\sqrt{2}} ||\varrho - \varrho_f||, \tag{3}$$

which we call the Fu distance [3]. We use the norm $||A|| = \sqrt{\mathrm{Tr}(A^\dagger A)}$. We have investigated whether the maximal Fu distance $\mathrm{d}_{\max}(\varrho)$,

$$\mathrm{d}_{\max}(\varrho) := \max_{\mathrm{loc.noneff.} U^B} \mathrm{d}(\varrho, U^B), \tag{4}$$

i.e. the maximal possible global distance achieved under any locally noneffective unitary operation, can indicate the existence of entanglement[4]. There is an upper bound [3] for the maximal Fu distance in the case of a classically correlated state ϱ_{cc}:

$$\mathrm{d}_{\max}(\varrho_{cc}) \leq \sqrt{\frac{2(M-1)(N-1)}{MN}} . \tag{5}$$

Thus, any Fu distance which is larger than this threshold reveals the existence of entanglement. A summary of our results [4] is as follows:

1) For a so-called *pseudopure state* ϱ, acting on $\mathcal{H}^M \otimes \mathcal{H}^N$, defined as

$$\varrho = \epsilon|\psi\rangle\langle\psi| + \frac{\mathbf{1} - \epsilon}{MN}I, \tag{6}$$

with $0 < \epsilon \leq 1$, the maximal Fu distance is

$$\mathrm{d}_{\max}(\varrho) = \begin{cases} \epsilon & \text{if } a_m^2 \leq \frac{1}{2}, \\ 2\epsilon\, a_m \sqrt{1 - a_m^2} & \text{otherwise.} \end{cases} \tag{7}$$

Here, a_m is the maximal coefficient in the Schmidt decomposition $|\psi\rangle = \sum_{k=0}^{N-1} a_k|k\rangle_A \otimes |k\rangle_B$, where a_k are non-negative real numbers (Schmidt coefficients), i.e. $a_m = \max_k a_k$. In connection with eq. (5) one finds a condition for a_m as function of ϵ that allows to detect entanglement of pseudopure states.

2) For a *Werner state* ϱ_W, acting on $\mathcal{H}^D \otimes \mathcal{H}^D$ with $D \geq 2$, defined as [2]:

$$\varrho_W = p\frac{2}{D^2 + D}P_{sym} + (1 - p)\frac{2}{D^2 - D}P_{as}, \tag{8}$$

with $P = \sum_{ij} |i\rangle\langle j| \otimes |j\rangle\langle i|$ and $P_{sym} = \frac{1}{2}(\mathbf{1}_{D^2} + P)$ and $P_{as} = \frac{1}{2}(\mathbf{1}_{D^2} - P)$, where $\mathbf{1}_{D^2}$ is the D^2-dimensional identity matrix and $0 \leq p \leq 1$, we find the maximal Fu distance

$$\mathrm{d}_{\max}(\varrho_W) = \frac{|2pD - D - 1|}{D^2 - 1}, \tag{9}$$

obtained using any traceless $D \times D$ choice of unitary U^B. Together with eq. (5) one can again find a condition for p that allows to detect the entanglement of a Werner state.

3) We have also studied three different types of *bound entangled states* (namely, two families of two-qutrit Horodecki states and bound entangled states from unextendible product bases); however, we did not find any possibility to detect bound entanglement via locally noneffective unitaries. But we did not find a proof that no bound entangled state can be detected in this way.

Regarding the search for an interpretation of the Fu distance, let us point out that it cannot be used as an entanglement measure (there exist entangled states with different degree of entanglement, leading to the same maximal Fu distance; in addition, some non-maximally entangled states can achieve a maximal Fu distance). It is also no non-locality measure, as purely classically correlated states can achieve a non-zero Fu distance. There does exist a connection of the Fu distance to the CHSH inequality [5] in the two-qubit case, namely both the CHSH inequality and the maximal Fu distance detect the same entangled Werner states of two qubits. A full understanding of the Fu distance is still missing at the moment.

References

1. Bruß, D., Leuchs, G. (eds.): Lectures on Quantum Information. Wiley-VCH, Weinheim (2007)
2. Werner, R.: Phys. Rev. A. 40, 4277 (1989)
3. Fu, L.: Europhys. Lett. 75, 1 (2006)
4. Gharibian, S., Kampermann, H., Bruß, D.: arXiv:0809.4469
5. Clauser, J., Horne, M., Shimony, A., Holt, R.: Phys. Rev. Lett. 23, 880 (1969)

Fractals and the Fock-Bargmann Representation of Coherent States

Giuseppe Vitiello

Dipartimento di Matematica e Informatica and INFN
Università di Salerno, I-84100 Salerno, Italy
vitiello@sa.infn.it
http://www.sa.infn.it/giuseppe.vitiello

Abstract. The self-similarity property of deterministic fractals is studied in the framework of the theory of entire analytical functions. The functional realization of fractals in terms of the q-deformed algebra of coherent states is presented. This sheds some light on the dynamical formation of fractals and provides some insight into the geometrical properties of coherent states. The global nature of fractals appears to emerge from coherent local deformation processes.

Much attention has been devoted to the study of fractals due to their relevance in science, from physics to biology, medical sciences, earth science, clustering of galaxies, etc. [1]. In this paper I focus my attention on self-similarity, which is a characterizing properties of a large class of fractals (in some sense self-similarity is the *most important property* of fractals (p. 150 in ref. [2])). In particular, I discuss the functional realization of fractals in the framework of the theory of the entire analytical functions and their relation with the deformed algebra of (Glauber) coherent states [3,4] (the existence of such a relation was conjectured in ref. [5]). This sheds some light on the dynamical formation of fractals and at the same time, from the perspective of coherent states, provides insights into the geometrical (fractal) properties of coherent states. The global nature of fractals appears to emerge from the coherence of local deformation processes.

In the present paper I will not discuss the measure of lengths in fractals, the Hausdorff measure, the fractal "mass", random fractals, and other fractal properties. My discussion is limited to the self-similarity property of fractals which are generated iteratively according to a prescribed recipe, the so-called deterministic fractals. In the following I will closely follow refs.[4].

Let me consider the example of the *Koch curve* (Fig. 1). In the step, or stage, of order $n = 0$, the one-dimensional ($d = 1$) segment u_0 of unit length L_0, called the *initiator* [1], is divided by the reducing factor $s = 3$, and the rescaled unit length $L_1 = \frac{1}{3}L_0$ is adopted to construct the new "deformed segment" u_1, called the *generator* [1], made of $\alpha = 4$ units L_1 (step of order $n = 1$). The "deformation" of the u_0 segment is only possible provided the one dimensional constraint $d = 1$ is relaxed. The u_1 segment "shape" lives in some $d \neq 1$ dimensions and thus we write $u_{1,q}(\alpha) \equiv q \, \alpha \, u_0$, $q = \frac{1}{3^d}$, $d \neq 1$ to be determined. The index q has been introduced in the notation of the deformed segment u_1.

P. Bruza et al. (Eds.): QI 2009, LNAI 5494, pp. 6–16, 2009.

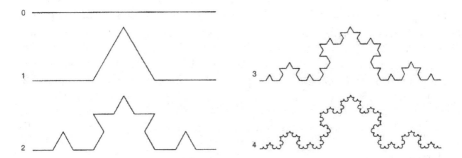

Fig. 1. The first five stages of Koch curve

In general, denoting by $\mathcal{H}(L_0)$ lengths, surfaces or volumes, one has

$$\mathcal{H}(\lambda L_0) = \lambda^d \mathcal{H}(L_0) \tag{1}$$

under the scale transformation: $L_0 \to \lambda L_0$. A square S of side L_0 scales to $\frac{1}{2^2}S$ when $L_0 \to \lambda L_0$ with $\lambda = \frac{1}{2}$. A cube V of same side with same rescaling of L_0 scales to $\frac{1}{2^3}V$. Thus $d = 2$ and $d = 3$ for surfaces and volumes, respectively. Note that $\frac{S(\frac{1}{2}L_0)}{S(L_0)} = p = \frac{1}{4}$ and $\frac{V(\frac{1}{2}L_0)}{V(L_0)} = p = \frac{1}{8}$, respectively, so that in both cases $p = \lambda^d$. For the length L_0 it is $p = \frac{1}{2}$; $\frac{1}{2^d} = \lambda^d$ and $p = \lambda^d$ gives $d = 1$.

In the case of any other "ipervolume" $\mathcal{H}$ one considers the ratio

$$\frac{\mathcal{H}(\lambda L_0)}{\mathcal{H}(L_0)} = p \,, \tag{2}$$

and Eq. (1) is assumed to be still valid. Then,

$$p\,\mathcal{H}(L_0) = \lambda^d \mathcal{H}(L_0) \,, \tag{3}$$

i.e. $p = \lambda^d$. For the Koch curve, setting $\alpha = \frac{1}{p} = 4$ and $q = \lambda^d = \frac{1}{3^d}$, the relation $p = \lambda^d$ gives

$$q\alpha = 1 \,, \quad \text{where} \quad \alpha = 4, \quad q = \frac{1}{3^d} \,, \tag{4}$$

i.e.

$$d = \frac{\ln 4}{\ln 3} \approx 1.2619 \,. \tag{5}$$

The non-integer d is called the *fractal dimension*, or the *self-similarity dimension* [2]. It is the dimension of the deformed space that ensures the existence of a solution of the relation $\frac{1}{\alpha} = \frac{1}{4} = \frac{1}{3^d} = q$. In this sense d is a measure of the "deformation" of the $u_{1,q}$-space with respect to the u_0-space. Note that the meaning of Eq. (4) is that the measure of the deformed segment $u_{1,q}$ with respect to the undeformed segment u_0 be 1: $\frac{u_{1,q}}{u_0} = 1$, i.e. $\alpha q = \frac{4}{3^d} = 1$. In some sense, it expresses the invariance of the lengths under the deformations we are performing. In the following, for brevity I will set $u_0 = 1$.

Since the deformation of u_0 into $u_{1,q}$ is performed by varying the number α of the component segments from 3 to 4, we expect that α and $\frac{d}{d\alpha}$ play a rôle in the fractal structure. We will see that $(\alpha, \frac{d}{d\alpha})$ play indeed the rôle of conjugate variables (cf. Eq. (10)).

Steps of higher order n, $n = 2, 3, 4, ..\infty$, can be obtained by iteration of the deformation process keeping $q = \frac{1}{3^d}$ and $\alpha = 4$. For example, in the step $n = 2$, $u_{2,q}(\alpha) \equiv q\,\alpha\,u_{1,q}(\alpha) = (q\,\alpha)^2\,u_0$, and so on. For the nth order deformation:

$$u_{n,q}(\alpha) \equiv (q\,\alpha)\,u_{n-1,q}(\alpha) , \quad n = 1, 2, 3, ... \tag{6}$$

i.e., for any n

$$u_{n,q}(\alpha) = (q\,\alpha)^n\,u_0 . \tag{7}$$

By iteration, or, equivalently, by requiring that $\frac{u_{n,q}(\alpha)}{u_0}$ be 1 for any n, gives $(q\alpha)^n = 1$ and Eq. (5) is again obtained. It should be stressed that the fractal is mathematically defined in the limit of infinite iterations of the deformation process, $n \to \infty$: in this sense, the fractal is the limit of the deformation process for $n \to \infty$. The definition of fractal dimension is indeed more rigorously given starting from $(q\alpha)^n = 1$ in the $n \to \infty$ limit [1,6]. As a matter of fact, self-similarity is defined only in the $n \to \infty$ limit (self-similarity does not hold when considering only a finite number n of iterations). Since $L_n \to 0$ for $n \to \infty$, the Koch fractal is a curve which is everywhere non-differentiable [2].

Eqs. (6) and (7) express, *in the $n \to \infty$ limit*, the *self-similarity* property of a large class of fractals (the Sierpinski gasket and carpet, the Cantor set, etc.) [1,6]. I also recall that invariance (always in the limit of $n \to \infty$ iterations) only under anisotropic magnification is called self-affinity. The discussion below can be extended to self-affine fractals.

I now observe that, by considering in full generality the complex α-plane, the functions

$$u_n(\alpha) = \frac{\alpha^n}{\sqrt{n!}} , \quad u_0(\alpha) = 1 , \quad\quad n \in \mathcal{N}_+ , \quad \alpha \in \mathbf{C} , \tag{8}$$

form in the space $\mathcal{F}$ of the entire analytic functions a basis which is orthonormal under the gaussian measure $d\mu(\alpha) = \frac{1}{\pi}e^{-|\alpha|^2}\,d\alpha d\bar{\alpha}$. In Eq. (8) the factor $\frac{1}{\sqrt{n!}}$ ensures the normalization condition with respect to the gaussian measure.

The functions $u_{n,q}(\alpha)|_{q \to 1}$ in Eq. (7) (for the factor $q \neq 1$ see the discussion below), are thus immediately recognized to be nothing but the restriction to real α of the functions in Eq. (8), apart the normalization factor $\frac{1}{\sqrt{n!}}$. The study of the fractal properties may be thus carried on in the space $\mathcal{F}$ of the entire analytic functions, by restricting, at the end, the conclusions to real α, $\alpha \to Re(\alpha)$. Since in Eq. (7) it is $q \neq 1$ ($q < 1$), actually one needs to consider the "q-deformed" algebraic structure of which the space $\mathcal{F}$ provides a representation.

To that aim, let me start by observing that the space $\mathcal{F}$ is a vector space which provides the so-called Fock-Bargmann representation (FBR) [7] of the Weyl–Heisenberg algebra

$$[a, a^\dagger] = 1 , \qquad [N, a^\dagger] = a^\dagger , \qquad [N, a] = -a , \qquad (9)$$

where $N \equiv a^\dagger a$, with the identification:

$$N \to \alpha \frac{d}{d\alpha} , \qquad a^\dagger \to \alpha , \qquad a \to \frac{d}{d\alpha} . \qquad (10)$$

The $u_n(\alpha)$ (Eq. (8)) are easily seen to be eigenkets of N with integer (positive and zero) eigenvalues. The FBR is the Hilbert space $\mathcal{K}$ generated by the $u_n(\alpha)$, i.e. the whole space $\mathcal{F}$ of entire analytic functions. Any vector $|\psi\rangle$ in $\mathcal{K}$ is associated, in a one-to-one correspondence, with a function $\psi(\alpha) \in \mathcal{F}$ and is thus described by the set $\{c_n; c_n \in \mathbf{C}, \sum_{n=0}^{\infty} |c_n|^2 = 1\}$ defined by its expansion in the complete orthonormal set of eigenkets $\{|n\rangle\}$ of N:

$$|\psi\rangle = \sum_{n=0}^{\infty} c_n |n\rangle \to \psi(\alpha) = \sum_{n=0}^{\infty} c_n u_n(\alpha), \qquad (11)$$

$$\langle \psi|\psi\rangle = \sum_{n=0}^{\infty} |c_n|^2 = \int |\psi(\alpha)|^2 d\mu(\alpha) = ||\psi||^2 = 1, \qquad (12)$$

$$|n\rangle = \frac{1}{\sqrt{n!}} (a^\dagger)^n |0\rangle , \qquad (13)$$

where $|0\rangle$ denotes the vacuum vector, $a|0\rangle = 0$, $\langle 0|0\rangle = 1$. The series expressing $\psi(\alpha)$ in Eq. (11) converges uniformly in any compact domain of the α-plane due to the condition $\sum_{n=0}^{\infty} |c_n|^2 = 1$ (cf. Eq. (12)), confirming that $\psi(\alpha)$ is an entire analytic function. In view of the correspondence $\mathcal{K} \to \mathcal{F}$ ($|n\rangle \to u_n(\alpha)$) we have

$$a^\dagger u_n(\alpha) = \sqrt{n+1}\, u_{n+1}(\alpha) , \qquad a\, u_n(\alpha) = \sqrt{n}\, u_{n-1}(\alpha) , \qquad (14)$$

$$N\, u_n(\alpha) = a^\dagger a\, u_n(\alpha) = \alpha \frac{d}{d\alpha} u_n(\alpha) = n\, u_n(\alpha) , \qquad (15)$$

which establish the mutual conjugation of a and $a^\dagger$ in the FBR, with respect to the measure $d\mu(z)$.

The Fock–Bargmann representation is known [7,8] to provide an useful frame to describe the (Glauber) coherent states (CS) $|\alpha\rangle$:

$$|\alpha\rangle = \exp\left(-\frac{|\alpha|^2}{2}\right) \sum_{n=0}^{\infty} \frac{\alpha^n}{\sqrt{n!}} |n\rangle = \exp\left(-\frac{|\alpha|^2}{2}\right) \sum_{n=0}^{\infty} u_n(\alpha) |n\rangle. \qquad (16)$$

These are generated by the action on the vacuum state $|0\rangle$ of the unitary displacement operator $\mathcal{D}(\alpha)$ given by:

$$\mathcal{D}(\alpha) = \exp(\alpha a^\dagger - \bar{\alpha} a) = \exp\left(-\frac{|\alpha|^2}{2}\right) \exp(\alpha a^\dagger) \exp(-\bar{\alpha}\, a) , \qquad (17)$$

$$|\alpha\rangle = \mathcal{D}(\alpha)|0\rangle , \qquad a|\alpha\rangle = \alpha|\alpha\rangle , \qquad \alpha \in \mathbf{C} . \qquad (18)$$

We also have

$$\mathcal{D}^{-1}(\alpha)\, a\, \mathcal{D}(\alpha) = a + \alpha . \qquad (19)$$

The explicit relation between the CS and the entire analytic function basis $\{u_n(\alpha)\}$ (Eq. (8)) is:

$$u_n(\alpha) = e^{\frac{1}{2}|\alpha|^2} \langle n|\alpha \rangle . \tag{20}$$

The operator $\mathcal{D}(\alpha)$ is a bounded operator defined on the whole $\mathcal{K}$. It provides a representation of the Weyl–Heisenberg group [7]. The set $\{|\alpha\rangle\}$ is an over-complete set of states. A complete set can be extracted by introducing in the complex α-plane a regular lattice L, called the von Neumann lattice [7].

I now introduce the deformation parameter $q = e^{\zeta}$, $\zeta \in \mathbf{C}$ and recall that the operator q^N acts on the whole $\mathcal{F}$ as [9]

$$q^N f(\alpha) = f(q\alpha) , \quad f(\alpha) \in \mathcal{F} . \tag{21}$$

which follows from the analysis of the q-deformation of the Weyl-Heisenberg algebra. See ref. [9] for details. Use of this relation gives $q^N u_n(\alpha) = u_n(q\alpha)$ and thus for the coherent state functional (16) we have

$$q^N |\alpha\rangle = |q\alpha\rangle = \exp\left(-\frac{|q\alpha|^2}{2}\right) \sum_{n=0}^{\infty} \frac{(q\alpha)^n}{\sqrt{n!}} |n\rangle . \tag{22}$$

Since $q\alpha \in \mathbf{C}$, from Eq. (18),

$$a |q\alpha\rangle = q\alpha |q\alpha\rangle , \quad q\alpha \in \mathbf{C} . \tag{23}$$

Eq. (7), with u_0 set equal to 1, is obtained by projecting out the nth component of $|q\alpha\rangle$ and restricting to real $q\alpha$, $q\alpha \to Re(q\alpha)$:

$$u_{n,q}(\alpha) = (q\alpha)^n = \sqrt{n!} \, \exp\left(\frac{|q\alpha|^2}{2}\right) \langle n|q\alpha\rangle, \quad \text{for any } n, \quad q\alpha \to Re(q\alpha), \tag{24}$$

which, taking into account that $\langle n| = \langle 0| \frac{(a)^n}{\sqrt{n!}}$, gives

$$u_{n,q}(\alpha) = (q\alpha)^n = \exp\left(\frac{|q\alpha|^2}{2}\right) \langle 0|(a)^n|q\alpha\rangle, \quad \text{for any } n, \quad q\alpha \to Re(q\alpha). \tag{25}$$

The operator $(a)^n$ thus acts as a "magnifying" lens [1]: the nth iteration of the fractal can be "seen" by applying $(a)^n$ to $|q\alpha\rangle$ and restricting to real $q\alpha$:

$$\langle q\alpha|(a)^n|q\alpha\rangle = (q\alpha)^n = u_{n,q}(\alpha), \quad q\alpha \to Re(q\alpha). \tag{26}$$

Summarizing, the nth fractal stage of iteration, with $n = 0, 1, 2, .., \infty$, is represented, in a one-to-one correspondence, by the nth term in the coherent state series Eq. (22). The operator q^N applied to $|\alpha\rangle$ (Eq. (22)) "produces" the fractal in the functional form of the coherent state $|q\alpha\rangle$. I call q^N *the fractal operator*.

Note that Eq. (23) expresses the invariance of the coherent state under the action of the operator $\frac{1}{q\alpha}a$ and allows to consider the coherent functional $\psi(q\alpha)$ as an "attractor" in $\mathcal{F}$. This reminds us of the fixed point equation $W(A) = A$, where W is the Hutchinson operator [1], characterizing the iteration process for the fractal A in the $n \to \infty$ limit.

The connection between fractals and the (q-deformed) algebra of the coherent states is formally established by Eqs. (24), (25) and (26).

I finally observe that the fractal operator q^N can be realized in $\mathcal{F}$ as:

$$q^N \psi(\alpha) = \frac{1}{\sqrt{q}} \, \exp\left(\frac{\zeta}{2}(c^2 - c^{\dagger^2}) \right) \psi(\alpha) \equiv \frac{1}{\sqrt{q}} \hat{S}(\zeta)\psi(\alpha) \equiv \frac{1}{\sqrt{q}} \psi_s(\alpha) \, , \quad (27)$$

where $q = e^\zeta$ (for simplicity, assumed to be real), $N = \alpha \frac{d}{d\alpha}$ and

$$c = \frac{1}{\sqrt{2}}\left(\alpha + \frac{d}{d\alpha}\right) \quad , \quad c^\dagger = \frac{1}{\sqrt{2}}\left(\alpha - \frac{d}{d\alpha}\right) \quad , \quad [c, c^\dagger] = 1 \, . \quad (28)$$

In $\mathcal{F}$, $c^\dagger$ is the conjugate of c [7,9]. It is convenient to set $\alpha \equiv x + iy$, x and y denoting the real and the imaginary part of α, respectively. In the limit $\alpha \to Re(\alpha)$, i.e. $y \to 0$, c and $c^\dagger$ turn into the conventional annihilation and creator operators associated with x and p_x in the canonical configuration representation, respectively.

Eq. (27) shows that q^N acts in $\mathcal{F}$, as well as in the configuration representation in the limit $y \to 0$, as the squeezing operator $\hat{S}(\zeta)$ (well known in quantum optics [9,10,11]) up to the numerical factor $\frac{1}{\sqrt{q}}$. $\zeta = \ln q$ is called the squeezing parameter. In (27) $\psi_s(\alpha)$ denotes the squeezed states in FBR. The q-deformation process, which we have seen is associated to the fractal generation process, is equivalent to the squeezing transformation.

In the $y \to 0$ limit, we have

$$\hat{S}^{-1}(\zeta) \, \alpha \, \hat{S}(\zeta) = \frac{1}{q}\alpha \to \frac{1}{q}x \, , \quad (29)$$

$$\hat{S}^{-1}(\zeta) \, p_\alpha \, \hat{S}(\zeta) = qp_\alpha \to qp_x \, , \quad (30)$$

where $p_\alpha \equiv -i\frac{d}{d\alpha}$. Eq. (29) shows that $\alpha \to \frac{1}{q}\alpha$ under squeezing transformation, which, in view of the fact that $q^{-1} = \alpha$ (cf. Eq. (4)), means that $\alpha \to \alpha^2$, i.e. under squeezing we proceed further in the fractal iteration process. Thus, the fractal iteration process can be described in terms of the coherent state squeezing transformation.

Due to the holomorphy conditions holding for $f(\alpha) \in \mathcal{F}$,

$$\frac{d}{d\alpha}f(\alpha) = \frac{d}{dx}f(\alpha) = -i\frac{d}{dy}f(\alpha) \, , \quad (31)$$

in the $y \to 0$ limit we get form (28)

$$c \to \frac{1}{\sqrt{2}}(x + ip_x) \equiv \hat{z} \, , \quad c^\dagger \to \frac{1}{\sqrt{2}}(x - ip_x) \equiv \hat{z}^\dagger \, , \quad [\hat{z}, \hat{z}^\dagger] = 1 \, , \quad (32)$$

where $p_x = -i\frac{d}{dx}$. $\hat{z}^\dagger$ and $\hat{z}$ are the usual creation and annihilation operators in the configuration representation. Under the action of the squeezing transformation, use of (29) and (30) leads to

$$\hat{z}_q = \frac{1}{q\sqrt{2}}(x + iq^2 p_x) , \quad \hat{z}_q^\dagger = \frac{1}{q\sqrt{2}}(x - iq^2 p_x), \quad [\hat{z}_q, \hat{z}_q^\dagger] = 1 . \tag{33}$$

Let $x_1 \equiv x$ and $x_2 \equiv q^2 p_x$ be the coordinates in the "deformed" phase space. Such coordinates do not commute:

$$[x_1, x_2] = iq^2 . \tag{34}$$

We thus recognize that q-deformation introduces non-commutative geometry in the (x_1, x_2)-space. The distance D in such a space is given by the noncommutative Pythagoras theorem:

$$D^2 = x_1^2 + x_2^2 = 2q^2 \left(\hat{z}_q^\dagger \hat{z}_q + \frac{1}{2}\right) . \tag{35}$$

Form the known properties of creation and annihilation operators we then get in $\mathcal{F}$, in the $y \to 0$ limit,

$$D_n^2 = 2q^2 \left(n + \frac{1}{2}\right) , \quad n = 0, 1, 2, 3... , \tag{36}$$

i.e. in the space (x_1, x_2) associated to the coherent state fractal representation, the (x_1, x_2)-distance is quantized according to the unit scale set by q. Eq. (36) shows that in the space (x_1, x_2) we have quantized "disks" of squared radius vector D_n^2. The "smallest" of such disks has non-zero radius given by the deformation parameter q (recall that $q = \frac{1}{3^d}$ when Koch fractal is considered). Recalling the expression of the energy spectrum of the harmonic oscillator, one could write Eq. (36) as $\frac{1}{2}D_n^2 = q^2\left(n + \frac{1}{2}\right) \equiv E_n$, $n = 0, 1, 2, 3...$, where E_n might be thought as the "energy" associated to the fractal n-stage.

Let me close the paper with a couple of comments. Certainly, connecting fractal self-similarity and coherent states opens the way to a series of possible theoretical developments and practical applications in a number of research sectors, including the field of complex dynamical systems. From the theoretical point of view, embedding the fractal study in the framework of the entire analytical functions may contribute to the treatment of fractal properties which present non-trivial difficulties due to the lack of a tractable mathematical representation; for example, the Koch curve is known to be everywhere non-differentiable [2], which makes it greatly interesting, but also difficult to deal with. From the standpoint of coherent states, there might be interesting applications in laser physics and quantum optics, where Glauber coherent states play a dominant role. Since coherent states always emerge as a result of boson condensation (see, e.g., Eq. (19), which describes indeed the (coherent) condensation transformation of the mode a in the vacuum $|0\rangle$), recognizing their fractal structure may contribute to the understanding of physical properties and behaviors in condensed matter systems endowed with a condensed ground state. Here, of course, the perspective opens towards the large class of phenomena based on the mechanism of spontaneous breakdown of symmetry. There the physical meaning of the process of coherent condensation of Nambu-Goldstone modes is the one of the appearance of long range correlation in the system, namely of the dynamical formation of ordered patterns in the ground state. Perhaps the common feature underlying fractal formation and ordered pattern formation is indeed in the dynamical emergence of long range correlation modes. We know that this is certainly the case in Quantum Field Theory systems characterized by ordered patterns in the ground state. The discussion presented above seems to suggest that this might be also the case for fractals. If so, this would provide an interesting example of interplay between the mesoscopic/macroscopic dimension of fractals and the microscopic dynamics out of which they are formed. It also provides an interesting example of a relation of global features (long range correlation at mesoscopic/macroscopic level) with local deformation processes (i.e. dynamical properties at microscopic level). On such a kind of relation is based our understanding of the formation of topologically non-trivial defects (also called extended objects), such as vortices, domain walls, boundary defects, and other soliton-like defects in condensed matter physics and high energy physics [3,11,12]. These extended objects behave as macroscopic classical objects, but are generated by the microscopic (quantum) dynamics. On the basis of our discussion, now one might suspect that perhaps the class of these extended objects might be enlarged so to include also fractals.

The relation between fractal self-similarity and squeezing is also interesting since it is known that squeezed coherent states are also related with dissipative systems [9,13,14]. Thus a link might be established between dissipation, which plays a relevant role in complex dynamical systems, and fractal studies.

Of course, a large field of applications might be offered by biological systems. An interesting example in such a direction is provided by brain studies [4,15]. Self-similarity is in fact observed to characterize the brain background activity. Measurements of the durations, recurrence intervals and diameters of neocortical EEG phase patterns have power-law distributions with no detectable minima. The power spectral densities in time and space of ECoGs from surface arrays conform to power-law distributions [16,17,18,19,20]. The activity patterns generated by neocortical neuropil appear to be scale-free [21,22] with self-similarity in ECoGs patterns over distances ranging from hypercolumns to an entire cerebral hemisphere (which might explain the similarity of neocortical dynamics in mammals, from mouse [23] to whale [24], differing in brain size by 4 orders of magnitude, which contrasts strikingly with the relatively small range of size of avian, reptilian and dinosaur brains lacking neocortex) [25]. In the dissipative model of brain squeezed coherent states describe the brain background activity [26]. According to the result of the present paper they provide the functional representation of self-similarity observed in neuro-phenomenological data. The dissipative model of brain thus accounts for the self-similarity in brain background activity. An application of the relation between fractal self-similarity and coherent squeezed states is thus found when studying the brain dynamics [4].

I recall that the Weyl-Heisenberg representations are labeled by the q-parameter by means of the squeezing transformations [14]. In the infinite volume limit (infinite degrees of freedom) representations labeled by different values of the q-deformation parameter are unitarily inequivalent representations [9]. Changes of the value of the q-parameter induce transitions through unitarily inequivalent representations. The trajectories so induced over the space of the representations can be shown to be, under quite general conditions, chaotic trajectories [5]. One might consider also phase parameters and translation parameters (which characterize generalized coherent states, such as $SU(2)$, $SU(1,1)$, etc. coherent states), besides the scale parameter, and relate them to the deformation q-parameter. By changing these parameters in a *deterministic iterated function process*, also called *multiple reproduction copy machine* process [2], the Koch curve may be then transformed into another fractal, e.g. into Barnsley's fern [2]. In the framework presented in this paper, these fractals are then described by corresponding unitarily inequivalent representations in the limit of infinitely many degrees of freedom (infinite volume limit) [3]. In this way one might recover the richness of the variety of "different" fractal shapes obtainable by changing the parameters of the fractal one starts with [2]. Work is in progress along such a direction.

In conclusion, by limiting my discussion to fractals generated iteratively according to a prescribed recipe (deterministic fractals), I have presented the functional realization of fractal self-similarity in terms of the q-deformed algebra of coherent states. Fractal study can be thus incorporated in the theory of entire analytical functions. From the discussion it appears that the reverse is also true: under convenient choice of the q-deformation parameter and by a suitable restriction to real α, coherent states exhibit fractal properties in the q-deformed space of the entire analytical functions.

The relation here established between fractals and coherent states introduces dynamical considerations in the study of fractals and of their origin, as well as geometrical insight into the coherent states properties. Fractals appear to be global systems arising from local deformation processes.

Acknowledgements

Partial financial support from INFN is acknowledged.

References

1. Bunde, A., Havlin, S. (eds.): Fractals in Science. Springer, Berlin (1995)
2. Peitgen, H.O., Jürgens, H., Saupe, D.: Chaos and fractals. New Frontiers of Science. Springer, Berlin (1986)
3. Vitiello, G.: Topological defects, fractals and the structure of quantum field theory. In: Licata, I., Sakaji, A. (eds.) The Nature Description in Quantum Field Theory. Open Problems and Epistemological Perspective. Springer, Berlin (in print, 2009)

4. Vitiello, G.: Coherent states, fractals and brain waves. New Mathematics and Natural Computation 5 (in print, 2009); Vitiello, G.: Self-similarity in fractals and coherent states (in preparation, 2009)
5. Vitiello, G.: Classical chaotic trajectories in quantum field theory. Int. J. Mod. Phys. B18, 785 (2004)
6. Bak, P., Creutz, M.: Fractals and self-organized criticality. In: Bunde, A., Havlin, S. (eds.) Fractals in Science. Springer, Berlin (1995)
7. Perelomov, A.: Generalized Coherent States and Their Applications. Springer, Berlin (1986)
8. Klauder, J.R., Skagerstam, B.: Coherent States. World Scientific, Singapore (1985)
9. Celeghini, E., De Martino, S., De Siena, S., Rasetti, M., Vitiello, G.: Quantum groups, coherent states, squeezing and lattice quantum mechanics. Annals of Physics 241, 50 (1995); Celeghini, E., Rasetti, M., Tarlini, M., Vitiello, G.: $SU(1,1)$ Squeezed States as Damped Oscillators. Mod. Phys. Lett. B3, 1213 (1989); Celeghini, E., Rasetti, M., Vitiello, G.: On squeezing and quantum groups. Phys. Rev. Lett. 66, 2056 (1991)
10. Yuen, H.P.: Two-photon coherent states of the radiation field. Phys. Rev. 13, 2226 (1976)
11. Umezawa, H.: Advanced Field Theory: Micro, Macro and Thermal Physics. American Institute of Physics (1993)
12. Vitiello, G.: Defect Formation Through Boson Condensation in Quantum Field Theory. In: Bunkov, Y.M., Godfrin, H. (eds.) Topological Defects and the Non-Equilibrium Dynamics of Symmetry Breaking Phase Transitions. Kluwer Academic Press, Dordrecht (2000)
13. Iorio, A., Vitiello, G.: Quantum dissipation and quantum groups. Annals of Physics 241, 496 (1995)
14. Iorio, A., Vitiello, G.: Quantum groups and von Neumann theorem. Mod. Phys. Lett. B8, 269 (1995)
15. Freeman, W.J., Vitiello, G.: Brain dynamics, dissipation and spontaneous breakdown of symmetry. J. Phys. A: Math. Theor. 41, 304042 (2008), http://Select.iop.org.q-bio.NC/0701053v1
16. Freeman, W.J.: Origin, structure, and role of background EEG activity. Part 1. Analytic amplitude. Clin. Neurophysiol. 115, 2077 (2004)
17. Freeman, W.J.: Origin, structure, and role of background EEG activity. Part 2. Analytic phase. Clin. Neurophysiol. 115, 2089 (2004)
18. Braitenberg, V., Schüz, A.: Anatomy of the Cortex: Statistics and Geometry. Springer, Berlin (1991)
19. Linkenkaer-Hansen, K., Nikouline, V.M., Palva, J.M., Iimoniemi, R.J.: Long-range temporal correlations and scaling behavior in human brain oscillations. J. Neurosci. 15, 1370 (2001)
20. Hwa, R.C., Ferree, T.: Scaling properties of fluctuations in the human electroencephalogram. Phys. Rev. E 66, 021901 (2002)
21. Wang, X.F., Chen, G.R.: Complex networks: small-world, scale-free and beyond. IEEE Circuits Syst. 31, 6 (2003)
22. Freeman, W.J.: A field-theoretic approach to understanding scale-free neocortical dynamics. Biol. Cybern. 92(6), 350 (2005)
23. Franken, P., Malafosse, A., Tafti, M.: Genetic variation in EEG activity during sleep in inbred mice. Am. J. Physiol. 275, R1127 (1998)

24. Lyamin, O.I., Mukhametov, I.M., Siegel, J.M., Nazarenko, E.A., Polyakova, I.G., Shpak, O.V.: Unihemispheric slow wave sleep and the state of the eyes in a white whale. Behav. Brain Res. 129, 125 (2002)
25. Freeman, W.J., Vitiello, G.: Nonlinear brain dynamics as macroscopic manifestation of underlying many-body dynamics. Phys. of Life Reviews 3, 93 (2006)
26. Vitiello, G.: Dissipation and memory capacity in the quantum brain model. Int. J. Mod. Phys. B9, 973 (1995)

Generalising Unitary Time Evolution

Kirsty Kitto[1], Peter Bruza[1], and Laurianne Sitbon[2]

[1] Faculty of Science and Technology, Queensland University of Technology
kirsty.kitto@qut.edu.au, p.bruza@qut.edu.au
[2] National ICT Centre, Australia
laurianne.sitbon@nicta.com.au

Abstract. In this third Quantum Interaction (QI) meeting it is time to examine our failures. One of the weakest elements of QI as a field, arises in its continuing lack of models displaying proper evolutionary dynamics. This paper presents an overview of the modern generalised approach to the derivation of time evolution equations in physics, showing how the notion of symmetry is essential to the extraction of operators in quantum theory. The form that symmetry might take in non-physical models is explored, with a number of viable avenues identified.

1 Quantum Interactions Are Not Evolving

As a field Quantum Interaction (QI) has progressed well in recent years [8, 10]. It is clear that something is to be gained from applying the quantum formalism to the description of systems not generally considered physical [1, 4, 14, 16, 23]. However, despite this initial promise, there are many elements of quantum theory that have yet to be properly applied within this framework. Perhaps most notably, it is clear that time evolution has yet to be properly implemented (i.e. derived) for any of these systems. This is a very significant weakness. Without an appreciation of how an entangled quantum-like system might come about it becomes rather difficult to justify the quantum collapse model that is very often leveraged in the quantum interaction community. This paper will explore the notion of time evolution in standard quantum theory (QT), sketching out the modern approach to extracting Hamiltonians and unitary operators. We shall then utilise this approach to suggest some interesting avenues that might be pursued in the future extraction of a fully-fledged quantum-like theory capable of evolving, entangling and then collapsing.

There is no *apriori* reason to expect that the Schrödinger equation is the only form of time evolution equation available in a quantum-like theory. This paper will discuss the reasons lying behind this, and propose ways in which the QI community might work to establish a new time dynamics, or to prove that the application of Schrödinger dynamics is appropriate. Even if some justification can be found for the application of the Schrödinger equation beyond the description of physical systems, it is highly unlikely that the common techniques used in the extraction of a quantum description will work. This is because the standard approach to constructing a quantum theory generally involves finding a description of the system of interest that bears resemblance to an existing quantum

P. Bruza et al. (Eds.): QI 2009, LNAI 5494, pp. 17–28, 2009.

description and then making use of a perturbative approach to extract the new quantum dynamics. Given that the systems modelled within the QI community are not necessarily physical in origin we might expect that this method will prove difficult to apply in this field.

It is worth emphasising at this point the necessity of these considerations. While the problem of describing composite quantum systems is well understood, there is no reason to expect that the systems described by the QI community will behave identically to physical systems. While entanglement and measurement are commonly used by QI models, almost none of them show how a quantum-like system might evolve to the point where it could be measured. One of the most commonly used techniques in the modelling of physical systems involves showing an approximate equivalence with a system already modelled and then applying that model to the new system. This may work for some QI models, but there is a very real possibility that not all QI systems will have direct physical analogues. This paper has been written in order to show those of the QI community who do not have a background in physics how they might proceed in constructing an evolving quantum-like theory if this becomes necessary.

2   Transformations in Quantum Theory

Time evolution is well understood in the standard quantum formalism, and the choices made in creating a model generally have very compelling reasons behind them. In this section we shall sketch out the modern approach to quantization, showing how this can be used to extract Schrödinger dynamics. The full approach can be found in any good modern text on QT [5, 21].

Physics has come a long way by assuming that the laws of nature are invariant under certain space-time transformations. These can include displacements, rotations and changes between frames of reference in uniform relative motion. In quantum theory, transformations of both states $|\psi\rangle \rightarrow |\psi'\rangle$ and observables $\hat{A} \rightarrow \hat{A}'$ must be considered together, and this places restrictions on the form that any transformation can take. Specifically, if $A|\phi_n\rangle = a_n|\phi_n\rangle$, then we must have $A'|\phi_n'\rangle = a_n|\phi_n'\rangle$ after transformation. Thus the eigenvalues of observable A cannot change under a transformation since the observable cannot be changed by the way we are looking at it. It is also essential that $|\langle\phi_n|\psi\rangle|^2 = |\langle\phi_n'|\psi'\rangle|^2$, which means that the probabilities for equivalent events in two different frames of reference should be equivalent. This requirement leads to Wigner's theorem [21], which places a strong restriction on the form that such a transformation can take, with only the above very minimal assumption about the nature of the inner product. This theorem shows that any mapping of a vector space onto itself that preserves the value of the inner product must be implemented by an operator U that is either unitary and linear or anti-unitary and anti-linear [25].[1] Unitary operators are very widely used in QT, as they are the only ones that can describe continuous transformations such as translations and rotations

---

[1] A unitary transformation is one such that $\langle\phi'|\psi'\rangle = \langle\phi|\psi\rangle$, whereas an anti-unitary transformation satisfies $\langle\phi'|\psi'\rangle = \langle\phi|\psi\rangle^*$.

(since every continuous transformation must have a square root [5]). However, anti-unitary transformations also play a part in the quantum formalism as they are used in the description of discrete time reversal symmetries.

Together, these very minimal requirements place strong constraints upon the form that transformation operators can take in a standard quantum theory. In the particular case of continuous transformations, we find that while states must transform according to

$$|\psi\rangle \rightarrow |\psi'\rangle = U|\psi\rangle, \tag{1}$$

observables must transform according to

$$\hat{A} \rightarrow A = UAU^{-1}. \tag{2}$$

Thus, with the assumption that the symmetries in quantum-like models will be continuous, we find ourselves to be looking for unitary operators satisfying equations (1) and (2).

In order to start sketching out the general form that such operators must take, we shall consider a set of unitary matrices $U(\alpha_1, \alpha_2, \dots)$ which depend upon the continuous parameters α_j. With a good choice of parameters, we find that these matrices are in a 1–1 correspondence with a continuous *group* of transformations, $\mathcal{G}_U$. That is, we find that the matrices satisfy:

Closure: for every $U_a, U_b \in \mathcal{G}_U$, the product of the two matrices is in the group, $U_a U_b \in \mathcal{G}_U$.

Associativity: for every $U_a, U_b, U_c \in \mathcal{G}_U$, $U_a(U_b U_c) = (U_a U_b)U_c$ (note that this property is automatically satisfied by matrices).

Identity element: there exists one, and only one, identity matrix in the group. We customarily define this matrix such that $U(0, 0, \dots) = \mathbb{1}$.

Inverse element: every matrix $U_a \in \mathcal{G}_U$ has a unique inverse also in the set. That is, there exists a matrix $U(\beta_1, \beta_2, \dots) \in \mathcal{G}_U$ such that:

$$U(\alpha_1, \alpha_2, \dots)U(\beta_1, \beta_2, \dots) = \mathbb{1}$$

Any set of matrices satisfying these properties forms a symmetry group. This is a remarkably important concept in modern physics. It is essential to realise that symmetry groups can take many different forms, the one sketched above for unitary matrices relies heavily upon multiplication, but the closure criterion could be just as easily framed for addition, or even some other operator. For example the Integers form a symmetry group under addition.

2.1 What Is a Symmetry?

The concept of symmetry has a very particular meaning in physics, where it applies to any physical or mathematical feature of a system that is preserved, or invariant, under some transformation. Thus, the concept is quite broad in physics, compared to the common lay usage which generally refers to properties of a more geometrical nature. Consider for example the following way in which a symmetry group can be constructed for motion in one dimension.

To appreciate the link between group theory and motion, imagine you are standing on a straight road that goes on forever both in front and behind you. Stand stock still; this is the identity of a group. Walk forwards a little, then a little more. But you are now where you would have been had you just walked further in the first place. So moving along a straight line exhibits the closure property. Associativity can be demonstrated by walking different distances forwards and backwards in different sequences and noting that the end result is always the same. Finally, if you walk forwards a bit then backwards to where you started you have discovered the inverse. [24]

Thus, a symmetry is not necessarily something that looks the same along an axis of view (like a mirror reflection symmetry), it has a much broader set of connotations.

Any transformation that satisfies the above group structure is a symmetry. Symmetries that commute with time evolution correspond to a conserved quantity in physics via Noether's theorem. This important theorem amounts to a statement that for every physical system exhibiting symmetry under time evolution there is some conserved physical property of that system, and conversely that each conserved physical quantity has a corresponding symmetry. Thus, symmetries can have physical consequences in their own right.

2.2   Symmetries, Operators, and Hamiltonians

It can be shown that any unitary transformation that depends upon a single parameter α (e.g. a rotation about a fixed axis by an angle $\alpha = \theta$) can be expressed as an exponential of a Hermitian *generator*, G, that is independent of α [5, 21]:

$$U(\alpha) = e^{-i\alpha G}. \tag{3}$$

The generators of transformations corresponding to symmetry properties often have simple physical meanings (such as energy, momentum, electric charge etc. in physics). It is important to realise that these symmetries often work together, forming larger groups which describe all allowable transformations within that space. Thus are the Gallilei, and Poincaré groups formed, as well as the larger groups used in The Standard Model of modern particle physics.

The Galilei group arises in non-relativistic quantum mechanics. It consists of a 10 dimensional representation of the symmetries of classical mechanics.[2] This group describes all of the rotations, displacements and transformations that can occur between uniformly (and slowly) moving frames of reference. Thus, this group describes all transformations of the form:

$$\mathbf{x} \to \mathbf{x}' = R\mathbf{x} + \mathbf{a} + \mathbf{v}t \tag{4}$$

$$t \to t' = t + s. \tag{5}$$

---

[2] Maxwell's equations do not satisfy this group and their inclusion in the group structure of modern physics led to the development of the Poincaré group which includes the Lorentz transformations of special relativity.

Here, R is a rotation (which can be thought of as a 3×3 matrix acting on a 3-vector $\mathbf{x}$), $\mathbf{a}$ is a space displacement, $\mathbf{v}$ is the velocity of a moving coordinate transformation and s is a small displacement of the time t.

Here, we are interested in the time evolution of a quantum system. In physics, time evolution is a symmetry of spacetime given by

$$t \to t + s, \; x \to x, \; y \to y, \; z \to z \tag{6}$$

with a conserved quantity that corresponds to the energy of the system. A system with more energy will move faster as time passes, so this conservation law is intuitively understandable. It is possible to derive Schrödinger's equation[3]

$$\frac{d}{dt}|\psi(x,t)\rangle = -iH(x,t)|(x,t)\rangle \tag{7}$$

from considerations of the dynamics of a free particle invariant under the full Galilei group of space-time transformations [5]. To do this, we make use of the properties of the Galilei group. We start by considering two sets of transformations, τ_1 followed by τ_2, and an equivalent single transformation τ_3. The equivalence means that $\tau_1\tau_2 = \tau_3$, and since these transformations are the same transformations we must require that $U(\tau_2)U(\tau_1)|\psi\rangle$ and $U(\tau_3)|\psi\rangle$ describe the same state. They do not necessarily have to be the same vector, they can differ up to a complex phase, which gives

$$U(\tau_3) = e^{i\omega(\tau_1,\tau_2)}U(\tau_2)U(\tau_1). \tag{8}$$

So symmetries must be relatable using some complex phase factor. Indeed, corresponding to the time displacement $t \to t' = t + s$, we find that the following vector space transformation holds [5]:

$$|\psi(t)\rangle \to e^{isH}|\psi(t)\rangle, \tag{9}$$

but if we consider figure 1 we quickly see that this can be written equivalently as $|\psi(t - s)\rangle$. We use this symmetry, by setting $s = t$, which gives $|\psi(x)\rangle = e^{-itH}|\psi(x)\rangle$. Finally, we note that only an equation of form (7) can generate this solution. $\qquad\qquad\qquad\qquad\qquad\qquad\qquad\qquad\qquad\qquad\qquad\qquad\quad\square$

While finding the Schrödinger equation through the application of symmetry information about time translations is the main point of this article, it will most likely prove useful to the QI community to see how this technique extends further. Indeed, it can be used to extract the full commutative structure of QT. We shall not perform that analysis here, the interested reader can refer to [5].

2.3   How Do Commutation Relations Relate to Symmetry?

As was mentioned above, symmetries that commute with time evolution correspond to conserved quantities via Noether's theorem. However, the commutation relations of QT have a wider set of relationships with the symmetry group of a physical theory.

---

[3] Here we have used natural units (which gives $\hbar = 1$).

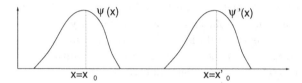

Fig. 1. A unitary time translation of the function $\psi(x)$, from a point around $x = x_0$ to a point around $x = x_0'$ is equivalent to a change in coordinate frame; there is an inverse relationhip between transformations on a function space and transformations on coordinates. Representing the change in coordinates as τ, we find that $\psi'(\tau x) = \psi(x)$, and hence that $U(\tau)\psi(x) = \psi(\tau^{-1}x)$ [5].

In extracting the generators of a Galileian group describing a QT it is necessary to couple the symmetry structure of the transformations in the group with the unitary requirements of (1) and (2). In doing this we find that the standard commutation relationships of QT must be satisfied [5].

Thus, it is possible to fully derive the structure of QT from a consideration of symmetry and unitarity, and this is the modern approach to quantization. It is likely that this approach will prove most effective in the construction of a fully-fledged quantum-like theory.

3   Symmetry Groups for Quantum-Like Theories?

There is no reason to believe that the symmetry groups of a quantum-like theory will be the same as for those of standard physics. Many of the relevant spaces considered in the field are of a very high dimension, and they do not need to satisfy the same set of physical conditions. Consider for example the very high dimensional cognitive spaces that are being modelled using QT [3, 9]. We would not immediately expect such systems to display the same symmetry behaviour as a standard QT. This raises an intriguing question; what form of symmetry could be satisfied by such models?

There are some early hints that we might explore in developing new symmetries, relevant to a much broader class of system. Some interesting avenues that we feel hold promise include:

- The use of Quantum Field Theory (QFT) in the modelling of biological systems [17, 23] and the use of symmetry breaking techniques in the modelling of dynamical emergence. This requires the identification of symmetry groups beyond those standard to physics, and it appears possible that complete groups might be identified as these theories develop; some of these might point towards a temporal symmetry that might be leveraged in deriving general time evolution equations in standard first quantized models.
- Some interesting work examining the concept of symmetry in object oriented programming languages has been performed [13, 27]. Here, the use of inheritance in the extraction of symmetry relations suggests that if a symmetry

group could be found for such systems then it should share some features with any biological models that make use of intergenerational symmetries (within the same species for example).

- A concept of *superfractals* has been coined [6] to describe the mathematics of natural imagery, art, and biology. Among the mathematics developed here, it is possible to make use of iterated function systems to generate complex landscape and biological images using a computer, which *look similar* to a human observer. This conception of similarity holds promise, and the mathematical nature of the theory leaves it ideal for extension to a theory of symmetry with respect to human cognition. This idea will be explored elsewhere.

- The different senses or meanings of a word might also be developed into a group theory. Such a theory would probably leverage the intuition that even when changing word senses you still have the same token. Thus, *bat* stands for "furry flying mammal" or "sporting implement". If a group denoting this could be found then it might even fit into a larger group structure of language, after all, the set of different languages still describes the same set of senses, at least approximately.

All of these different avenues are currently under investigation, but the problem of finding proper formalisations of what are generally quite vague arguments is very difficult. We might wonder if perhaps there is a new generalised mathematics of symmetry groups waiting to be found.

3.1 Towards a New Mathematics?

Group theory as it currently stands is concerned with relatively simple structures and behaviour. It has been developed primarily for physical systems, and we might wonder if the behaviour of quantum-like systems can be described by the same sets of groups developed for physics. Some reasons to believe that this is probably not the case will be briefly discussed in this section.

Many of the systems described by the QI community display complex behaviour [18], and as such they will have features such as internal structure, hierarchical organisation, contingent dependency upon historical events, and an evolving dynamics. This would lead us to suggest that their symmetries will be far more difficult to extract, and in themselves far more complex, than those of physical systems.

4 Towards Time Evolution in Quantum-Like Theories

In this section we shall summarise some recently developed ideas that we feel hold sufficient promise for the future creation of a fully-fledged quantum-like theory. Both of them have been generated through attempts to develop the idea of a symmetry group of the system of interest, and in particular to find properties of that system that are conserved under time evolution and so might be used to generate some sort of quantum time evolution dynamics.

4.1 Quantum Models of Biological Development

Symmetries play a vital role in models of biological development. In fact, it is the breaking of symmetries that generates actual outcomes in terms of axial orientation, and through this eventual cell differentiation. It will be instructive to consider some of the issues involved in constructing a full description of the dynamics involved in this process.

Let us consider a perfectly spherical egg. It is symmetrical under all rotations and translations in space, and as such could be represented by the O(3) group. Differentiation of the cell starts from the moment it is impregnated by a sperm cell; a new axis of symmetry arises from the line joining the site of sperm penetration with the centre of the egg. Once this event occurs the O(3) rotational symmetry of the egg is lost, and developmental events will quickly lead to a loss of more and more symmetry. However, over time, there is a sense of conservation; the organism remains the same organism, even if it gradually becomes very different in form.

It is hoped that this idea might be leveraged in order to develop a quantum-like model of biological development. Here we would see a situation where the environment in which the fertilised egg is developing influences the eventual form of the egg itself, however, there is every reason to suppose that a QFT would prove most appropriate for such systems. This is because QFT's allow for the existence of unitarily inequivalent ground states [19, 22, 23], which allows for a model of development that sees the organism as growing through a number of different stable states. The alternative picture supplied by a first quantized theory (such as is discussed in this paper) would see the developing through a process of excitation, this is not feasible, after all, such a model would open up the possibility that a fully developed organism might de-excite back to the ground state!

Can we find a situation where a first quantized model is the most appropriate approach?

4.2 Quantum Models of Semantic Structure

Cognitive scientists have produced a collection of models which have an encouraging, and at times impressive, track record of replicating human information processing, such as word association norms. These are generally referred to as *semantic space* models. As used here, the term "semantic" derives from the intuition that the meaning of a word derives from the "company it keeps", as the linguist J.R. Firth (1890-1960) famously remarked. For example, the words "mobile" and "cellular" would exhibit a strong association in semantic space as the distribution of words with which they co-occur tends to be similar, even though the two words almost never co-occur themselves. Although the details of the various semantic space models differ, they all process a corpus of text and "learn" representations of words in a high dimensional space.

There is already an existing body of work linking semantic space theory to QT [2, 7, 11, 12, 26]. In one set of examples [11, 12], a semantic space S_w surrounding word w is constructed by collecting a corpus of traces centred around w. Such matrices are square symmetric matrices and hence self-adjoint. For a given set of words u, v and w we shall represent the corresponding matrices as S_u, S_v and S_w.

We shall now sketch how a symmetry group might be developed for a semantic space model, and perhaps eventually used to generate a model of semantic dynamics.

The product $S_u S_v$ can be interpreted as the the effect on the semantic representation of u when seen in the context of word v. That is, how much of u's semantic representation project onto that of v. This product satisfies closure, since the word itself is still in the combined semantic space. Combining semantic representations using such a product is also associative: $S_u(S_v S_w) = (S_u S_v)S_w$. The identity operator can be easily identified as the word itself, it has the same semantic representation as itself.

The question of an inverse S_w^{-1} for S_w for an arbitrary word w in not a straightforward issue. Intuitively we might expect the inverse to be something that "undoes" the projection, hence removing a word from its context. However, a word removed from its context is a highly artificial thing, and this is not necessarily the best way to proceed. Perhaps instead a notion of inverse might be developed that would produce a representation that is "orthogonal" to the meaning of w. One possible candidate is the Householder reflection:

$$ S_w^{-1} = I - 2|w\rangle\langle w| $$

This formula exploits the complementary representations S_w and $|w\rangle$ noted in [7]. S_w is a matrix representation for the word w, but also the unit vector $|w\rangle$ is a prominent column vector in S_w. The above formula produces a self adjoint matrix S_w^{-1} which is a reflection in the hyperplane perpendicular to the vector $|w\rangle$. This problem of the inverse is something that will be investigated in future work.

Obviously there are significant details to be worked out in such an approach. Firstly the product $S_u S_v$ is not guaranteed to be self adjoint. This is not necessarily a problem, but it would be much cleaner if a product operation could be defined which resulted in a self adjoint matrix. In addition, the above definition of an inverse only covers the inverses corresponding to individual words, not compound representations, e.g., $S_u S_w$ which are also elements of the group. There is a whole avenue of research in relation to forming the semantic representations of compounds, indeed, there has been some speculation that concepts are entangled [3, 9] Finally, there is the question about what the interpretation of invariance should be in relation to the a semantic representation. As semantic space models are derived directly from an underlying corpus, the semantic representations of the words change accordingly. That is, the meaning of the words changes according how the company around them evolves [20]. As a consequence, the strength of semantic association between words varies as the corpus evolves. However, there will be a point where the semantic representations stabilise and semantic associations will stabilise. The stabilisation of semantic association was demonstrated recently in relation to the BEAGLE model [15]. In BEAGLE, representations are primed initially by random vectors. So each time the model is run over a given corpus the actual semantic representations of words will be different. However the strength of semantic association is largely invariant across different runs of the model.

5   Conclusions, and a Question for the Future

This article is obviously of a very exploratory nature. Here we shall ask a question in the hope that others might be interested in considering it.

The symmetry groups of modern physics are, in a number of ways, boring. The requirement to satisfy space-time symmetries is a very strong one, which leads to some very profound restrictions upon the nature of physical reality. Such restrictions do not necessarily apply in the high dimensional conceptual spaces often considered in QI. This actually makes the derivation of group structures in this field much more challenging as there are no clear restrictions to incorporate into our models. However, as we have seen in sections 3 and 4.2, there are some early intuitive ideas that might be investigated. More generally, there are many of mathematically interesting ideas that could be considered. For example, the structure of the unitary operators must be taken into account, many interesting systems take a hierarchical form, and while the Standard Model does have something of a nested structure, it was not necessary to consider any truly hierarchical behaviour in the construction of this model. However, we can ask if there might be a way of constructing a set of more general tensorial operators, ones that could incorporate the complex and interrelated hierarchical symmetries of biological systems. This is a problem that will be investigated in future work.

If QI is to truly come of age then it must start to develop complete theories. These must include both time evolution and symmetry considerations. The entanglement so often relied upon in the field must emerge from a truly evolving quantum model, not just be assumed to exist at the outset. This paper has presented some ideas about how such models might be constructed, and pointed at some of the possible avenues that might be pursued in the future. We hope that these ideas might prove fruitful to any future investigations of dynamics in the new field of quantum interaction..

Acknowledgements. This project was supported in part by the Australian Research Council Discovery grant DP0773341 (KK and PB). NICTA is funded by the Australian Government as represented by the Department of Broadband, Communications and the Digital Economy and the Australian Research Council through the ICT Centre of Excellence program.

References

[1] Aerts, D., Aerts, S., Broekaert, J., Gabora, L.: The Violation of Bell Inequalities in the Macroworld. Foundations of Physics 30, 1387–1414 (2000)
[2] Aerts, D., Czachor, M.: Quantum Aspects of Semantic Analysis and Symbolic Artificial Intelligence. Journal of Physics A-Mathematical and General 37, L123–L132 (2004), http://uk.arxiv.org/abs/quant-ph/0309022

[3] Aerts, D., Gabora, L.: A Theory of Concepts and Their Combinations II: A Hilbert Space Representation. Kybernetes 34, 176–205 (2005)

[4] Baaquie, B.E.: Quantum Finance: Path Integrals and Hamiltonians for Options and Interest Rates. Cambridge University Press, Cambridge (2004)

[5] Ballentine, L.: Quantum Mechanics: A modern development. World Scientific, Singapore (1998)

[6] Barnsley, M.F.: Superfractals: Patterns of nature. Cambridge University Press, Cambridge (2006)

[7] Bruza, P., Cole, R.: Quantum Logic of Semantic Space: An Exploratory Investigation of Context Effects in Practical Reasoning. In: Artemov, S., Barringer, H., d'Avila Garcez, A.S., Lamb, L.C., Woods, J. (eds.) We Will Show Them: Essays in Honour of Dov Gabbay, vol. 1, pp. 339–361. College Publications, London (2005)

[8] Bruza, P.D., Lawless, W., van Rijsbergen, C.J., Sofge, D. (eds.): Proceedings of the second conference on Quantum Interaction, March 26-28. College Publications, London (2008)

[9] Bruza, P.D., Kitto, K., Nelson, D., McEvoy, C.L.: Entangling words and meaning. In: Proceedings of the Second Quantum Interaction Symposium (QI 2008), pp. 118–124. College Publications (2008)

[10] Bruza, P.D., Lawless, W., van Rijsbergen, C.J., Sofge, D. (eds.): Quantum Interaction. Association for the Advancement of Artificial Intelligence. AAAI Press, Menlo Park (2007)

[11] Bruza, P.D., Widdows, D., Woods, J.A.: Quantum Logic of Down Below. In: Engesser, K., Gabbay, D., Lehmann, D. (eds.) Handbook of Quantum Logic and Quantum Structures, vol. 2. Elsevier, Amsterdam (2007)

[12] Bruza, P.D., Woods, J.H.: Quantum collapse in semantic space: interpreting natural language argumentation. In: Proceedings of the Second Quantum Interaction Symposium (QI 2008), pp. 141–147. College Publications (2008)

[13] Coplien, J.O., Zhao, L.: Symmetry breaking in software patterns. In: Butler, G., Jarzabek, S. (eds.) GCSE 2000. LNCS, vol. 2177, p. 37. Springer, Heidelberg (2001)

[14] Gabora, L., Rosch, E., Aerts, D.: Toward an ecological theory of concepts. Ecological Psychology 20, 84–116 (2008)

[15] Jones, M.N., Mewhort, D.J.K.: Representing Word Meaning and Order Information in a Composite Holographic Lexicon. Psychological Review 114(1), 1–37 (2007)

[16] Khrennikov, A.: Quantum-like Probabilistic Models outside Physics (2007), arXiv:0702250

[17] Kitto, K.: Modelling and Generating Complex Emergent Behaviour. Ph.D thesis, School of Chemistry Physics and Earth Sciences, The Flinders University of South Australia (2006)

[18] Kitto, K.: Why quantum theory? In: Proceedings of the Second Quantum Interaction Symposium, pp. 11–18. College Publications (2008)

[19] Kronz, F.M., Lupher, T.A.: Unitarily Inequivalent Representations in Algebraic Quantum Theory. International Journal of Theoretical Physics 44(8), 1239–1258 (2005)

[20] McArthur, R.M., Bruza, P.D.: Discovery of tacit knowledge and topical ebbs and flows within the utterances of online community. In: Ohsawa, Y., McBurney, P. (eds.) Chance Discovery - Foundations and Applications. Advances in Information Processing, vol. XVI. Springer, Heidelberg (2003)

[21] Peres, A.: Quantum theory: concepts and methods. Fundamental theories of physics, vol. 57. Kluwer academic publishers, Dordrecht (1993)

[22] Umezawa, H.: Advanced Field Theory: Micro, macro, and thermal physics. American Institute of Physics, New York (1993)
[23] Vitiello, G.: My Double Unveiled. John Benjamins Publishing Company, Amsterdam (2001)
[24] Watson, A.: The mathematics of symmetry. New Scientist 1740 (1990)
[25] Weinberg, S.: The Quantum Theory of Fields, vol. 1. Cambridge University Press, Cambridge (1995)
[26] Widdows, D.: Geometry and Meaning. CSLI Publications, Stanford (2004)
[27] Zhao, L., Coplien, J.: Symmetry in Class and Type Hierarchy. In: Noble, J., Potter, J. (eds.) Proceedings Fortieth International Conference on Technology of Objects, Languages and Systems (TOOLS Pacific 2002), Sydney, Australia, pp. 181–190. Australian Computer Society (2002)

Comparison of Quantum and Bayesian Inference Models

Jerome R. Busemeyer and Jennifer Trueblood

Cognitive Science, Indiana University,
1101 E. 10th Street, Bloomington Indiana, 47405
jbusemey@indiana.edu

Abstract. The mathematical principles of quantum theory provide a general foundation for assigning probabilities to events. This paper examines the application of these principles to the probabilistic inference problem in which hypotheses are evaluated on the basis of a sequence of evidence (observations). The probabilistic inference problem is usually addressed using Bayesian updating rules. Here we derive a quantum inference rule and compare it to the Bayesian rule. The primary difference between these two inference principles arises when evidence is provided by incompatible measures. Incompatibility refers to the case where one measure interferes or disturbs another measure, and so the order of measurement affects the probability of the observations. It is argued that incompatibility often occurs when evidence is obtained from human judgments.

1 Introduction

Quantum theory was originally invented by physicists to explain findings that seemed paradoxical from the classical physical view point. Later Dirac (1930) and Von Neumann (1932) provided an axiomatic foundation for quantum theory, and by doing so, they discovered that it implied a new type of logic and probability theory. Consequently, there are now at least two general theories for assigning probabilities to events: classic theory and quantum theory [7], [10], [16].

An important application that should be addressed by any general probability theory is the problem of inference – that is, the evaluation of hypotheses on the basis of evidence. Inference is a general problem that arises in many applications. For example, a detective must infer the person who committed a crime on the basis of facts collected from the crime scene and testimony of witnesses. A physician must infer the cause of an illness based on medical symptoms and medical tests. A commander must infer the location of an enemy on the basis of sensory data and intelligence reports. According to classic probability theory, Bayes rule is used to model this kind of probabilistic inference. Quantum probability theory provides an alternative model, and the purpose of this paper is to compare models of probabilistic inference based on Bayesian versus quantum principles.

The evidence used to make an inference is based on observations and measurements. It is well known that the key point upon which Bayesian and quantum models differ is the concept of compatibility of the measures. If all the measures are compatible, then Bayesian and quantum models always agree and the two models assign exactly the same probabilities. Differences only arise when incompatible measures are

P. Bruza et al. (Eds.): QI 2009, LNAI 5494, pp. 29–43, 2009.
© Springer-Verlag Berlin Heidelberg 2009

involved. It is argued that incompatibility can arise when measurements are based on human judgments which interfere with each other and change depending on order of presentation [3].

There already exists a mature literature on hypothesis testing in the area of quantum information theory [8], [9], [10]. This literature is concerned with the problem of determining which of several quantum states describe the true state of a quantum system by performing measurements. The main concern of this literature is designing tests and analyzing the probability of incorrectly deciding which quantum state is the true state. This differs from the present goal which is to describe the revision of a quantum state on the basis of new evidence.

There is also a growing literature concerned with quantum networks that are comparable to Bayesian networks [13], [15], [17]. This literature is concerned with efficient algorithms for computing quantum probabilities from graphical representations of relations. However, the present paper examines more directly the effect of employing sequences of compatible and incompatible measurements on revision of quantum probabilities.

New research is beginning to appear on quantum models of human probability judgments [2], [6], [11]. This work focuses on explaining some paradoxical findings about the way humans make judgments. In contrast, the goal of this paper is to develop a general model that uses human judgments as sources of evidence for making coherent and rational inferences.

2 Probabilistic Inference Task

To begin, we limit our discussion to finite sets (although the number of elements can be very large). The ideas can be extended to infinite sets, but the latter requires more careful handling of convergence and so it is simpler to start with the finite case.

It is assumed that there is a finite set of m hypotheses labeled $\{h_1, ..., h_i, ..., h_m\}$. For example, these might be suspects for a crime, or causes of an illness, or possible locations of an enemy, or intentions of an opponent.

Evidence is obtained from a sequence of measurements that are taken across time, $t = 1, 2, ..., T$. Different *types* of measures may be selected at each time step. The notation $X(t) = x_t$ symbolizes that the measure selected at time t produced outcome x_t. For example, a physician may first measure the patient's temperature (producing a degree on a digital thermometer), then ask the patient to judge how much pain he or she experiences (providing a rating on a one to ten scale), and finally ask the patient how long the pain lasts (evaluated in minutes). Each measure is assumed to produce one of a finite set of outcomes.

The task is to determine the probability of each hypothesis after observing a sequence of outcomes:

$$p(h_i \mid X(1) = x_1, X(2) = x_2, ..., X(t) = x_t) \text{ for } i = 1,...,m; t = 1, ..., T. \tag{1}$$

2.1 Classic Probability

Classic probability theory [12] assigns probabilities to events defined as subsets of a sample space, S, which is the universal set. Suppose the cardinality of S equals N.

Then we can define N elementary events $S = \{z_1, z_2, ..., z_N\}$. Two elementary events can be joined by union to form a new set. Joining elementary events this way, one can generate a family of 2^N sets (including the empty set). This forms a Boolean algebra of sets.

Classic probability postulates the existence of a probability function p that assigns a probability, $0 \leq p(z_i) \leq 1$ to each elementary event. The probability of an arbitrary A event is then defined by $p(A) = \sum_{i \in A} p(z_i)$. Classic probabilities must satisfy $p(S) = \sum_{i \in S} p(z_i) = 1$ for the universal event and $p(\varnothing) = 0$ for the null event.

For purposes of comparison, it is worthwhile to describe classic probability theory using vectors and projection operations. First, we can define an $N \times 1$ vector $|z_j\rangle$ corresponding to elementary event z_j that has all zeros except for a one in row j. Then we can define a projector for event z_j as the outer product $\mathbf{P}_j = |z_j\rangle\langle z_j|$, which is an $N \times N$ matrix full of zeros except a one on the diagonal for row j. The projectors corresponding to different elementary events are orthogonal, $\mathbf{P}_i \cdot \mathbf{P}_j = \mathbf{0}$ for $i \neq j$, and they are complete in the sense that $\sum_j \mathbf{P}_j = \mathbf{I}_N$, where $\mathbf{I}_N$ is the identity matrix.

The projector for an arbitrary event A then equals $\mathbf{P}(A) = \sum_{j \in A} \mathbf{P}_j$. We can represent the probability function by an $N \times 1$ vector of probabilities

$$\pi = \sum_{j \in S} p(z_i) \cdot |z_j\rangle . \tag{2}$$

This vector π can be interpreted as the state of the classic probability system. This is called a mixed state. The probability of an event A is determined by the projection of the mixed state followed by a sum of the projection:

$$p(A) = \mathbf{1} \cdot \mathbf{P}(A) \cdot \pi, \text{ with } \mathbf{1} = [\ 1\ 1\ ...\ 1\ ] . \tag{3}$$

In particular, the probability of the event corresponding to elementary event $|z_j\rangle$ is simply

$$p(z_j) = \mathbf{1} \cdot \mathbf{P}_j \pi . \tag{4}$$

Also note that $p(S) = \mathbf{1} \cdot \mathbf{P}(S) \cdot \pi = \mathbf{1} \cdot \mathbf{I} \cdot \pi = 1.0$ and $p(\varnothing) = \mathbf{1} \cdot \mathbf{P}(0) \cdot \pi = \mathbf{1} \cdot \mathbf{0} \cdot \pi = 0$.

2.2 Quantum Probability

Quantum probability theory [5] assigns probabilities to events defined as subspaces of a Hilbert space, H, which is the universal space. Suppose the dimensionality of H is N. Then we can define N orthonormal basis vectors, $\{|z_1\rangle, ..., |z_i\rangle, ..., |z_N\rangle\}$ where each basis vector represents a one dimensional subspace (corresponding to an elementary event). Two basis vectors can be joined to form a subspace that spans the two vectors. Joining basis vectors this way, one can form a family of 2^N subspaces (including the zero point). As discussed in the concluding section, this forms a partial Boolean algebra of events.

In quantum theory, each basis vector $|z_j\rangle$ corresponds to a projector $\mathbf{P}_j = |z_j\rangle\langle z_j|$ that projects unit length vectors in H onto this basis vector. This forms a complete set of orthogonal projectors $\mathbf{P}_i \cdot \mathbf{P}_j = \mathbf{0}$ for $i \neq j$, and $\sum_j \mathbf{P}_j = \mathbf{I}_N$. The projector corresponding to the join of basis vector $|z_i\rangle$ with basis vector $|z_j\rangle$ equals $\mathbf{P}_i + \mathbf{P}_j$. This implies that each event A is also defined by a projector $\mathbf{P}(A) = \sum_{j \in A} \mathbf{P}_j$. Note that the projector for

H is $P(H) = \sum_{j \in H} P_j = I_N$ (the identity operator) and the projector for the null event is $P(\varnothing) = 0$ (the zero operator).

Quantum probability postulates the existence of a state vector, denoted here as $|Z\rangle$, which is a unit length vector in the Hilbert space. This state vector can be expressed in terms of coordinates of the basis states as follows:

$$|Z\rangle = I_N \cdot |Z\rangle = (\sum_{j \in H} |z_j\rangle\langle z_j|) \cdot |Z\rangle = \sum \langle z_j|Z\rangle|z_j\rangle = \sum \alpha_j \cdot |z_j\rangle . \tag{5}$$

The coefficient, $\alpha_j = \langle z_j|Z\rangle$ is called probability amplitude corresponding to basis state $|z_j\rangle$. The state vector $|Z\rangle$ is a superposition of the basis states.

The probability of an event A is determined by the squared length of the projection of the state vector onto the subspace that defines the event: $q(A) = \| P(A)|Z\rangle\|^2$. In particular, the probability of the event corresponding to basis state $|z_j\rangle$ is simply

$$q(z_j) = \| P_j|Z\rangle\|^2 = \|(|z_j\rangle\langle z_j| \cdot |Z\rangle)\|^2 = \|\alpha_j \cdot |z_j\rangle\|^2 = \|\alpha_j\|^2 . \tag{6}$$

The analogy between classic and quantum theory can be made even clearer if we work directly with the coordinates of the superposition state. According to quantum theory, the coordinates of the state vector $|Z\rangle$ with respect to the $|z_i\rangle$ basis forms an $N \times 1$ complex vector α. Also with respect to this basis, the projector $P_j = |z_j\rangle\langle z_j|$ is simply an $N \times N$ matrix full of zeros except a one on the diagonal for row j. Finally, the probability for the event A is simply

$$q(A) = \| P(A)|Z\rangle\|^2 = \sum_{j \in A} \|\alpha_j\|^2 . \tag{7}$$

2.3 Conditional Probabilities

Both classic and quantum probabilities revise (collapse, reduce) the state after observing an event. First consider state revision given by the conditional probability rule of classical probability theory. Suppose the vector π describes the probability distribution across the elementary events prior to a measurement. The probability distribution across the elementary events following an observation of event A equals the projection onto event A followed by normalization:

$$\pi|A = (P(A) \cdot \pi) / (1 \cdot P(A) \cdot \pi) . \tag{8}$$

Normalization guarantees that the elements of the new state vector sum to unity. All subsequent conditional probabilities are then computed from projections on the new mixed state $\pi|A$. Specifically, the probability of a new event B given A is

$$p(B|A) = 1 \cdot P(B) \cdot (\pi|A) = 1 \cdot (P(B) \cdot P(A) \cdot \pi)/(1 \cdot P(A) \cdot \pi) . \tag{9}$$

Next consider the state revision given by Luder's rule [14] of quantum probability theory. Suppose the vector $|Z\rangle$ describes the superposition state prior to a measurement. The state following an observation of event A equals the projection onto event A followed by normalization:

$$|Z|A\rangle = P(A)|Z\rangle / \|P(A)|Z\rangle\| . \tag{10}$$

Normalization guarantees that the new state vector has length equal to one. Probabilities of new events, conditioned on already observing event A, are then computed from projections on the new state vector $|Z|A\rangle$. The coefficients for this conditional state vector, with respect to the $|z_j\rangle$ basis, are defined by

$$\alpha|A = \mathbf{P}(A)\cdot\alpha / \|\mathbf{P}(A)\cdot\alpha\| . \tag{11}$$

For example the probability of a new event B given A is

$$q(B|A) = \|\mathbf{P}(B)\cdot(\alpha|A)\|^2 = \|\mathbf{P}(B)\cdot\mathbf{P}(A)\alpha\|^2/\|\mathbf{P}(A)\cdot\alpha\|^2 . \tag{12}$$

2.4 Inference Based on a Single Measurement

Suppose $T=1$ and one wishes to make an inference based on a single measure denoted X. Classic and quantum theories provide the same answers to this problem. First we present the classic Bayesian inference model (using projection operators), followed by the quantum inference model.

Suppose the measure X can take n different values, with x representing an arbitrary outcome. When we combine each possible outcome x of X with each of the m possible hypotheses, we obtain $N = m\cdot n$ unique elementary joint events such as $(h_i \wedge x)$. (Later this number will change.) The classic inference process starts with initial distribution over these N events represented by the $N \times 1$ vector π. The prior probability is given by

$$p(h_i) = \mathbf{1}\cdot\mathbf{P}(h_i)\cdot\pi , \tag{13}$$

where $\mathbf{P}(h_i)$ is a $N \times N$ matrix with ones on the diagonal corresponding to the rows matching hypothesis h_i.

The marginal probability equals

$$p(X(1) = x) = \mathbf{1}\cdot\mathbf{P}(X(1) = x)\cdot\pi , \tag{14}$$

where $\mathbf{P}(X(1)=x)$ is a $N \times N$ matrix with ones on the diagonal corresponding to the rows matching $X(1) = x$. The new state after observing $X(1) = x$ becomes

$$\pi|x = \mathbf{P}(X(1) = x)\cdot\pi / (\mathbf{1}\cdot\mathbf{P}(X(1) = x)\cdot\pi) . \tag{15}$$

If hypothesis h_i is known to be true then the new state is

$$\pi|h_i = \mathbf{P}(h_i)\cdot\pi / (\mathbf{1}\cdot\mathbf{P}(h_i)\cdot\pi) . \tag{16}$$

Finally, the likelihood is then given by

$$p(X(1) = x| h_i) = \mathbf{1}\cdot\mathbf{P}(X(1) = x)\cdot(\pi|h_i) . \tag{17}$$

Bayes inference rule follows from the definition of conditional probability:

$$p(h_i | X(1) = x_1) = \mathbf{1}\cdot\mathbf{P}(h_i)\cdot(\pi|x)$$

$$= \mathbf{1}\cdot\mathbf{P}(h_i)\cdot\mathbf{P}(X(1)=x)\cdot\pi / (\mathbf{1}\cdot\mathbf{P}(X(1)=x)\cdot\pi)$$

$$= \mathbf{1}\cdot\mathbf{P}(X(1)=x)\cdot\mathbf{P}(h_i)\cdot\pi / (\mathbf{1}\cdot\mathbf{P}(X(1)=x)\cdot\pi) . \tag{18}$$

Substituting $\pi|h_i = \mathbf{P}(h_i)\cdot\pi / (\mathbf{1}\cdot\mathbf{P}(h_i)\cdot\pi)$ into the above

$$p(h_i | X(1) = x_1) = (\mathbf{1}\cdot\mathbf{P}(h_i)\cdot\pi)\cdot(\mathbf{1}\cdot\mathbf{P}(X(1)=x)\cdot(\pi|h_i)) / (\mathbf{1}\cdot\mathbf{P}(X(1)=x)\cdot\pi)$$

$$= p(h_i) \cdot [\, p(X(1) = x \mid h_i) \, / \, p(X(1)=x) \,] \,. \tag{19}$$

Next we examine the quantum inference model for a single measure. We begin with an initial state defined on an N dimensional Hilbert space. This Hilbert space can be represented by $N = m \cdot n$ basis vectors such as $|h_i \wedge x\rangle$ representing the elementary event $(h_i \wedge x)$. (Later this dimension will change.) The initial state can be represented by the $N \times 1$ coordinate vector α with respect to this basis.

The prior probability of hypothesis h_i equals

$$q(h_i) = \|\mathbf{P}(h_i) \cdot \alpha\|^2 \,. \tag{20}$$

The marginal probability of event $X(1) = x$ is

$$q(X(1) = x) = \|\mathbf{P}(X(1) = x) \cdot \alpha\|^2 \,. \tag{21}$$

After observing first observation, $X(1)=x$, the initial state α changes to a new state

$$\alpha|x = \mathbf{P}(X(1)=x) \cdot \alpha \, / \, \|\mathbf{P}(X(1) = x) \cdot \alpha\|, \text{ and } \|\alpha|x\| = 1 \,. \tag{22}$$

Suppose we assume that h_i is true. Then the conditional state given h_i equals

$$\alpha|h_i = \mathbf{P}(h_i) \cdot \alpha \, / \|\mathbf{P}(h_i) \cdot \alpha\| \text{ and } \|\alpha|h_i\| = 1 \,. \tag{23}$$

If we assume that h_i is true, then the conditional probability of observing $X(1) = x$ equals

$$q(X(1) = x \mid h_i) = \|\mathbf{P}(X(1) = x) \cdot (\alpha|h_i)\|^2 \,. \tag{24}$$

Finally, quantum inference follows from Luder's rule [14]

$$\begin{aligned} q(h_i \mid x) &= \|\mathbf{P}(h_i) \cdot (\alpha|x)\|^2 \\ &= \|\mathbf{P}(h_i)\mathbf{P}(X(1) = x) \cdot \alpha\|^2 \, / \, \|\mathbf{P}(X(1) = x) \cdot \alpha\|^2 \\ &= \|\mathbf{P}(X(1) = x)\mathbf{P}(h_i) \cdot \alpha\|^2 \, / \, \|\mathbf{P}(X(1) = x) \cdot \alpha\|^2 \,. \end{aligned} \tag{25}$$

Substituting $\mathbf{P}(h_i) \cdot \alpha = (\alpha|h_i) \cdot \|\mathbf{P}(h_i) \cdot \alpha\|$ yields

$$\begin{aligned} q(h_i \mid x) &= \|\mathbf{P}(h_i) \cdot \alpha\|^2 \cdot \|\mathbf{P}(X(1)=x) \cdot (\alpha|h_i)\|^2 \, / \|\mathbf{P}(X(1)=x) \cdot \alpha\|^2 \\ &= q(h_i) \cdot [\, q(\, X(1) = x \mid h_i) \, / \, q(\, X(1) = x) \,] \,. \end{aligned} \tag{26}$$

This is identical to Bayes rule if the classic probability function p replaces the quantum probability function q.

3 Representation of Measurements

3.1 Change of Basis Vectors

A key issue arises from the idea of using a Hilbert space representation of events. One can choose different sets of basis vectors for spanning a Hilbert space. Two different sets of basis vectors are related by a unitary transformation, denoted $\mathbf{U}$ with $\mathbf{U}\mathbf{U}^\dagger = \mathbf{U}^\dagger\mathbf{U} = \mathbf{I}_N$:

$$\{|z_1\rangle\,,\,...,\,|z_i\rangle\,,\,...,\,|z_N\rangle\} = \{U|z'_1\rangle\,,\,...,U|z'_i\rangle\,,\,...,\,U|z'_N\rangle\}\,,$$

$$\{|z'_1\rangle\,,\,...,\,|z'_i\rangle\,,\,...,\,|z'_N\rangle\} = \{U^\dagger|z_1\rangle\,,\,...,U^\dagger|z_i\rangle\,,\,...,\,U^\dagger|z_N\rangle\}\,. \tag{27}$$

A state vector can be expressed with respect to either one of these two sets of basis vectors:

$$|Z\rangle = I_N \cdot |Z\rangle = (\textstyle\sum_{j\in H} |z_j\rangle\langle z_j|) \cdot |Z\rangle = \sum \langle z_j|Z\rangle|z_j\rangle = \sum \alpha_j \cdot |z_j\rangle\,,$$

$$= I_N \cdot |Z\rangle = (\textstyle\sum_{j\in H} |z'_j\rangle\langle z'_j|) \cdot |Z\rangle = \sum \langle z'_j|Z\rangle|z'_j\rangle = \sum \beta_j \cdot |z'_j\rangle\,. \tag{28}$$

The coordinates of the state vector $|Z\rangle$ with respect to the $|z_i\rangle$ basis forms a $N \times 1$ complex vector α, and the probability of elementary event z_i is $\|\alpha_i\|^2$. The coordinates of $|Z\rangle$ with respect to the $|z'_i\rangle$ basis forms a $N \times 1$ complex vector denoted β, and the probability of elementary event z_i is $\|\beta_i\|^2$. These coordinates are related by a unitary matrix $U = [u_{ij}] = [\langle z'_i|z_j\rangle] = [\langle z_i|U|z_j\rangle]$: $\beta = U \cdot \alpha$, and $\alpha = U^\dagger \cdot \beta$.

By changing basis vectors, we change the nature of the set of elementary events under consideration. In particular, the $|z_i\rangle$ basis is needed to define elementary events $\{z_1, z_2, ..., z_N\}$, and the projector $|z_j\rangle\langle z_j|$ defines event z_j; but $|z'_i\rangle$ is needed to define elementary events $\{z'_1, z'_2, ..., z'_N\}$, and the projector $|z'_j\rangle\langle z'_j|$ defines event z'_j. In other words, if we ask questions about the elementary events $\{z_1, z_2, ..., z_N\}$, then we need to use the α coordinates to compute probabilities, and $\|\alpha_i\|^2$ determines the probability of elementary event z_i; but if we ask questions about the elementary events $\{z'_1, z'_2, ..., z'_N\}$, then we need to use the β coordinates to compute probabilities and $\|\beta_i\|^2$ determines the probability of elementary event z'_i.

3.2 Compatibility of Measures

The concept of compatibility is unique to quantum theory. It is concerns the possible disturbing effect of one measure, say X, on another measure, say Y. We could take these measurements in two different orders: X first followed by Y, or Y first followed by X. If the probability of the two events produced by the two measurements does not depend on the order, then these two measures are compatible; otherwise they are incompatible [5]. Human judgments frequently exhibit order effects, hence the concern for compatibility.

One measure is labeled X and it yields an event such as A = $(X=x)$ where x is one of the n possible outcomes produced by X. We assume that these $n(X)$ outcomes correspond to a set of $n(X)$ orthogonal projectors $\{P(X = x_1),...,P(X = x),...,P(X = x_{n(X)})\}$ operating on the N dimensional Hilbert space that forms a complete set so that

$$P(X = x) \cdot P(X=y) = 0 \text{ for } x \neq y, \textstyle\sum_x P(X = x) = I_N\,. \tag{29}$$

Suppose the A = $(X=x)$ event uses the $|z_j\rangle$ set of basis vectors for its definition and corresponds to the projector

$$P(X = x) = \textstyle\sum_{j\in A} |z_j\rangle\langle z_j|\,. \tag{30}$$

The other measure is labeled Y and it yields an event such as B = $(Y = y)$ where y is one of the $n(Y)$ possible outcomes produced by Y. Again we assume that these $n(Y)$

outcomes correspond to a set of $n(Y)$ orthogonal projectors $\{P(Y=y_1),...,P(Y = y),...,P(Y = y_{n(Y)})\}$ that forms a complete set so that

$$P(Y = x) \cdot P(Y=y) = 0 \text{ for } x \neq y, \sum_y P(Y = y) = I_N. \tag{31}$$

The $B = (Y=y)$ event requires the $|z'_j\rangle$ set of basis vectors for its definition and corresponds to the projector

$$P(Y = y) = \sum_{j \in B} |z'_j\rangle\langle z'_j| = \sum_{j \in B} U^\dagger |z_j\rangle\langle z_j| U. \tag{32}$$

The probability of the A followed by B sequence is

$$\| P(Y = y) \cdot P(X = x) \cdot |Z\rangle \|^2. \tag{33}$$

The probability of the opposite sequence of events is

$$\| P(X = x) \cdot P(Y = y) \cdot |Z\rangle \|^2. \tag{34}$$

Two measures are said to be compatible if the probability distribution over the joint outcomes from the two measures does not depend on the order of measurement; otherwise they are incompatible.

These two sequences give the same probability for any arbitrary state vector $|Z\rangle$ and any pair of outcomes x and y if and only if the commutator is always zero:

$$[P(X = x) \cdot P(Y = y) - P(Y = y) \cdot P(X = x)] = 0. \tag{35}$$

If the commutator is zero for all pairs of values that can be observed on the two measures, then the two measures are compatible; otherwise they are incompatible. The commutator is always zero when

$$(|z_j\rangle\langle z_j|) \cdot (|z'_k\rangle\langle z'_k|) - (|z'_k\rangle\langle z'_k|) \cdot (|z_j\rangle\langle z_j|) = 0 \tag{36}$$

for all pairs, which holds when $\langle z_j | z'_k \rangle = 0$ for $j \neq k$ and $\langle z_j | z'_j \rangle = 1$. This implies that $U = I_N$, or in other words, the basis set $|z_j\rangle = U \cdot |z'_j\rangle$ used for X is identical to the basis set $|z'_j\rangle$ used for Y. If $U \neq I_N$ then the measure X will be incompatible with the measure Y.

If the measures are compatible, then we can define the joint event

$$P(X=x \wedge Y=y) = P(X=x) \cdot P(Y=y) = P(Y=y) \cdot P(X=x). \tag{37}$$

This forms a new complete set of $n(X) \cdot n(Y)$ orthogonal projectors $\{..., P(X=x \wedge Y=y), ...\}$ so that $P(X=x \wedge Y=y) \cdot P(X=u \wedge Y=v) = 0$, $x \neq u$ or $y \neq v$, $\sum_x \sum_y P(X=x \wedge Y=y) = I_N$. These projectors are then used to define the joint probabilities for these two measures

$$q(X=x \wedge Y=y) = \| P(X=x \wedge Y=y)|Z\rangle \|^2. \tag{38}$$

Classic probability theory assumes that it is always possible to define these joint probabilities between measures. However, in quantum theory, this joint probability does not exist for incompatible measures.

When two measures are compatible, then the first measure does not disturb or affect the second measure, order of measurement does not matter, and both measures can be determined simultaneously. However, when two measures are incompatible, then determining the value of one measure necessarily makes the values of the other measure uncertain. To see how this uncertainty principle arises with incompatible measures, suppose the inference state $|Z\rangle$ is placed at the following point after a measurement:

$$|Z|x\rangle = \mathbf{P}(X = x)|Z\rangle / \|\mathbf{P}(X = x)|Z\rangle\| . \tag{39}$$

We can express this state using the coordinates defined by $|z_j\rangle$ as follows:

$$(\alpha|x) = \mathbf{P}(X = x)\cdot\alpha / \|\mathbf{P}(X = x)\cdot\alpha\|, \text{ and so } \|\alpha|x\|^2 = 1 . \tag{40}$$

Here $\mathbf{P}(X = x)$ is the matrix representation of the projector with respect to the $|z_j\rangle$ basis (it is simply a matrix with zeros everywhere except for ones on the diagonal in the rows corresponding to combinations that satisfy $X = x$). Given this state, the outcome x is certain to occur again with measure X:

$$q(X = x) = \|\mathbf{P}(X = x)\cdot(\alpha|x)\|^2$$
$$= \|\mathbf{P}(X = x)^2\cdot\alpha\|^2 / \|\mathbf{P}(X = x)\cdot\alpha\|^2 = 1 . \tag{41}$$

Now let us examine this same state using the coordinates defined by $|z'_j\rangle$:
$(\beta|x) = U\cdot(\alpha|x)$, and note that $\|(\beta|x)\|^2 = \|U\cdot(\alpha|x)\|^2 = 1$ because U is unitary. The probability of the outcome y for the measure Y is determined by

$$q(Y = y) = \|\mathbf{P}(Y = y)\cdot(\beta|x)\|^2 \tag{42}$$

where $\mathbf{P}(Y = y)$ is the matrix representation with respect to the $|z'_j\rangle$ basis. That is, it is a matrix with zeros everywhere except ones on the diagonal for rows that satisfy $Y = y$. Now we find that $q(Y = y) = \|\mathbf{P}(Y=y)\cdot(\beta|x)\|^2 < 1$ because $\|(\beta|x)\|^2 = 1$ and $\mathbf{P}(Y=y)$ is a projection on a subspace of H.

In other words, if X and Y are incompatible, and if we are certain about the outcome that X will produce, then we must be uncertain about the outcome produced by Y.

3.3 Constructing a Hilbert Space Representation

We construct our Hilbert space using the principles initially described by Dirac [5]. The dimension, N, of the Hilbert space used to represent all T measures is determined by a maximum number $K \le T$ of mutually compatible measures. Incompatible measures, being unitary transformations of a set of compatible measures, remain within the same space, and so they do not increase the dimensionality of the Hilbert space.

If all of the measures are compatible with each other, and $K = T$, then we can use the same set of basis vectors to represent events for all T of the measures. This is exactly the key assumption of classical probability theory. In fact, quantum probability assigns the same probabilities to all of the events as classical probability when all of the measures are compatible.

Hereafter we will assume that there are only $K \le T$ compatible measures labeled $\{X_1,...,X_k,...,X_K\}$. As described earlier, an incompatible measure Y_k can be constructed from one of the compatible set X_k by a unitary transformation of the basis.

Any given measure from the compatible set, say X_k, has $n(k)$ possible outcomes $\{x_1,x_2,...,x_k,...,x_{n(k)}\}$. This produces a total of $n = [n(1)\cdot n(2)\cdot...\cdot n(k)\cdot...\cdot n(K)]$ combinations of possible outcomes from all K compatible measures, such as $z_j = (X_1=x_1)\wedge...\wedge(X_k=x_k)\wedge...\wedge(X_K=x_K)$.

To model inference, we also need to include the m possible hypotheses, $\{h_1,\ldots,h_m\}$. We assume the hypotheses are compatible with all of the measures. Combining each hypothesis with each combination of measurement outcomes defines an elementary event, and so this produces a set of $N = m \cdot n$ elementary events, with typical element $(h_i \wedge z_j) = h_i \wedge (X_1=x_1) \wedge \ldots \wedge (X_K=x_K)$. These events are represented as subspaces (rays) of an N dimensional Hilbert space.

The $n(k)$ outcomes produced by X_k are represented by a complete set of $n(k)$ orthogonal projectors $\{\ldots,\mathbf{P}(X_k=x),\ldots\}$, $\mathbf{P}(X_k=x)\cdot\mathbf{P}(X_k=y)=\mathbf{0}$, $x\neq y$, $\sum_x \mathbf{P}(X_k=x) = \mathbf{I}_N$. There is one such set of projectors from each measure, and all K compatible measures are defined by the same basis. Finally, each hypothesis is represented by a set of orthogonal projectors $\{\ldots,\mathbf{P}(h_i),\ldots\}$, $\mathbf{P}(h_i)\cdot\mathbf{P}(h_j) = \mathbf{0}$, $i \neq j$, $\sum_i \mathbf{P}(h_i) = \mathbf{I}_N$. The projectors for the hypotheses are defined by the same basis as the projectors for the compatible measures.

The projectors for the outcomes of each measure can be combined to form a projector for each combination of outcomes. The projector for an elementary event $(h_i \wedge z_j)$ is $\mathbf{P}_{ij} = \mathbf{P}(h_i)\cdot\mathbf{P}(X_1=x_1)\cdot\ldots\cdot\mathbf{P}(X_k=x_k)\cdot\ldots\cdot\mathbf{P}(X_K=x_K)$. This forms a complete set of N orthogonal projectors: $\{\ldots,\mathbf{P}_{ij}\ldots\}$, $\mathbf{P}_{ij}\cdot\mathbf{P}_{i'j'} = \mathbf{0}$, $ij \neq i'j'$, $\sum_i \sum_j \mathbf{P}_{ij} = \mathbf{I}_N$. Each projector $\mathbf{P}_{ij}$ for an elementary event has a single and unique (unit length) eigenvector

$$|h_i \wedge z_i\rangle = |h_i \wedge (X_1=x_1)\wedge\ldots\wedge(X_k=x_k)\wedge\ldots\wedge(X_K=x_K)\rangle . \tag{43}$$

These basis vectors span the $N \times 1$ dimensional Hilbert space denoted H.

It is useful to construct each basis vector $|h_i \wedge z_j\rangle$ of H from a tensor product of vectors representing the K compatible measures and the hypotheses as follows. Consider the measure X_k that has $n(k)$ possible values. For this measure, we can define a set of $n(k)$ orthonormal vectors $\{|X_k=x_1\rangle,\ldots,|X_k=x\rangle,\ldots,|X_k=x_{n(k)}\rangle\}$, where each possible outcome, say x, from the measure X_k corresponds to a normalized vector $|X_k=x\rangle$. The set of orthonormal vectors for X_k spans an $n(k) \times 1$ subspace H_k. Then the tensor product of the individual measurement vectors produces a vector representing a combination of measurement outcomes:

$$|z_j\rangle = |X_1=x_1\rangle\otimes\ldots\otimes|X_k=x_k\rangle\otimes\ldots\otimes |X_K=x_K\rangle$$

$$= | (X_1=x_1)\wedge\ldots\wedge(X_k=x_k)\wedge\ldots\wedge(X_K=x_K)\rangle . \tag{44}$$

There are total of n such vectors which form an orthonormal set that spans an n dimensional tensor product space $H_1\otimes H_2\otimes\ldots\otimes H_k\otimes\ldots\otimes H_K$.

The hypotheses are represented by a set of orthonormal vectors $\{|h_1\rangle, \ldots,|h_i\rangle,\ldots,|h_m\rangle\}$. This orthonormal set forms an m-dimensional Hilbert space denoted H_m. Then each basis vector of H can be constructed from a tensor product of vectors from each measure and hypothesis:

$$|h_i \wedge z_j\rangle = |h_i\rangle\otimes|z_j\rangle$$

$$= |h_i\rangle\otimes|X_1=x_1\rangle\otimes\ldots\otimes|X_k=x_k\rangle\otimes\ldots\otimes |X_K=x_K\rangle$$

$$= |h_i \wedge (X_1=x_1)\wedge\ldots\wedge(X_k=x_k)\wedge\ldots\wedge(X_K=x_K)\rangle . \tag{45}$$

This spans an N dimensional tensor product space $H = H_m \otimes H_1 \otimes ... \otimes H_k \otimes ... \otimes H_K$. Finally the state of the quantum system is defined as

$$|\psi(t)\rangle = \sum \sum \psi_{ij}(t) \cdot |h_i\rangle \otimes |z_j\rangle . \tag{46}$$

The $N \times 1$ vector of coefficients, $\psi(t)$, represents the state with respect to the basis formed by $|h_i\rangle \otimes |z_j\rangle$. So if we wish to compute the joint probability that hypothesis h_i is true and that elementary event z_j occurs, then this is given by

$$\|\mathbf{P}(h_i \wedge z_j) \cdot \psi(t)\|^2 = \|\psi_{ij}(t)\|^2 , \tag{47}$$

where $\mathbf{P}(h_j \wedge z_j)$ is a $N \times N$ matrix with zeros everywhere except a one on the diagonal in the row corresponding to the event $(h_j \wedge z_j)$.

4 Quantum Inference

4.1 Quantum Inference after the First Measurement

The first measure to be selected is denoted $X(1)$. If this is one of the measures in the compatible set $\{X_1,...,X_k,...,X_K\}$ then we proceed by using the basis with vectors $|z_j(1)\rangle = |X_1=x_1\rangle \otimes ... \otimes |X_k=x_k\rangle \otimes ... \otimes |X_K=x_K\rangle$.

If the first measure is incompatible with one of the measures in compatible the set $\{X_1,...,X_k,...,X_K\}$, then we need to change the basis by applying a unitary transformation. Suppose the first measure is Y_k which is incompatible with X_k. Then the basis vectors for these elementary events are defined by

$$|z_j(1)\rangle = |X_1=x_1\rangle \otimes ... \otimes |Y_k=y_k\rangle \otimes ... \otimes |X_K=x_K\rangle$$
$$= |X_1=x_1\rangle \otimes ... \otimes U^{\dagger}_k|X_k=x_k\rangle \otimes ... \otimes |X_K=x_K\rangle . \tag{48}$$

where we have replaced $(X_k=x_k)$ with $(Y_k=y_k)$. In other words, we start with a basis $|z_j(1)\rangle$ that is defined by the first measure. The initial state is represented by:

$$|\psi(0)\rangle = \sum \sum \psi_{ij}(0) \cdot |h_i\rangle \otimes |z_j(1)\rangle . \tag{49}$$

The $m \cdot N$ vector of coordinates $\psi(0)$ represents the state with respect to this initial basis.

In section 2.4, we calculated the quantum inference based on a single measurement. By letting $\alpha = \psi(0)$ in these calculations we have the inference after the first observation for the initial state $\psi(0)$.

4.2 Quantum Inference after a Second Measurement

Suppose we take another measurement $X(2) = x_2$ which defines our second event. If it is compatible with the first measure, $X(1)$, then we simply continue working with the same basis by setting $\psi(1|x_1) = \alpha(1|x_1)$, where $\alpha(1|x_1)$ is the new state defined by Equation 22. If it is incompatible with the first measure, then we need to change coordinates.

Suppose the first measure was chosen to be $X(1) = X_k$ and the coefficients for $X(1)$ are initially expressed in terms of the $|z_j(1)\rangle$ basis with coordinates $\alpha(1|x_1)$ given by Equation 22. Now suppose the second measure is $X(2) = Y_k$, which is incompatible with X_k. So we need to change the coordinates from α which are defined by the $|z_j(1)\rangle$ basis to β which are defined with respect to the new basis: $\beta(1|x_1) = (I\otimes...\otimes I\otimes U_k\otimes I\otimes...I)\cdot\alpha(1|x_1)$. Finally we set $\psi(1|x_1) = \beta(1|x_1)$ and continue as before. The projector for $P(X(2) = x_2)$ is simply an $N \times N$ matrix with zeros everywhere except ones on the diagonals of the rows that correspond to the event $X(2) = x_2$. $P(h_i)$ is simply an $N \times N$ matrix with zeros everywhere except ones on the diagonals of the rows that correspond to hypothesis h_i.

The prior after the first measure but before the second observation equals

$$q(h_i|x_1) = \|P(h_i)\cdot\psi(1|x_1)\|^2 . \tag{50}$$

The probability of event $X(2) = x_2$ given $X(1)=x_1$ is

$$q(X(2)=x_2|X(1)=x_1) = \|P(X(2) = x_2)\psi(1|x_1)\|^2 . \tag{51}$$

After observing the second observation, $X(2)=x_2$, the state $\psi(1|x_1)$ changes to a new state

$$\alpha(2|x_1, x_2) = P(X(2)=x_2)\psi(1|x_1)/\|P(X(2) = x_2)\psi(1|x_1)\| \tag{52}$$

and $\|\alpha(2| x_1, x_2)\| = 1$.

Suppose we assume that h_i is true. Then the conditional state given h_i and $X(1)=x_1$ equals

$$\psi(1|x_1, h_i) = P(h_i)\psi(1|x_1)/\|P(h_i)\psi(1|x_1)\| \tag{53}$$

and $\|\psi(1|x_1, h_i)\| = 1$.

If h_i is true, and we already observed $X(1) = x_1$, then the conditional probability given of observing $X(2) = x_2$ equals

$$q(X(2)=x_2 |x_1, h_i) = \|P(X(2) = x_2)\psi(1|x_1, h_i)\|^2 . \tag{54}$$

Finally, our inference after the second observation equals

$$\begin{aligned}
q(h_i \mid x_1, x_2) &= \|P(h_i)\alpha(2|x_1, x_2)\|^2 \\
&= \|P(h_i)P(X(2)=x_2)\psi(1|x_1)\|^2 / \|P(X(2)=x_2)\psi(1|x_1)\|^2 \\
&= \|P(X(2)=x_2)P(h_i)\psi(1|x_1)\|^2 / \|P(X(2)=x_2)\psi(1|x_1)\|^2 \\
&= \|P(h_i)\psi(1|x_1)\|^2 \times \|P(X(2)=x_2)\psi(1|x_1, h_i)\|^2 \div \|P(X(2)=x_2)\psi(1|x_1)\|^2 \\
&= q(h_i|x_1) \cdot [q(X(2) = x_2 |x_1, h_i) / q(X(2) = x_2 | x_1)] .
\end{aligned} \tag{55}$$

Once again, this corresponds with Bayes rule.

4.3 Quantum Inference after Several Observations

A new measure is denoted $X(t+1)$. The process continues with the coefficients produced by the last measurement, $\alpha(t|x_1, ...,x_t)$. For example, if $t+1=3$, then we would start with Equation 52. If $X(t+1)$ is compatible with $X(t)$ then we use the same basis states as used for the last measurement. That is, we continue using the same coordinates $\psi(t|x_1, ...,x_t) = \alpha(t|x_1, ...,x_t)$. If $X(t+1) = Y_k$ is incompatible with $X(t)$, then we transform the coordinates to the new basis for the new measure:

$\beta(t|x_1, ...,x_t) = (\mathbf{I}\otimes...\otimes U_j\otimes...\otimes\mathbf{I})\cdot\alpha(t|x_1, ...,x_t)$. In this case, we express the coordinates of the current inference state as $\psi(t|x_1, ...,x_t) = \beta(t|x_1, ...,x_t)$.

The prior after the first measure but before the second observation equals

$$q(h_i|x_1,...,x_t) = \|\mathbf{P}(h_i)\cdot\psi(t|x_1,...,x_t)\|^2 . \tag{56}$$

The probability of event $X(t+1) = x_{t+1}$ given the previous history is

$$q(X(t+1)=x_{t+1}|x_1,...,x_t) = \|\mathbf{P}(X(2) = x_2)\psi(t|x_1,...,x_t)\|^2 . \tag{57}$$

After observing the next observation, $X(t+1)=x_{t+1}$, the state $\psi(t|x_1,...,x_t)$ changes to a new state

$$\alpha(t+1|x_1, ...,x_{t+1}) = \mathbf{P}(X(t+1)=x_{t+1})\psi(t|x_1,...,x_t) \div \|\mathbf{P}(X(t+1)=x_{t+1})\psi(t|x_1,...,x_t)\| \tag{58}$$

and $\|\alpha(t+1| x_1,..., x_{t+1})\| = 1$.

Suppose we assume that h_i is true. Then the conditional state given h_i and the past history equals

$$\psi(t|x_1, ...,x_t, h_i) = \mathbf{P}(h_i)\psi(t|x_1,...,x_t)/\|\mathbf{P}(h_i)\psi(t|x_1,...,x_t)\| \tag{59}$$

and $\|\psi(t|x_1,...,x_t, h_i)\| = 1$. If h_i is true, and we already observed a history of events, then the conditional probability of $X(t+1) = x_{t+1}$ given h_i and this history equals

$$q(X(t+1)=x_{t+1}|x_1,...,x_t,h_i) = \|\mathbf{P}(X(t+1)=x_{t+1})\cdot\psi(t|x_1,...,x_t, h_i)\|^2 . \tag{60}$$

Finally, our inference after the next observation equals

$$q(h_i \mid x_1, ...,x_{t+1}) = \|\mathbf{P}(h_i)\alpha(t+1|x_1,..., x_{t+1})\|^2$$

$$= \|\mathbf{P}(h_i)\mathbf{P}(X(t+1)=x_{t+1})\cdot\psi(t|x_1,...,x_t)\|^2 \div \|\mathbf{P}(X(t+1)=x_{t+1})\psi(t|x_1,...,x_t)\|^2$$

$$= \|\mathbf{P}(X(t+1)=x_{t+1})\mathbf{P}(h_i)\cdot\psi(t|x_1,...,x_t)\|^2 \div \|\mathbf{P}(X(t+1)=x_{t+1})\cdot\psi(t|x_1,...,x_t)\|^2$$

$$= \|\mathbf{P}(h_i)\psi(t|x_1,...,x_t)\|^2 \times \|\mathbf{P}(X(t+1)=x_{t+1})\psi(t|x_1,...,x_t,h_i)\|^2$$
$$\div \|\mathbf{P}(X(t+1)=x_{t+1})\cdot\psi(t|x_1,...,x_t)\|^2$$

$$= q(h_i|x_1,...,x_t) \times [q(X(t+1)=x_{t+1}|x_1,...,x_t,h_i) \div q(X(t+1)=x_{t+1}| x_1,...,x_t)] . \tag{61}$$

Again, this corresponds with Bayes rule if the classic probability function p replaces the quantum probability function q.

5 Summary and Concluding Comments

This paper began with the assumption that the abstract mathematical basis of quantum theory is not tied to physics per se, but rather it can be used as a generalized probability theory with meaningful applications outside of physics. If so, it should be applicable to probabilistic inference problems.

If we do this, we find that quantum inferences are updated in manner that correspond exactly to Bayesian updating except that the coordinates of the state must be transformed by unitary matrices to coordinates of a different a basis for changes between incompatible measurements. Determining the unitary matrices that transform from one set of coordinates to another is a critical step that remains to be achieved for applications outside of physics.

Quantum inference is identical to Bayesian inference when only compatible measures are involved. But quantum inference can depart dramatically from Bayesian inference when incompatible measurements are involved. In particular, one can start out certain about a particular value of a measure, but if this is followed later by an incompatible measure, then one will become uncertain again about the value of the earlier (certain) measure. This results from the disturbance of one incompatible measure on another.

When all the measures are compatible, we have one set of elementary events and this forms a single Boolean algebra of events. When incompatible measures are involved, we need to define different incompatible sets of elementary events, which correspond to different sets of basis vectors within the same Hilbert space. These sets of events cannot be combined into a single comprehensive set of events using Boolean logic.[1] Thus we are forced to work with a partial Boolean algebra of events.

So the most crucial question is whether incompatible measurements occur outside of physics. There is clear evidence that one type of human judgment can disturb another and the order of human judgments changes the probabilities [3]. This suggests that it may be fruitful to employ quantum probabilities when human judgments are involved.

Cognitive psychologists have attempted to describe the disturbing effect of one judgment on another by building cognitive models that describe a separate probability distribution for each order. However, they have implicitly assumed a partial Boolean algebra to formulate these models, and thus these models are not really consistent with classic probability theory either. An important empirical question is whether simpler yet more generalizable probabilistic models can be found using quantum probabilities. The success in physics suggests this may be the case. This remains to be seen outside of physics (but see, [1], [11]).

Acknowledgments. This research was supported by NSF SES-0753164 and SES-0753168. Thanks for the IU quantum group (Amr Sabry, Larry Moss, Andrew Hanson, Gerardo Ortiz, Michael Dunn) for discussions of the ideas described in this paper.

References

1. Aerts, D., Broekaert, J., Gabora, L.: A Case for Applying an Abstracted Quantum Formalism to Cognition. In: Campbell, R. (ed.) Mind in Interaction. John Benjamin, Amsterdam (2003)
2. Bordley, R.F.: Quantum Mechanical and Human Violations of Compound Probability Principles: Toward a Generalized Heisenberg Uncertainty Principle. Operations Research 46, 923–926 (1998)
3. Busemeyer, J.R., Wang, Z., Townsend, J.T.: Quantum Dynamics of Human Decision Making. Journal of Mathematical Psychology 50(3), 220–241 (2006)

---

[1] The events from incompatible measures follow quantum logic (not discussed here), which obeys all of the rules of Boolean logic except for the distributive axiom.

4. Degroot, M.H.: Optimal Statistical Decisions. McGraw-Hill, New York (1970)
5. Dirac, P.A.M.: The Principles of Quantum Mechanics. Oxford University Press, Oxford (1930, 2001)
6. Franco, R.: The Conjunction Fallacy and Interference Effects. Journal of Mathematical Psychology (2007) (to appear), arXiv:0708.3948v1 [physics.gen-ph]
7. Guder, S.P.: Stochastic Methods in Quantum Mechanics. Dover Press (1979)
8. Hayashi, M.: Quantum Information Theory: An Introduction. Springer, Berlin (2006)
9. Helstrom, C.W.: Quantum Detection and Estimation Theory. Academic Press, New York (1976)
10. Holevo, A.S.: Statistical Structure of Quantum Mechanics. Springer, Berlin (2001)
11. Khrennikov, A.: Can Quantum Information be Processed by Macroscopic Systems? Quantum Information Theory (in press, 2007)
12. Kolmogorov, A.N.: Foundations of the Theory of Probability. Chelsea Publishing Co., New York (1950)
13. Laskey, K.B.: Quantum Causal Networks. In: Bruza, P.D., Lawless, W., van Rijsbergen, C.J., Sofge, D. (eds.) Proceedings of the AAAI Spring Symposium on Quantum Interaction, Stanford University, March 27-29. AAAI Press, Menlo Park (2007)
14. Luders, G.: Uber die Zustandsanderung durch den Messprozess. Annalen der Physik 8, 322–328 (1951)
15. La Mura, P., Swiatczak, L.: Markov Entanglement Networks. In: Bruza, P.D., Lawless, W., van Rijsbergen, C.J., Sofge, D. (eds.) Proceedings of the AAAI Spring Symposium on Quantum Interaction, Stanford University, March 27-29. AAAI Press, Menlo Park (2007)
16. Pitowski, I.: Quantum Probability, Quantum logic. Lecture Notes in Physics, vol. 321. Springer, Heidelberg (1989)
17. Tucci, R.R.: Quantum Bayesian Nets (1997) quantph/9706039
18. Von Neumann, J.: Mathematical Foundations of Quantum Theory. Princeton University Press, Princeton (1932)

Quantum-Like Representation of Macroscopic Configurations

Andrei Khrennikov*

International Center for Mathematical Modeling
in Physics and Cognitive Sciences
University of Växjö, S-35195, Sweden
Andrei.Khrennikov@vxu.se

Abstract. The aim of this paper is to use a contextual probabilistic model (in the spirit of Mackey, Gudder, Ballentine) to represent and to generalize some results of quantum logic on macroscopic quantum-like (QL) behaviour. The crucial point is that our model provides the QL-representation of macroscopic configurations in terms of complex probability amplitudes – wave functions of such configurations. Thus, instead of the language of propositions which is common in quantum logic, we use the language of wave functions which common in the conventional presentation of QM.

Keywords: Contextual probabilistic model, complex Hilbert space, hyperbolic Hilbert space, quantum-like representation.

1   Introduction

One should sharply distinguish QM as a physical theory and the mathematical formalism of QM. In the same way as one should distinguish classical Newtonian mechanics and its mathematical formalism. Nobody is surprised that the differential and integral calculi which are basic in Newtonian mechanics can be fruitfully applied in other domains of science. Unfortunately, the situation with the mathematical formalism of QM is essentially more complicated – some purely mathematical features of QM are identified with features of quantum physical systems. Although already Niels Bohr pointed to the possibility to apply the mathematical formalism of QM outside of physics, prejudice based on the identification of mathematics and physics still survives, but cf., e.g., Accardi, Aerts et al, Ballentine, Coecke, Czachor, D'Hooghe, De Muynck, Grib et al., Gudder, Landé, Mackey, Pykacz et al., detailed bibliography can be found in new author's book [1]. We can point out to a few applications of the quantum formalism outside of physics, [2]. Here we discuss not *reductionist models* in that the quantum description appears as a consequence of the evident fact that any physical system, even living (for example, the brain), is composed of quantum particles, but

---

* This paper was written during author's visit to Technical University of Copenhagen, Lungby, May 2008.

P. Bruza et al. (Eds.): QI 2009, LNAI 5494, pp. 44–58, 2009.
© Springer-Verlag Berlin Heidelberg 2009

really the possibility to use the mathematical formalism of QM without direct coupling with quantum physics.

We remark that importance of mentioned separation between quantum physics and quantum mathematics has been already well recognized in *quantum logic*. In particular, an exiting possibility to apply quantum mathematics to macroscopic systems is not surprising for quantum logicians.

Recently I developed so called contextual probability theory [1] which was inspired essentially by quantum logic and quantum probability. The main distinguishing feature of the theory of contextual probabilities is a possibility to derive the complex probability amplitude, the wave function, from probabilistic data. Such an algorithm for transfer of probabilistic data into the complex probability amplitude is presented in [1], *quantum-like representation algorithm* – QLRA.

The aim of this paper is to use our contextual probabilistic model, the *Växjö model,* to represent and to generalize some results of quantum logic on macroscopic quantum-like (QL) behaviour in terms of complex probability amplitudes. On the one hand, it may be interesting for physicists, since some rather mystical quantum features will be illustrated on the basis of behavior of macroscopic systems. On the other hand, the approach developed in this paper may be used e.g. in biology, sociology, or psychology. Our example of a QL-representation of hidden macroscopic configurations can find natural applications in those domains of science.

The basic example which we would like to generalize in the contextual probabilistic framework is well known in quantum logic. This is "firefly in the box". It was proposed by Foulis who wanted to show that a macroscopic system, firefly, can exhibit a QL-behavior which can be naturally represented in terms of quantum logics, [3].

We can mention some consequences of our QL-representation of macroscopic configurations for foundations of quantum physics. For such macroscopic models the QL-description is not complete. Thus hidden variables exist, but they could not be observed on the basis of available observables. Nevertheless, a kind of "Einstein demon" can observe behavior of hidden variables.

2   Firefly in the Box

We recall the well known example [3] by emphasizing its probabilistic structure. Let us consider a box which is divided into four sub-boxes. These small boxes which are denoted by $\omega_1, \omega_2, \omega_3, \omega_4$ provides internal description. These elements are available for a "Einstein demon", but they are not available for some external observable.

We consider the Kolmogorov probability space: $\Omega = \{\omega_1, \omega_2, \omega_3, \omega_4\}$, the algebra of all finite subsets $\mathcal{F}$ of Ω and a probability measure determined by probabilities $\mathbf{P}(\omega_j) = p_j$, where $0 < p_j < 1, p_1 + ... + p_4 = 1$.

We now consider two different disjoint partitions of the set Ω :

$$C_{\alpha_1} = \{\omega_1, \omega_2\}, C_{\alpha_2} = \{\omega_3, \omega_4\},$$

Fig. 1. Internal description

$$C_{\beta_1} = \{\omega_1, \omega_3\}, C_{\beta_2} = \{\omega_2, \omega_4\}.$$

We can obtain such partitions by dividing the box: a) into two equal parts by the vertical line: the left-hand part gives C_{α_1} and the right-hand part C_{α_2}; b) into two equal parts by the horizontal line: the top part gives C_{β_1} and the bottom part C_{β_2}.

We introduce two random variables corresponding to these partitions: $\xi_a(\omega) = \alpha_i$, if $\omega \in C_{\alpha_i}$ and $\xi_b(\omega) = \beta_i \in$ if $\omega \in C_{\beta_i}$. Here α_i and β_i are arbitrary labels. Suppose now that the external observer is able to measure only these two variables, denote the corresponding observables by the symbols a and b. We remark that there exist other random variables, they are available for the "Einstein demon", but not for the external observer. Roughly speaking elements ω_j are not visible for the latter observer. They are "hidden variables."

Such a model was illustrated (from the viewpoint of quantum logic) by the following example [3]. Let us consider a firefly in the box. It has definite position in space. The firefly position can be seen by a "Einstein demon" living inside this box.

Now we consider an external observer who has only two possibilities to observe the firefly in the box:

Fig. 2. The a-observable

Fig. 3. The b-observable

1) to open a small window at the point a which is located in such a way (the bold dot in the middle of the bottom side of the box, Figure 2) that it is possible to determine only either the firefly is in the section C_{α_1} or in the section C_{α_2} of the box;

2) to open a small window at the point b which is located in such a way (the bold dot in the middle of the right-hand side of the box, Figure 3) that it is possible to determine only either the firefly is in the section C_{β_1} or in the section C_{β_2} of the box.

In the first case such an external observer can determine in which part, C_{α_1} or C_{α_2}, the firefly is located. In the second case he can only determine in which part, C_{β_1} or C_{β_2}, the firefly is located. But he is not able to look into both windows simultaneously. In such a situation the observables a and b are the only source of information about the firefly ("reference observables"). The Kolmogorov description is meaningless for the external observer (although it is present in the latent form).

Can one apply in such a situation the QL-description? The answer is to be positive.

3   Contextual Probability

A general statistical model for observables based on the contextual viewpoint to probability will be presented. It will be shown that classical as well as quantum probabilistic models can be obtained as particular cases of our general contextual model, the *Växjö model*, [1].

A physical, biological, social, mental, genetic, economic, or financial *context* C is a complex of corresponding conditions.

Construction of any model M should be started with fixing the collection of contexts of this model. Denote the collection of contexts by the symbol $\mathcal{C}$ (so the family of contexts $\mathcal{C}$ is determined by the model M under consideration). In the mathematical formalism $\mathcal{C}$ is an abstract set (of "labels" of contexts).

We remark that in some models it is possible to construct a set-theoretic representation of contexts – as some family of subsets of a set Ω. For example, Ω can be the set of all possible parameters (e.g., physical, or mental, or economic) of the model. However, in general we *do not assume the possibility to construct a set-theoretic representation of contexts*.

Another fundamental element of any contextual probabilistic model M is a set of observables $\mathcal{O}$: each observable $a \in \mathcal{O}$ can be measured under each complex of conditions $C \in \mathcal{C}$. For an observable $a \in \mathcal{O}$, we denote the set of its possible values ("spectrum") by the symbol X_a.

Axiom 1. *For any observable $a \in \mathcal{O}$ and its value $\alpha \in X_a$, there are defined contexts, say C_α, corresponding to $[a = \alpha]$-selections: if we perform a measurement of the observable a under the complex of physical conditions C_α, then we obtain the value $a = \alpha$ with probability 1. We assume that the set of contexts $\mathcal{C}$ contains C_α-selection contexts for all observables $a \in \mathcal{O}$ and $\alpha \in X_a$.*

For example, let a be the observable corresponding to some question: $a = +$ (the answer "yes") and $a = -$ (the answer "no"). Then the C_+-selection context is the selection of those participants of the experiment who answering "yes" to this question; in the same way we define the C_--selection context.

Axiom 2. *There are defined contextual (conditional) probabilities* $p_C^a(\alpha) \equiv \mathbf{P}$ $(a = \alpha|C)$ *for any context* $C \in \mathcal{C}$ *and any observable* $a \in O$.

Thus, for any context $C \in \mathcal{C}$ and any observable $a \in O$, there is defined the probability to observe the fixed value $a = \alpha$ under the complex of conditions C.

Especially important role will be played by "transition probabilities" $p^{a|b}$ $(\alpha|\beta) \equiv \mathbf{P}(a = \alpha|C_\beta), a, b \in \mathcal{O}, \alpha \in X_a, \beta \in X_b$, where C_β is the $[b = \beta]$-selection context. By axiom 2 for any context $C \in \mathcal{C}$, there is defined the set of probabilities: $\{p_C^a : a \in \mathcal{O}\}$. We complete this probabilistic data for the context C by transition probabilities. The corresponding collection of data $D(\mathcal{O}, C)$ consists of contextual probabilities: $p^{a|b}(\alpha|\beta), p_C^b(\beta), p^{b|a}(\beta|\alpha), p_C^a(\alpha)...$, where $a, b, ... \in \mathcal{O}$. Finally, we denote the family of probabilistic data $D(\mathcal{O}, C)$ for all contexts $C \in \mathcal{C}$ by the symbol $\mathcal{D}(\mathcal{O}, \mathcal{C})(\equiv \cup_{C \in \mathcal{C}} D(\mathcal{O}, C))$.

Definition 1. (Växjö Model) *A contextual probabilistic model of reality is a triple* $M = (\mathcal{C}, \mathcal{O}, \mathcal{D}(\mathcal{O}, \mathcal{C}))$, *where* $\mathcal{C}$ *is a set of contexts and* $\mathcal{O}$ *is a set of observables which satisfy to axioms 1,2, and* $\mathcal{D}(\mathcal{O}, \mathcal{C})$ *is probabilistic data about contexts* $\mathcal{C}$ *obtained with the aid of observables belonging* $\mathcal{O}$.

We call observables belonging the set $\mathcal{O} \equiv \mathcal{O}(M)$ *reference of observables.* Inside of a model M observables belonging to the set $\mathcal{O}$ give the only possible references about a context $C \in \mathcal{C}$. In the definition of the Växjö Model we speak about "reality." In our approach it is reality of contexts.

In what follows we shall consider Växjö models with two dichotomous reference observables.

4   Quantum-Like Representation Algorithm – QLRA

In [1] we derived the following formula for interference of probabilities:

$$p_C^b(\beta) = \sum_\alpha p_C^a(\alpha)p^{b|a}(\beta|\alpha) + 2\lambda(\beta|\alpha, C)\sqrt{\prod_\alpha p_C^a(\alpha)p^{b|a}(\beta|\alpha)}, \qquad (1)$$

where the coefficient of interference

$$\lambda(\beta|a, C) = \frac{p_C^b(\beta) - \sum_\alpha p_C^a(\alpha)p^{b|a}(\beta|\alpha)}{2\sqrt{\prod_\alpha p_C^a(\alpha)p^{b|a}(\beta|\alpha)}}. \qquad (2)$$

A similar representation we have for the a-probabilities. Such interference formulas are valid for any collection of contextual probabilistic data satisfying the conditions:

R1). Observables a and b are symmetrically conditioned[1]: $p^{b|a}(\beta|\alpha) = p^{a|b}(\alpha|\beta)$.

---

[1] This condition will induce symmetry of the scalar product.

R2). Observables a and b are mutually nondegenerate[2]: $p^{a|b}(\alpha|\beta) > 0,;$ $p^{b|a}(\beta|\alpha) > 0$.

R2a). Context C is nondegenerate with respect to both observables a and b : $p_C^b(\beta) > 0$, $p_C^a(\alpha) > 0$.

Suppose that also the following conditions hold:

R3). Coefficients of interference are bounded by one[3]:

$$\left| \frac{p_C^b(\beta) - \sum_\alpha p_C^a(\alpha) p^{b|a}(\beta|\alpha)}{2\sqrt{\prod_\alpha p_C^a(\alpha) p^{b|a}(\beta|\alpha)}} \right| \le 1,$$

$$\left| \frac{p_C^a(\alpha) - \sum_\beta p_C^b(\beta) p^{a|b}(\alpha|\beta)}{2\sqrt{\prod_\alpha p_C^b(\beta) p^{a|b}(\alpha|\beta)}} \right| \le 1,$$

A context C such that R3) holds is called trigonometric, because in this case we have the conventional formula of trigonometric interference:

$$p_C^b(\beta) = \sum_\alpha p_C^a(\alpha) p^{b|a}(\beta|\alpha) + 2\cos\theta(\beta|\alpha, C)\sqrt{\prod_\alpha p_C^a(\alpha) p^{b|a}(\beta|\alpha)}, \qquad (3)$$

where $\lambda(\beta|a, C) = \cos\theta(\beta|a, C)$. Parameters $\theta(\beta|\alpha, C)$ are said to be $b|a$-relative phases with respect to the context C. We defined these phases purely on the basis of probabilities. We have not started with any linear space; in contrast we shall define geometry from probability.[4]

We denote the collection of all trigonometric contexts by the symbol $\mathcal{C}^{tr}$.

By using the elementary formula: $D = A + B + 2\sqrt{AB}\cos\theta = |\sqrt{A} + e^{i\theta}\sqrt{B}|^2$, for real numbers $A, B > 0, \theta \in [0, 2\pi]$, we can represent the probability $p_C^b(\beta)$ as the square of the complex amplitude (Born's rule):

$$p_C^b(\beta) = |\psi_C(\beta)|^2 . \qquad (4)$$

Here

$$\psi(\beta) \equiv \psi_C(\beta) = \sqrt{p_C^a(\alpha_1) p^{b|a}(\beta|\alpha_1)} + e^{i\theta_C(\beta)}\sqrt{p_C^a(\alpha_2) p^{b|a}(\beta|\alpha_2)}, \quad \beta \in X_b, \quad (5)$$

where $\theta_C(\beta) \equiv \theta(\beta|\alpha, C)$.

The formula (5) gives the quantum-like representation algorithm – QLRA. For any trigonometric context C by starting with the probabilistic data

$$p_C^b(\beta), p_C^a(\alpha), p^{b|a}(\beta|\alpha)$$

---

[2] This condition will induce noncommutativity of operators $\hat{a}$ and $\hat{b}$ representing these observables.

[3] This condition will induce representation of the context C in the complex Hilbert space. Thus complex numbers appear due to this condition.

[4] We remark that conditions R1) and R3) are also necessary.

– QLRA produces the complex amplitude ψ_C. This algorithm can be used in any domain of science to create the QL-representation of probabilistic data (for a special class of contexts).

We point out that QLRA contains the reference observables as parameters. Hence the complex amplitude give by (5) depends on $a, b : \psi_C \equiv \psi_C^{b|a}$.

We denote the space of functions: $\varphi : X_b \to \mathbf{C}$ by the symbol $\Phi = \Phi(X_b, \mathbf{C})$. Since $X = \{\beta_1, \beta_2\}$, the Φ is the two dimensional complex linear space. By using QLRA we construct the map

$$J^{b|a} : \mathcal{C}^{\text{tr}} \to \Phi(X, \mathbf{C}) \tag{6}$$

which maps contexts (complexes of, e.g., physical conditions) into complex amplitudes. The complex amplitude $\psi_C(x)$ can be called a **wave function** of the complex of physical conditions (context) C or a (pure) *state*. We set $e_\beta^b(\cdot) = \delta(\beta - \cdot)$ – Dirac delta-functions concentrated in points $\beta = \beta_1, \beta_2$. The Born's rule for complex amplitudes (4) can be rewritten in the following form: $p_C^b(\beta) = |\langle \psi_C, e_\beta^b \rangle|^2$, where the scalar product in the space $\Phi(X_b, C)$ is defined by the standard formula: $\langle \phi, \psi \rangle = \sum_{\beta \in X_b} \phi(\beta) \bar{\psi}(\beta)$. The system of functions $\{e_\beta^b\}_{\beta \in X_b}$ is an orthonormal basis in the Hilbert space $H_{ab} = (\Phi, \langle \cdot, \cdot \rangle)$.

Let $X_b \subset \mathbf{R}$. By using the Hilbert space representation of the Born's rule we obtain the Hilbert space representation of the expectation of the observable b: $E(b|C) = \sum_{\beta \in X_b} \beta |\psi_C(\beta)|^2 = \sum_{\beta \in X_b} \beta \langle \psi_C, e_\beta^b \rangle \langle \psi_C, e_\beta^b \rangle = \langle \hat{b} \psi_C, \psi_C \rangle$, where the (self-adjoint) operator $\hat{b} : H_{ab} \to H_{ab}$ is determined by its eigenvectors: $\hat{b} e_\beta^b = \beta e_\beta^b, \beta \in X_b$. This is the multiplication operator in the space of complex functions $\Phi(X_b, \mathbf{C}) : \hat{b}\psi(\beta) = \beta\psi(\beta)$. It is natural to represent the b-observable (in the Hilbert space model) by the operator $\hat{b}$.

We would like to have Born's rule not only for the b-variable, but also for the a-variable: $p_C^a(\alpha) = |\langle \varphi, e_\alpha^a \rangle|^2, \alpha \in X_a$.

How can we define the basis $\{e_\alpha^a\}$ corresponding to the a-observable? Such a basis can be found starting with interference of probabilities. We set $u_j^a = \sqrt{p_C^a(\alpha_j)}, p_{ij} = p(\beta_j|\alpha_i), u_{ij} = \sqrt{p_{ij}}, \theta_j = \theta_C(\beta_j)$. We have:

$$\varphi = u_1^a e_{\alpha_1}^a + u_2^a e_{\alpha_2}^a, \tag{7}$$

where

$$e_{\alpha_1}^a = (u_{11}, \ u_{12}), \ e_{\alpha_2}^a = (e^{i\theta_1} u_{21}, \ e^{i\theta_2} u_{22}) \tag{8}$$

The condition R1) implies that the system $\{e_{\alpha_i}^a\}$ is an orthonormal basis iff the probabilistic phases satisfy the constraint: $\theta_2 - \theta_1 = \pi \mod 2\pi$, but, as we have seen [1], we can always choose such phases (under the condition R1).

In this case the a-observable is represented by the operator $\hat{a}$ which is diagonal with eigenvalues α_1, α_2 in the basis $\{e_{\alpha_i}^a\}$. The conditional average of the observable a coincides with the quantum Hilbert space average: $E(a|C) = \sum_{\alpha \in X_a} \alpha p_C^a(\alpha) = \langle \hat{a} \psi_C, \psi_C \rangle$.

If condition R3) is violated, non conventional QL-representations of probabilistic data arise: for example, in the hyperbolic analogue of complex Hilbert space [1].

5 Flies in a Packet

We consider a metal box. Food (which is attractive for flies) is distributed at different points inside this box. This distribution is not uniform: at some points food's concentration is higher, there are even domains without food. An external observer (who is outside this box) has no idea about the real distribution of food in the box. But the "Einstein demon" living inside this box knows well this distribution.

We put a population of flies, say Ω, inside this box. After a while they will be distributed in space inside the box concentrating at sites with food.

The "Einstein demon" knows the probability distribution $\mathbf{P}(x, y, z)$ to find a fly at the point with coordinates (x, y, z). It is assumed to be stationary (at least for a while). (In principle, some flies can move between attractive points, but statistically the number of flies at each site with food is stable.)

As in the example "firefly in the box", one can divide this box in two ways: a) by the vertical wall – a, see Figure 2; b) by horizontal wall – b, see Figure 3. Here $a(\omega) = \alpha_1$ if the "Einstein demon" finds a fly ω in the left-hand part and $a(\omega) = \alpha_2$ if he finds a fly ω in the right-hand part (e.g. $\alpha = \pm 1$). We define b in a similar way: $b(\omega) = \beta_1$ if the "Einstein demon" finds a fly ω in the top part and $b(\omega) = \beta_2$ if he finds a fly ω in the bottom part (e.g. $\beta = \pm 1$). The "Einstein demon" sets:

$$\Omega_\alpha = \{\omega \in \Omega : a(\omega) = \alpha\}, \quad \Omega_\beta = \{\omega \in \Omega : a(\omega) = \beta\}.$$

By assuming that $\mathbf{P}(\Omega_\alpha), \mathbf{P}(\Omega_\beta) > 0$, he can define transition probabilities:

$$p^{b|a}(\beta|\alpha) = \mathbf{P}(\Omega_\beta|\Omega_\alpha) \equiv \frac{\mathbf{P}(\Omega_\beta \cap \Omega_\alpha)}{\mathbf{P}(A_\alpha)}$$

and in the same way probabilities $p^{a|b}(\alpha|\beta)$.

Let C be some domain inside the box. We shall consider it as a geometric-context. The "Einstein demon" can be find (by using the Bayes' formula) conditional probability distribution:

$$\mathbf{P}_C(U) = \frac{\mathbf{P}(\Omega_U \cap \Omega_C)}{\mathbf{P}(\Omega_C)}, \tag{9}$$

for any subset U of box (if we consider general distribution $\mathbf{P}$ (i.e., not discrete), then we should take Borel sets). Here $\Omega_C = \{\omega \in \Omega : (x_\omega, y_\omega, z_\omega) \in C\}$ and Ω_U is defined in the same way.

This probability distribution $\mathbf{P}_C$ provides the probabilistic representation of the domain C. The "Einstein demon" encoded geometry by probability. Of course, probability provides only rough images of geometric structures, since the map:

$$C \rightarrow \mathbf{P}_C$$

is not one-to-one. Denote now by $\mathcal{F}$ some σ-algebra of subsets of the box such that the probability $\mathbf{P}$ – flies' distribution – can be defined on it. Denote also the set of all probability measures on the $\mathcal{F}$ by the symbol $\mathcal{P}$. Then we have the map:

$$J : \mathcal{F} \rightarrow \mathcal{P}. \tag{10}$$

This is the classical probabilistic representation of geometry (of distribution of food). It is available for any internal observer ("Einstein demon") who lives inside this box. In this mapping a lot of geometric information is neglected. However, the whole probabilistic information is taken into account. This is the end of the classical story!

Remark 5.1. (Food and flies version of fields and particles) This representation has one interesting feature. Geometry of food distribution is represented by ensembles of flies. We can make the following analogy: electromagnetic field can be represented by photons. One can compare the food distribution with a kind of a "food-field" and flies with particles representing this field. If we put another type of insects into the box, they may be not interested in this sort of food. They would not reproduce the distribution $\mathbf{P}(x, y, z)$. Thus we may speak about various food-fields which are represented by corresponding types of insects-particles. In some sense this picture reminds Bohmian mechanics.

Now we modify the previous framework. We have the same box with the same distribution of fly-attractive food. But flies are put not directly in the box, but in a plastic packet, say C. The geometric configuration is unknown for us – external observers. Moreover, we are not able to find its configuration directly (even by making a hole in the box), because packet's surface is covered by a "B2-bomber type" material. Thus we look inside the box, but we see nothing.[5] Nevertheless, we (external observers) would like to get at least partial information about this packet configuration by using flies distribution. The problem is even more complicated, because we shall proceed under the assumption that any attempt to open the metal box will induce destruction of the packet which in its turn induces redistribution of flies in the space.

We do the following. As in the firefly-example we introduce fuzzy coordinates a and b. We measure them in the following way. We assume that we can put very quickly either vertical or horizontal wall into the box. Such a moving wall divides (practically adulating, at least in comparing with fly's velocity) the box into two sub-boxes, but at the same time it destroys (of course) the plastic packet. It is assumed that after this act we can open each sub-box and find numbers of flies in each part of the box.

At the moment we consider *nondisturbing measurements:* walls do not change food distributions in corresponding parts of the box (those walls are negligibly thin and destruction of the packet does not change the distribution of food). However, opening of any box induces a strong disturbing effect, flies are essentially redistributed.

Thus first we do the a-measuring by using the vertical wall. It divides the box into two parts, say C_{α_1} and C_{α_2}. In this way we get probabilities $p_C^a(\alpha)$ that a fly was located in the α-side of the box. Since the vertical wall moves quickly relatively to fly's velocity, the number of flies which were able to change the left-hand part of the box to the right-hand part or vice versa is statistically negligible. In principle, we might try to use the classical formula:

---

[5] The "Einstein demon" also gets a problem, but he can still investigate packet's geometry just by moving over its surface. Of course, if the packet is disconnected, so it has a few components, a few "Einstein demons" should be employed.

$$p_C^a(\alpha) = \mathbf{P}_C(\Omega_\alpha) \equiv \frac{\mathbf{P}(\Omega_\alpha \cap \Omega_C)}{\mathbf{P}(\Omega_C)},$$

However, it is not useful for us, because we do not know the configuration C and hence $\mathbf{P}_C$.

We point out that if we do not open sub-boxes C_α and if after while the corresponding "Einstein demons" measure the b-coordinate of flies in each part C_α of the box they will obtain the original transition probabilities $p^{b|a}(\beta|\alpha)$, since flies will again redistribute in the domain C_α according to the food-field.[6] However, the original distribution of flies in the domain $C \cap C_\alpha$ has been lost for ever even for the "Einstein demons." We (external observers) are not able to find transition probabilities in this way, since opening of a box produces redistribution of flies in it.

We also remark that trivially $a(\omega) = \alpha$ on the α-part of the box.

Remark 5.2. (Reaction of "food-field" to space reconfiguration) At the moment we proceed under the assumption that the "food-field" is not sensitive to the disturbing effect of the moving wall (separating the box into two sub-boxes). Moreover, the "food-field" is *not sensitive to changes of the geometry of space* ("boundary conditions"). In principle, we can imagine the following situation. The appearance of a separating wall does not induce a disturbing effect which could move food in space. However, the wall by itself can have some physical properties influencing the food distribution. For example, food is electrically charged (in some way) and walls of the box (including walls used in separation experiments) also carry electric charges. Thus even "mechanically peaceful appearance" of a separating wall will induce (after a while) redistribution of food in the sub-box.

To construct the QL-representation of the context C by a complex probability amplitude, we need also probabilities:

$$p_C^b(\beta) = \mathbf{P}_C(\Omega_\beta) \equiv \frac{\mathbf{P}(\Omega_\beta \cap \Omega_C)}{\mathbf{P}(\Omega_C)},$$

However, since we do not know the configuration C, we are not able to apply Bayes' formula directly. We should repeat previous considerations, but by using now the horizontal wall which separates quickly the box into top and bottom parts, C_{β_1} and C_{β_2}. Then by opening these sub-boxes and counting flies in each of them we find the probabilities $p_C^b(\beta)$.

Of course, we should have two boxes with the same configuration C, because each falling wall destroys this configuration. Thus we should be able to make such a preparation a few times. Moreover, if one wants to exclude effects of interaction between flies (as one does in QM), there should be created an ensemble of boxes, each box containing just one fly. It is assumed that flies would reproduce the food distribution.

In particular, for $C = C_\alpha$, i.e., the configuration C which coincides with the α-part of the box we get: $\Omega_{C_\alpha} = \Omega_\alpha$ and

$$p_{C_\alpha}^b(\beta) = p^{b|a}(\beta|\alpha).$$

---

[6] Two "Einstein demons" should be involved – one for each sub-box.

However, we do not know from the very beginning that a hidden geometric configuration is the half-box C_α. Therefore this is not an experimental way to find transition probabilities.

To find transition probabilities, we assume that each half-box C_α can be divided by the horizontal wall (as in the original b-measurement in the whole box) in two parts, say $C_{\beta|\alpha}, \beta = \beta_1, \beta_2$. By counting flies in each of these boxes we find the transition probabilities. At the moment we proceed under the same assumptions as before: by putting the horizontal walls in the box C_α we do not change the distribution of food in it.

Now everything is prepared for application of QLRA. A necessary condition is given by R2), since in QM matrices of transition probabilities are symmetrically conditioned. Thus from the very beginning one should assume that the distribution of attracting sites in the box induces this condition. This happens iff $\mathbf{P}(\Omega_\alpha) = \mathbf{P}(\Omega_\beta) = 1/2$.

The next condition is that variables are statistically conjugate, i.e., $\mathbf{P}(\Omega_\alpha \cap \Omega_\beta) \neq 0$ for all α and β.

Finally, the context C should be "large enough" with respect to both variables: $\mathbf{P}(\Omega_C \cap \Omega_\beta), \mathbf{P}(\Omega_C \cap \Omega_\alpha) > 0$.

We also know that, besides a complex probability amplitude, some contexts can be represented by hyperbolic amplitudes, thus to guarantee real QM-like representation we should have $|\lambda| \leq 1$ for the coefficient of interference.

Thus we represent all "trigonometric configurations" C by complex vectors and the observables a and b by self-adjoint operators. The map:

$$J^{b|a} : C^{\mathrm{tr}} \to H$$

is a QL-analogue of the classical map J given by (10). Of course, the former map is "better". However, we are not able to use it in the situation with invisible configuration C.

As was shown [1], for some contexts, hyperbolic ones, $|\lambda| > 1$. They are mapped into hyperbolic amplitudes:

$$J^{b|a} : C^{\mathrm{hyp}} \to H_{\mathrm{hyp}},$$

where H_{hyp} is the hyperbolic analogue of Hilbert space [1].

6 Quantum-Like Representation of Kolmogorov's Model

In spite of the presence of the underlying Kolmogorov space, we constructed the QL-representation of probabilistic data for macroscopic configurations (essentially incomplete representation) which has all distinguishing features of the conventional quantum representation of probabilistic data for a pair of incompatible observables: interference formula for probabilities, Born's rule, representation of these observables by self-adjoint operators. As was mentioned, the map $J^{b|a}$ is not injective. We no ask: Is it surjective? Can one get any quantum state ψ and any pair of quantum observables $\hat{a}$ and $\hat{b}$ in such a way? The answer is no. This is a consequence of Bell's type inequality for transition probabilities.

To apply conditional Bell's inequality to our macroscopic situation, it is better to consider a ball bounded by the metal sphere, instead of the box. We now can divide this ball into parts with the aid of central planes. To simplify considerations, we can consider a bundle of planes which are enumerated by the angle ϕ. Then we shall obtain a family of observables a_ϕ, say taking values $\pm$. Parts of the ball obtained by the ϕ-separation are $C_{\phi,+} = \{\theta : \phi \leq \theta < \phi + \pi\}$ and $C_{\phi,-} = \{\theta : \phi + \pi \leq \theta < \phi\}$, respectively.

For each pair of them we find transition probabilities $p^{\phi_1|\phi_2}(\epsilon_1|\epsilon_2)$. For each context C (a plastic packet with flies inside it; this packet is placed inside the metal ball; any attempt to open the ball would destroy this packet) and any ϕ-section, we find probabilities $p_C^\phi(\epsilon), \epsilon = \pm$. If we choose a context C such that $p_C^\phi(+1) = p_C^\phi(-1) = 1/2$ for all ϕ, then we can apply arguments of appendix and we see that some types of transition probabilities could not be obtained from a single Kolmogorov model.

One Kolmogorov space is too small to generate all quantum (or better to say quantum-like) states and observables.

7 Disturbing Measurements

However, we can easily modify our example to destroy the (hidden) Kolmogorov structure of the model. Suppose now that everything is as it was before with only one difference: destruction of the packet by a wall (encoded by some ϕ-plane) induces not only the possibility for flies to move outside the packet, but also induces a redistribution of food sites, cf. Remark 5.2. The latter is determined by the wall. Thus after e.g. the ϕ-plane separation of the ball the distribution of sites with food in its parts $C_{\phi,+1}$ and $C_{\phi,-1}$ is not such as it was before this separation. Therefore, for any successive ϕ'-separation of the sectors $C_{\phi,\epsilon}$ (which were produced by the previous ϕ-separation), the transition probabilities $p^{\phi'|\phi}(\epsilon'|\epsilon)$ obtained by an external observer do not coincide with the transition probabilities which would be obtained by the "Einstein demon" on the basis of the original ensemble. Hence Bell's type inequality for transition probabilities, see appendix and [1], cannot be applied.

In fact, by using random generators we can simulate probabilities for any complex probability amplitude and any pair of self-adjoint operators in the two dimensional Hilbert space.

For example, suppose that we would like to simulate the transition probabilities for successive measurement of spin projections as well as the uniform probability distribution for the a and b measurements for the original context C (state ψ_C). To provide the latter condition, we start with the uniform distribution of food. It would induce probabilities $p_C^a(\alpha) = p_C^b(\beta) = 1/2$.

Now to simplify considerations, we consider not three dimensional configurations, but just two dimensional, in particular, we consider a circle, instead of a ball, and sections by central lines, instead of planes.

The disturbance induced by the a_{ϕ_0}-measurement, $0 \leq \phi_0 < \pi$, induces redistribution of food in the sectors and finally generates e.g. in the sector $C_{\phi_0,+}$ the density of flies:

$$\rho_{\phi_0}^{+}(r, \theta) = \sin(\theta - \phi_0). \tag{11}$$

(We assume that the circle has unit radius). Then we separate the sector $C_{\phi_0, +1}$ by the ϕ-plane, say $\phi > \phi_0$. Then the probability

$$p^{\phi|\phi_0}(+|+) = \int_0^1 rdr \int_\phi^{\phi_0+\pi} \sin(\theta - \phi_0)d\theta = \cos^2 \frac{\phi - \phi_0}{2},$$

$$p^{\phi|\phi_0}(-|+) = \int_0^1 rdr \int_{\phi_0}^{\phi} \sin(\theta - \phi_0)d\theta = \sin^2 \frac{\phi - \phi_0}{2}.$$

For the sector $C_{\phi_0, -}$, we choose the probability distribution

$$\rho_{\phi_0}^{-}(r, \theta) = -\sin(\theta - \phi_0). \tag{12}$$

Here transition probabilities are given by

$$p^{\phi|\phi_0}(+|-) = -\int_0^1 rdr \int_{\phi_0+\pi}^{\phi+\pi} \sin(\theta - \phi_0)d\theta = \sin^2 \frac{\phi - \phi_0}{2},$$

$$p^{\phi|\phi_0}(-|-) = -\int_0^1 rdr \int_{\phi+\pi}^{\phi_0+2\pi} \sin(\theta - \phi_0)d\theta = \cos^2 \frac{\phi - \phi_0}{2}.$$

Remark 7.1. (Complementarity or supplementarity?) Since we consider disturbing measurements, we (external observers) are not able to measure two observables, a_{ϕ_1} and a_{ϕ_2}, simultaneously. Thus these are *incompatible observables*. However, such measurement incompatibility does not exclude that an element of reality can be assigned to each fly – the pair $a_{\phi_1}(\omega), a_{\phi_2}(\omega)$. We recall that we consider such separations that they do not induce redistribution of flies between sectors: the ϕ-plane moves so quickly that flies are not able to change sectors (or at least only statistically negligible number of flies could make such changes). Moreover, only negligible number of flies can be killed by a moving-separating plane. Thus the values of $a_{\phi_1}(\omega)$ and $a_{\phi_2}(\omega)$ which would be obtained by an external observer coincide with the values which have been known by the "Einstein demon" before measurements. Therefore complementarity (in the sense of mutual exclusivity) is only external observer's complementarity. The "Einstein demon" still has supplementarity, in the sense on additional information (of course, fuzzy) about fly's location.

8   Classical Probabilistic Structure and Disturbance Effects

By Bell's inequality for transition probabilities, see appendix, it is impossible to find a single underlying classical probabilistic space which would reproduce all possible wave functions and pairs of self-adjoint noncommutative operators in the contextual probabilistic framework. One can not find such a Kolmogorov probability space that by choosing different pairs of reference observables a, b

and corresponding families of trigonometric contexts $\mathcal{C}^{\mathrm{tr}}(a, b)$ (represented by sets from the σ-algebra of the Kolmogorov space) he would (by applying QLRA) cover the whole unit sphere of Hilbert state space as well as obtain all pairs of noncommutative self-adjoint operators. We showed that by considering disturbing measurements we can reproduce all quantum structures. Can one approach the same result without disturbance? In principle, yes!

8.1   Ensemble Fluctuations

Another important point is that we proceeded by using counterfactual arguments. To be on really realistic ground, we should consider at least three different balls and perform on them conditional measurements for pairs of observables a_{ϕ_i}, a_{ϕ_j}. In principle, we cannot guarantee that we would be able to reproduce statistically identical distributions of food in balls and identical hidden configurations. As was emphasized, the map, see (6), from the collection of trigonometric contexts into complex probability amplitudes is not injection, various contexts can be mapped in the same complex probability amplitude. Even if we are sure that we have the same QL-state given by the same complex probability amplitude, ψ, we could never be sure that contexts in different balls are the same. Therefore we should work in multi-Kolmogorovian framework and the Bell's inequality for conditional probabilities can also be violated without any disturbance.

9   ERP-Bohm Type Experiments with Flies

We have considered in very detail measurements (in fact, position-type measurements) for ensembles of single flies. In principle, we could consider the real EPR-Bohm type experiment for pairs of "entangled flies" which we put into different metal balls. One of technological problems is to produce such pairs of flies. However, this is not the main point. The main point is that in the macroscopic framework such experiments would not give so much more than experiments with single flies. In contrast to photons or electrons, we have no doubts that flies have objective properties, in particular, the position. Therefore the only consequence of the EPR-Bohm type experiment with flies would be that disturbing effects should be excluded.[7]

---

[7] We remind that we consider not only mechanical disturbance by moving planes, but also the field type disturbance. To exclude the latter type of disturbance, one should be sure that the effect of the "food-field" (e.g. smell) from one ball would be not able to propagate to another ball. If balls have small windows (or produced not of metal, but of some less isolating material), then smell can propagate from one ball to another. We recall that insects can find smell-traces on huge distances. Thus to exclude completely disturbing effects, we should either isolate balls completely or to make measurements on balls with a time-window such that a signal from one ball would not be able to approach another during this time window.

Thus as well as in the case of a single system we have tree choices: a) unfair sampling; b) ensemble fluctuations; c) nonrelativistic communications between flies.

The last condition cannot be completely rejected even for human beings, but the EPR-type experiment could not be used to provide the crucial argument in its favor.

Conclusion. *We shown that macroscopic configurations can be naturally represented in the QL-way – by complex probability amplitudes. Classical probabilistic structure can be violated. In particular, Bell's type inequality can be violated.*

References

1. Khrennikov, A.Y.: Contextual Approach to Quantum Formalism. Springer, Berlin (2009)
2. Khrennikov, A.Y.: Ubiquitous Quantum Structure: from Psychology to Finances. Springer, Berlin (2009)
3. Foulis, D.J.: A Half-Century of Quantum-Logic. What Have We Learned? In: Quantum Structures and the Nature of Reality. Einstein Meets Magritte, vol. 7, pp. 1–36. Kluwer, Dordrecht (1990)

Appendix: Bell's Inequality for Transition Probabilities

Theorem. *Let $a, b, c = \pm 1$ be dichotomous uniformly distributed random variables on a single Kolmogorov space. Then the following inequality holds true:*

$$\mathbf{P}(a = +1|b = +1) + \mathbf{P}(c = +1|b = -1) \geq \mathbf{P}(a = +1|c = +1) \qquad (13)$$

We underline again that the main distinguishing feature of (13) is the presence of only transition probabilities. Transition probabilities can always be calculated by using quantum formalism for noncomposite systems. In fact, we need not consider pairs of particles.

Experimental Evidence for Quantum Structure in Cognition

Diederik Aerts[1], Sven Aerts[1], and Liane Gabora[2]

[1] Leo Apostel Center, Brussels Free University, Brussels, Belgium
diraerts@vub.ac.be, saerts@vub.ac.be
http://www.vub.ac.be/CLEA/aerts/
[2] Psychology and Computer Science, University of British Columbia, Canada
liane.gabora@ubc.ca
http://www.vub.ac.be/CLEA/liane/

Abstract. We prove a theorem that shows that a collection of experimental data of membership weights of items with respect to a pair of concepts and its conjunction cannot be modeled within a classical measure theoretic weight structure in case the experimental data contain the effect called overextension. Since the effect of overextension, analogue to the well-known guppy effect for concept combinations, is abundant in all experiments testing weights of items with respect to pairs of concepts and their conjunctions, our theorem constitutes a no-go theorem for classical measure structure for common data of membership weights of items with respect to concepts and their combinations. We put forward a simple geometric criterion that reveals the non classicality of the membership weight structure and use experimentally measured membership weights estimated by subjects in experiments from [26] to illustrate our geometrical criterion. The violation of the classical weight structure is similar to the violation of the well-known Bell inequalities studied in quantum mechanics, and hence suggests that the quantum formalism and hence the modeling by quantum membership weights, as for example in [17], can accomplish what classical membership weights cannot do.

1 Introduction

Many branches of mathematics, such as geometry, complexity theory, and even number theory, were originally conceived not as domains of mathematics, but as describing a particular domain of physical reality. It was only much later that they were conceived more abstractly, and their applicability to a wide range of phenomena was realized. We believe this is also proving to be the case for the mathematical formalisms originally developed to describe events observed in the microworld: quantum mechanics.

Meanwhile the mathematical formalism of quantum mechanics has indeed been used successfully to model situations pertaining to domains different from the micro-world, for example, in economics [1, 2, 3], operations research and management sciences [4, 5], psychology and cognition [6, 7, 8, 9, 10, 11, 12, 13, 14, 15, 16, 17], and language and artificial intelligence [18, 19, 20, 21, 22, 23].

P. Bruza et al. (Eds.): QI 2009, LNAI 5494, pp. 59–70, 2009.
© Springer-Verlag Berlin Heidelberg 2009

More specifically, in [17] a quantum mechanical representation of experimental data corresponding to membership weights of items with respect to pairs of concepts and their conjunctions was elaborated. It was proven that these data cannot be modelled by a classical theory of membership weights, i.e. a theory where membership weights are represented within a measure theoretic structure (see theorems 1, 2 and 3).

In the present paper we introduce a very simple geometrical criterion that allows the identification of the classical or non-classical nature of membership weight data gathered for pairs of concepts and their conjunctions, or more generally, collections of concepts and conjunctions of some of the pairs in these collections. More specifically we determine for such a collection of concepts and some of the conjunctions of these concepts a geometrical figure called a polytope (which is the higher dimensional generalization of a polygon in a real vector space) and a geometrical way of representing the measured membership weights of this collection and its conjunctions by means of a vector in this real vector space called a correlation vector. We prove that if this correlation vector is located inside the polytope a classical measure theoretic model exists for these data, while if the correlation vector is located outside of the polytope then such a model does not exist.

2   Membership Weights on Pairs of Concepts and Their Conjunctions

It has been shown that *Guppy* is neither a very typical example of *Pet* nor *Fish* but is a very typical example of *Pet-Fish* [24]. Hence, the typicality of a specific item with respect to the conjunction of concepts can behave in an unexpected way. The problem is often referred to as the 'pet-fish problem' and the effect is usually called the 'guppy effect'. The guppy effect is abundant; it appears almost in every situation where concepts combine. Meanwhile many experiments and analyses of this effect and related to the problem of combining concepts have been conducted [24, 25, 26, 27, 28, 29, 30, 31, 32, 33, 34, 35].

The guppy effect was not only identified for the typicality of items with respect to concepts and their conjunctions but also for the membership weights of items with respect to concepts and their conjunction [26]. For example,

Table 1. The list of pairs of concepts and their conjunction used in [26]

A_1	A_2	A_1 and A_2
Furniture	*Household Appliances*	*Furniture and Household Appliances*
Food	*Plant*	*Food and Plant*
Weapon	*Tool*	*Weapon and Tool*
Building	*Dwelling*	*Building and Dwelling*
Machine	*Vehicle*	*Machine and Vehicle*
Bird	*Pet*	*Bird and Pet*

Table 2. Three of the pairs of concepts and items of experiment 4 in [26]. The non classical items are labeled by q and the classical items by c.

		$\mu(A_1)$	$\mu(A_2)$	$\mu(A_1 \text{and} A_2)$			$\mu(A_1)$	$\mu(A_2)$	$\mu(A_1 \text{and} A_2)$
A_1=Furniture, A_2=Household Appliances					A_1=Building, A_2=Dwelling				
Filing Cabinet	q	0.9744	0.3077	0.5263	Castle	c	1	1	1
Clothes Washer	q	0.15	1	0.725	Cave	c	0.2821	0.95	0.2821
Vacuum Cleaner	q	0.075	1	0.3846	Phone box	c	0.2308	0.0526	0.02778
Hifi	q	0.5789	0.7895	0.7895	Apartment Block	q	0.9231	0.8718	0.9231
Heated Water Bed	q	1	0.4872	0.775	Library	q	0.95	0.175	0.3077
Sewing Chest	q	0.8718	0.5	0.55	Trailer	q	0.35	1	0.6154
Floor Mat	q	0.5641	0.15	0.2051	Jeep	q	0	0.05	0.05
Coffee Table	q	1	0.15	0.3846	Palena	q	0.975	1	1
Piano	q	0.95	0.1282	0.3333	Igloo	q	0.875	1	0.9
Rug	q	0.5897	0.05128	0.1842	Synagogue	c	0.925	0.4872	0.4474
Painting	q	0.6154	0.0513	0.1053	Tent	q	0.5	0.9	0.55
Chair	q	0.975	0.175	0.3590	Bown	q	0.9487	0.8205	0.8974
Fridge	q	0.4103	1	0.775	Theatre	q	0.95	0.1282	0.2821
Desk Lamp	q	0.725	0.825	0.825	LogCabin	c	1	1	1
Cooking Stove	q	0.3333	1	0.825	House	c	1	1	1
TV	q	0.7	0.9	0.925	Tree House	q	0.7692	0.8462	0.85
A=Food, B=Plant					A=Machine, B=Vehicle				
Garlic	q	0.9487	0.7105	0.8514	Dog Sled	q	0.1795	0.925	0.275
Toadstool	q	0.1429	0.6061	0.2727	Dishwasher	q	1	0.025	0
Steak	c	1	0	0	Backpack	c	0	0	0
Peppercorn	q	0.875	0.6207	0.7586	Bicycle	q	0.85	0.975	0.95
Potato	q	1	0.7436	0.9	Sailboat	c	0.5641	0.8	0.4211
Raisin	q	1	0.3846	0.775	Roadroller	q	0.9375	0.9063	0.9091
Mint	q	0.8718	0.8056	0.8974	Raft	c	0.2051	0.725	0.2
Sunflower	q	0.7692	1	0.775	Elevator	q	0.9744	0.6	0.7949
Seaweed	q	0.825	0.9744	0.8684	Course liner	q	0.875	0.875	0.95
Sponge	q	0.0263	0.3421	0.0882	Automobile	c	1	1	1
Bread	q	1	0.0769	0.2051	Horsecart	q	0.3846	0.95	0.2895
Cabbage	q	1	0.9	1	Skateboard	q	0.2821	0.8421	0.3421
Eucalyptus	q	0.1622	0.8974	0.3243	Bus	c	1	1	1
Poppy	q	0.3784	0.8947	0.5405	Bulldozer	q	1	0.925	0.95
Mushroom	q	1	0.6667	0.9	Lawn-mower	q	0.975	0.1053	0.2632
Lettuce	q	1	0.925	1	Ski Lift	q	1	0.5897	0.875
A=Weapon, B=Tool					A=Bird, B=Pet				
Ruler	q	0.05	0.9	0.1538	Dog	c	0	1	0
Toothbrush	c	0	0.55	0	Cuckoo	q	1	0.575	0.8421
Chisel	q	0.4	0.975	0.6410	Parakeet	c	1	1	1
Axe	q	0.875	1	0.975	Cat	c	0	1	0
Screwdriver	q	0.3	1	0.625	Lark	q	1	0.275	0.4872
Arrow	q	1	0.225	0.575	Heron	q	0.9412	0.1515	0.2581
Knife	c	1	0.975	0.975	Peacock	q	1	0.4	0.5789
Rifle	q	1	0.35	0.5	Cow	q	0	0.425	0.025
Whip	q	0.875	0.2632	0.625	Toucan	q	1	0.6154	0.8026
Hammer	q	0.575	1	0.8	Parrot	c	1	1	1
Scissors	q	0.6053	0.9744	0.7692	Mynah Bird	q	1	0.8710	0.8438
Spoon	q	0	0.752	0.075	Raven	q	1	0.2368	0.4
Spear	q	1	0.275	0.7179	Elephant	c	0	0.25	0
Chain-saw	q	0.55	1	0.75	Goldfish	c	0	1	0
Club	q	1	0.3590	0.775	Homing Pigeon	q	1	0.775	0.8974
Razor	q	0.625	0.775	0.825	Canary	c	1	1	1

subjects rate *Cuckoo* a better member of the conjunction '*Bird and Pet*' than of the concept *Pet* on its own. This is a strange effect; if the conjunction of concepts behaved like a conjunction of logical propositions the second should be at least as great as the first. This deviation from what one would expect of a standard classical interpretation of conjunctions of concepts is referred to as 'overextension' [26]. Table 1 gives the list of six pairs of concepts and their conjunction for which in [26] the membership weights were measured with respect to different items, and in Table 2 the outcomes of these measurements are given for each of the items.

3   Classical and Non Classical Membership Weights

The behavior of a standard classical weight for a conjunction is described mathematically for the case of one pair of concepts and their conjunction in section 3 of [17]. Consider weights $\mu(A_1)$, $\mu(A_2)$ and $\mu(A_1$ and $A_2)$ of an item X with respect to a pair of concepts A_1 and A_2 and their conjunction 'A_1 and A_2'. We say that they are 'classical membership weights' if and only if there exists a normed measure space $(\Omega, \sigma(\Omega), P)$ and events $E_{A_1}, E_{A_2} \in \sigma(\Omega)$ of the events algebra $\sigma(\Omega)$ such that

$$P(E_{A_1}) = \mu(A_1) \quad P(E_{A_2}) = \mu(A_2) \quad \text{and} \quad P(E_{A_1} \cap E_{A_2}) = \mu(A_1 \text{ and } A_2) \quad (1)$$

A normed measure P is a function defined on a σ-algebra $\sigma(\Omega)$ over a set Ω and taking values in the interval $[0, 1]$ such that the following properties are satisfied: (i) The empty set has measure zero, i.e. $P(\emptyset) = 0$; (ii) Countable additivity or σ-additivity: if E_1, E_2, E_3, ... is a countable sequence of pairwise disjoint sets in $\sigma(\Omega)$, the measure of the union of all the E_i is equal to the sum of the measures of each E_i, i.e. $P(\bigcup_{i=1}^{\infty} E_i) = \sum_{i=1}^{\infty} P(E_i)$; (iii) The total measure is one, i.e. $P(\Omega) = 1$. The triple $(\Omega, \sigma(\Omega), P)$ is called a normed measure space, and the members of $\sigma(\Omega)$ are called measurable sets. A σ-algebra over a set Ω is a nonempty collection $\sigma(\Omega)$ of subsets of Ω that is closed under complementation and countable unions of its members. Measure spaces are the most general structures devised by mathematicians and physicists to represent weights.

We generalize this definition to the case of n concepts $A_1, A_2, \ldots, A_n$ with weights $\mu(A_i)$ for each concept A_i, and weights $\mu(A_i$ and $A_j)$ for the conjunction of concepts A_i and A_j. It is not necessary that weights are measured with respect to each one of the possible pairs of concepts. Hence, to describe this situation formally, we consider a set S of pairs of indices $S \subseteq \{(i,j) \mid i < j; i, j = 1, 2, \ldots, n\}$ corresponding to those pairs of concepts for which the weights have been measured with respect to the conjunction of these pairs. As a consequence, the following set of weights have been experimentally determined

$$p_i = \mu(A_i) \quad i = 1, 2, \ldots, n \quad p_{ij} = \mu(A_i \text{ and } A_j) \quad (i,j) \in S \quad (2)$$

We say that the set of weights in (2) is a 'classical set of membership weights' if it has a normed measure representation, hence if there exists a normed measure space $(\Omega, \sigma(\Omega), P)$ with $E_{A_1}, E_{A_2}, \ldots, E_{A_n} \in \sigma(\Omega)$ elements of the event algebra, such that

$$p_i = P(E_{A_i}) \quad i = 1, 2, \ldots, n \quad p_{ij} = P(E_{A_i} \cap E_{A_j}) \quad (i, j) \in S \qquad (3)$$

4 Geometrical Characterization of Membership Weights

We now introduce a geometric language that makes it possible to verify the existence of a normed measure representation for the set of weights in (2), much like the characterization of Kolmogorovian probability models in [36]. Following [36], we first define an $n + |S|$-tuple, called the $n + |S|$-dimensional correlation vector,

$$\overrightarrow{p} = (p_1, p_2, \ldots, p_n, \ldots, p_{ij}, \ldots) \qquad (4)$$

where $|S|$ is the cardinality of S. Denote $R(n, S) = \mathbb{R}^{n+|S|}$ the $n+|S|$ dimensional vector space over the real numbers. Let $\epsilon \in \{0, 1\}^n$ be an arbitrary n-dimensional vector consisting of $0's$ and $1's$. For each ϵ we construct the following vector $\overrightarrow{u}^\epsilon \in R(n, S)$

$$u_i^\epsilon = \epsilon_i \quad i = 1, 2, \ldots, n \quad u_{ij}^\epsilon = \epsilon_i \epsilon_j \quad (i, j) \in S \qquad (5)$$

The set of convex linear combinations of the $u's$ is called the classical correlation polytope

$$c(n, S) = \{\overrightarrow{f} \in R(n, S) \mid \overrightarrow{f} = \sum_{\epsilon \in \{0,1\}^n} \lambda_\epsilon \overrightarrow{u}^\epsilon; \ \lambda_\epsilon \geq 0; \ \sum_{\epsilon \in \{0,1\}^n} \lambda_\epsilon = 1\} \qquad (6)$$

The following theorem can now be proven similar to what was done in [36] for the case of Kolmogorovian probabilities

Theorem 1. *The set of weights*

$$p_i = \mu(A_i) \quad i = 1, 2, \ldots, n \quad p_{ij} = \mu(A_i \ and \ A_j) \quad (i, j) \in S \qquad (7)$$

admits a normed measure space, and hence is a classical set of membership weights, if and only if its correlation vector $\overrightarrow{p}$ belongs to the correlation polytope $c(n, S)$.

Proof: Suppose that (7) is a classical set of weights, and hence we have a normed measure space $(\Omega, \sigma(\Omega), P)$ and events $E_{A_i} \in \sigma(\Omega)$ such that (2) are satisfied. Let us show that in this case $\overrightarrow{p} \in c(n, S)$. For an arbitrary subset $X \subset \Omega$ we define $X^1 = X$ and $X^0 = \Omega \backslash X$. Consider $\epsilon = (\epsilon_1, \ldots, \epsilon_n) \in \{0, 1\}^n$ and define $A(\epsilon) = \cap_i A_i^{\epsilon_i}$. Then we have that $A(\epsilon) \cap A(\epsilon') = \emptyset$ for $\epsilon \neq \epsilon'$, $\cup_\epsilon A(\epsilon) = \Omega$, and $\cup_{\epsilon, \epsilon_j=1} A(\epsilon) = A_j$. We put now $\lambda_\epsilon = P(A(\epsilon))$. Then we have $\lambda_\epsilon \geq 0$ and $\sum_\epsilon \lambda_\epsilon = 1$, and $p_i = P(A_i) = \sum_{\epsilon, \epsilon_i=1} \lambda_\epsilon = \sum_\epsilon \lambda_\epsilon \epsilon_i$. We also have $p_{ij} = P(A_i \cap A_j) = \sum_{\epsilon, \epsilon_i=1, \epsilon_j=1} \lambda_\epsilon = \sum_\epsilon \lambda_\epsilon \epsilon_i \epsilon_j$. This means that $\overrightarrow{p} = \sum_\epsilon \lambda_\epsilon u^\epsilon$, which shows that $\overrightarrow{p} \in c(n, S)$. Conversely, suppose that $\overrightarrow{p} \in c(n, S)$. Then there exists numbers $\lambda_\epsilon \geq 0$ such that $\sum_\epsilon \lambda_\epsilon = 1$ and $\overrightarrow{p} = \sum_\epsilon \lambda_\epsilon u^\epsilon$. We define $\Omega = \{0, 1\}^n$ and $\sigma(\Omega)$ the power set of Ω. For $X \subset \Omega$ we define then $P(X) = \sum_{\epsilon \in X} \lambda_\epsilon$. Then we choose $A_i = \{\epsilon, \epsilon_i = 1\}$ which gives that $P(A_i) = \sum_\epsilon \lambda_\epsilon \epsilon_i = \sum_\epsilon \lambda_\epsilon u_i^\epsilon = p_i$ and $P(A_i \cap A_j) = \sum_\epsilon \lambda_\epsilon \epsilon_i \epsilon_j = \sum_\epsilon \lambda_\epsilon u_{ij}^\epsilon = p_{ij}$. This shows that we have a classical set of weights.

5   The Correlation Polytopes for Pairs of Concepts and Their Conjunctions

In the case of two concepts A_1, A_2 and their conjunction 'A_1 and A_2' the set of indices is $S = \{(1,2)\}$ and the correlation polytope $c(2,S)$ is contained in the $2 + |S| = 3$ dimensional euclidean space, i.e. $R(2, \{1,2\}) = \mathbb{R}^3$. Further we have four vectors $\epsilon \in \{0,1\}^n$, namely $(0,0), (0,1), (1,0)$ and $(1,1)$, and hence the four vectors $\overrightarrow{u}^\epsilon \in \mathbb{R}^3$ which are the following

$$\overrightarrow{u}^{(0,0)} = (0,0,0) \quad \overrightarrow{u}^{(1,0)} = (1,0,0) \quad \overrightarrow{u}^{(0,1)} = (0,1,0) \quad \overrightarrow{u}^{(1,1)} = (1,1,1) \quad (8)$$

This means that the correlation polytope $c(2, \{1,2\})$ is the convex region spanned by the convex combinations of the vectors $(0,0,0), (1,0,0), (0,1,0)$ and $(1,1,1)$, and the correlation vector is given by $\overrightarrow{p} = (\mu(A_1), \mu(A_2), \mu(A_1 \text{ and } A_2))$. It is well-known that every polytope admits two dual descriptions: one in terms of convex combinations of its vertices, and one in terms of the inequalities that define its boundaries [37]. For the polytope $c(2, \{1,2\})$ the inequalities defining its boundaries are $0 \leq \mu(A_1 \text{ and } A_2)$; $\mu(A_1 \text{ and } A_2) \leq \mu(A_1)$; $\mu(A_1 \text{ and } A_2) \leq \mu(A_2)$ and $\mu(A_1) + \mu(A_2) - \mu(A_1 \text{ and } A_2) \leq 1$.

In Figures 1, 2 and 3 we have represented this correlation polytope $c(2, \{1,2\})$ and all the correlation vectors $\overrightarrow{p}$ for the different items (we have presented the vectors as points not to overload the figure). If the point of the correlation vector corresponding to the data of a specific item lies inside the polytope spanned by $(0,0,0), (1,0,0), (0,1,0)$ and $(1,1,1)$, it is a classical item, for which the membership weights can be represented within a normed measure space. If the point

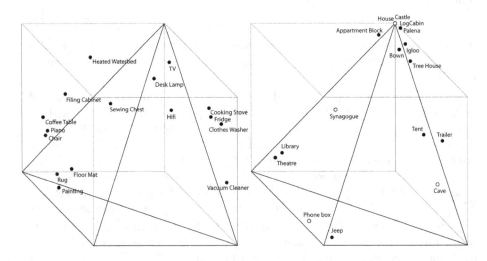

Fig. 1. The polytopes for the concepts *Furniture* and *Household Appliances* and the concepts *Building* and *Dwelling*. The classical items are *Castle, Cave, Phone Box, Synagogue, Log Cabin* and *House*. The other items are non classical.

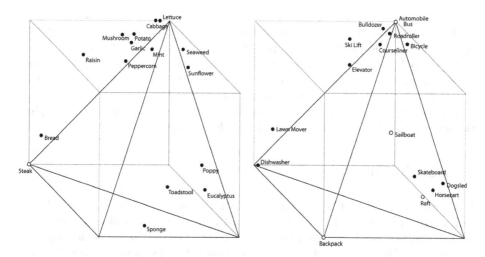

Fig. 2. The polytopes for the concepts *Food* and *Plant* and the concepts *Machine* and *Vehicle*. The classical item are *Steak*, *Backpack*, *Automobile*, *Bus*, *Sailboat* and *Raft*. The other items are non classical.

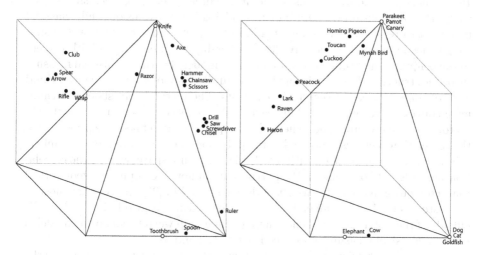

Fig. 3. The polytopes for the concepts *Weapon* and *Tool* and the concepts *Bird* and *Pet*. The classical item are *Knife*, *Toothbrush*, *Elephant*, *Dog*, *Cat*, *Goldfish*, *Parakeet*, *Parrot* and *Canary*. The other items are non classical.

does not lie inside the polytope, the corresponding item is non-classical, indicating that perhaps a quantum representation, for example the one presented in [17], can be elaborated for its weights. Since the polytope is also given by the inequalities defining its boundaries, points lying inside (or outside) the polytope can be characterized by their coordinates satisfying (or violating) these inequalities. The inequalities that define the boundaries of polytope $c(2, \{1, 2\})$ are a lower dimensional variant ([37] and [36]) of the well-known Bell inequalities, studied in the foundations of quantum mechanics. This means that the violation of these inequalities, such as it happens by the data corresponding to items for which the points lie outside the polytope, has from a probabilistic perspective an analogous meaning as the violation of Bell inequalities. Hence these violations may indicate the presence of quantum structures in the domain where the data is collected, which makes it plausible that a quantum model, such as for example the one proposed in [17], can be used to model the data.

6   Conclusion

If two concepts are combined to form a conjunction we can measure the membership weights of items with respect to each of these concepts and also with respect to their conjunction. If the 'conjunction of concepts' behaved like a classical logical conjunction of propositions does, we would expect that the membership weight of an item with respect to the conjunction would never be bigger than the membership weight of this item with respect to one of the concepts. Experiments show that this is not the case, and this counterintuitive effect is referred to as both the guppy effect [24] and overextension [26] in the literature. It has been shown elsewhere that a quantum description can model this overextension while classical measure theoretic structures cannot [17]. In this paper we have elaborated a simple geometric method to identify the membership weights of items with respect to the conjunction of concepts that cannot be modeled within a classical measure theoretic structure. We do this for the general situation of a set of n concepts and a set of conjunctions between these concepts. The method consists of determining a convex polytope, and making each of the items correspond to a correlation vector in the real vector space where also the polytope is defined. We prove that if the correlation vector is contained in the polytope, the considered set of membership weights can be modeled within a classical measure space, while if the correlation vector is not contained in the polytope it cannot. We apply this geometrical characterization method to the set of data collected in [26] and see that most of the tested items have membership weights for which the correlation vector falls outside of the polytope, and hence these membership weights cannot be modeled within a classical measure space (see Figures 1, 2 and 3).

The experimental data for which we show in the present article by means of the polytope criterion that they are non classical, i.e. cannot be represented within a classical measure structure, are all 'conjunction data', meaning that they are membership weights of items with respect to the conjunction of two concepts. The phenomenon of structural non classicality that we put into evidence in this article is however much more general and does not only appear with membership weights of conjunctions. For example, it appears in a very analogous way for disjunctions of concepts. as experimentally shown in [27] and theoretically analyzed in [17], theorems 4, 5 and 6. Following our contextual theory of concepts developed in [9, 10], we have good reason to believe that the effect appears whenever concepts are combined. Unfortunately, the non-classical effect is difficult to identify for an arbitrary combination of concepts, since we do not have a simple mathematical characterization, (like we have for conjunction and disjunction) of what the classical structure of such an arbitrary combination would be. In [27] next to the disjunction also the negation is investigated, and also there significant deviations of what one would expect classically from the logical structure of a negation are measured. We have not yet analyzed these effects on the negation with respect to the quantum models developed in [13, 14, 17], but plan to do so in future work. It would be interesting to make this type of experiments with the remaining not yet tested simple logical connective which is called 'the implication'. The appearance of this type of non-classical weights is however not limited to the domain of concepts and their combinations. In decision theory and in economics different types of situations have been studied entailing a very similar type of non classical weight structure than the one we consider in the present paper [38]. The disjunction effect [39] and the conjunction fallacy [40, 41] are the best known ones, and the disjunction effect has been modeled by a quantum mechanical description in [12], while the conjunction fallacy was analyzed with respect to quantum mechanical modeling in [15]. In [42] a 'disjunction fallacy' is experimentally identified, by considering disjunctions that are not combinations of concepts but already received a name of themselves, such as for example *Natural Sciences* being the disjunction of *Astronomy, Physics, Chemistry, Biology* and *Earth Sciences*. The deviation from how classically this type of disjunction should behave is shown in [42] to be very big. Hence this demonstrates that the non classicality does not find its origin in a kind of 'wrong application of the combination rules', a hypothesis put forward in [43]. It shows that the effect is also present for disjunctions and conjunctions that are single concepts and not combinations. From the perspective of the quantum model developed in [13, 14, 17] we have put forward an explanation for the non classicality, due to the fact that for combinations of concepts equally the single new emergent concept as well as the combination as a logical connective play a role in the influence provoked by the conceptual landscape surrounding the decision situation [38]. A similar simple criterion with polytopes can be worked out for situations of non classicality of other combinations of concepts, and also for the non classical phenomena identified in decision theory, such as the disjunction effect and the conjunction fallacy. We plan to elaborate this in future work.

We note that have formulated all hypothesis and claims in the present article by considering the notion of a 'normed measure space' and its elements representing the membership weights of the items with respect to concepts. Alternatively, we could equally well have considered 'the probability for a specific subject to choose in favor of membership' in replacement of 'the membership weight' as central element. If we would have done so our theorem would become a theorem on probability models instead of a theorem on unitary measure spaces, and it would be mathematically completely equivalent with Pitowsky's main theorem in [36].

The geometrical identification presented here gives rise to a demarcation similar to the violation of the well-known Bell inequalities in physics, which is generally regarded as experimental evidence for the need of a fundamental change in the classical paradigm to describe the process under consideration. Hence,

Hampton's membership weight data giving rise to a situation equivalent to the violation of Bell inequalities, constitutes a strong argument in favor of the fact that quantum structure would be at work within the mechanism giving rise to these data, hence within human cognition. If so, then these experiments constitute a pioneering example of experimentally tested quantum structure in cognition performed by a psychologist in tempore non suspecto. It also would mean that only a non classical description, for example one based on quantum mechanics eventually as the one elaborated in [17], is able to model the mechanism giving rise to the data.

Acknowledgments

This work was supported by grant G.040508N from the FWO-Flanders, Belgium, and a grant from the Social Sciences and Humanities Research Council of Canada.

References

1. Schaden, M.: Quantum finance: A quantum approach to stock price fluctuations. Physica A 316, 511–538 (2002)
2. Baaquie, B.E.: Quantum Finance: Path Integrals and Hamiltonians for Options and Interest Rates. Cambridge University Press, Cambridge (2004)
3. Bagarello, F.: An operational approach to stock markets. Journal of Physics A 39, 6823–6840 (2006)
4. Bordley, R.F.: Quantum mechanical and human violations of compound probability principles: Toward a generalized Heisenberg uncertainty principle. Operations Research 46, 923–926 (1998)
5. Bordley, R.F., Kadane, J.B.: Experiment dependent priors in psychology and physics. Theory and Decision 47, 213–227 (1999)
6. Aerts, D., Aerts, S.: Applications of quantum statistics in psychological studies of decision processes. Foundations of Science 1, 85–97 (1994); Reprinted in: Van Fraassen, B. (ed.): Topics in the Foundation of Statistics, pp. 111–122. Kluwer Academic, Dordrecht (1997)
7. Gabora, L., Aerts, D.: Contextualizing concepts. In: Proceedings of the 15th International FLAIRS Conference. Special track: Categorization and Concept Representation: Models and Implications, Pensacola Florida, May 14-17, pp. 148–152. American Association for Artificial Intelligence (2002)
8. Gabora, L., Aerts, D.: Contextualizing concepts using a mathematical generalization of the quantum formalism. Journal of Experimental and Theoretical Artificial Intelligence 14, 327–358 (2002)
9. Aerts, D., Gabora, L.: A theory of concepts and their combinations I: The structure of the sets of contexts and properties. Kybernetes 34, 167–191 (2005)
10. Aerts, D., Gabora, L.: A theory of concepts and their combinations II: A Hilbert space representation. Kybernetes 34, 192–221 (2005)
11. Busemeyer, J.R., Wang, Z., Townsend, J.T.: Quantum dynamics of human decision making. Journal of Mathematical Psychology 50, 220–241 (2006)

12. Busemeyer, J.R., Matthew, M., Wang, Z.: A quantum information processing theory explanation of disjunction effects. Proceedings of the Cognitive Science Society (2006)
13. Aerts, D.: Quantum interference and superposition in cognition: Development of a theory for the disjunction of concepts (2007), Archive address and link: http://arxiv.org/abs/0705.0975
14. Aerts, D.: General quantum modeling of combining concepts: A quantum field model in Fock space (2007), Archive address and link: http://arxiv.org/abs/0705.1740
15. Franco, R.: The conjunction fallacy and interference effects (2007), Archive address and link: http://arxiv.org/abs/0708.3948
16. Franco, R.: Quantum mechanics and rational ignorance (2007), Archive address and link: http://arxiv.org/abs/physics/0702163
17. Aerts, D.: Quantum Structure in Cognition. Journal of Mathematical Psychology (2009), archive reference and link: http://uk.arxiv.org/abs/0805.3850
18. Aerts, D., Czachor, M.: Quantum aspects of semantic analysis and symbolic artificial intelligence. Journal of Physics A, Mathematical and Theoretical 37, L123-L132 (2004)
19. Widdows, D.: Orthogonal negation in vector spaces for modelling word-meanings and document retrieval. In: Proceedings of the 41st Annual Meeting of the Association for Computational Linguistics, Sapporo, Japan, July 7-12, 2003, pp. 136–143 (2003)
20. Widdows, D., Peters, S.: Word vectors and quantum logic: Experiments with negation and disjunction. In: Mathematics of Language, vol. 8, pp. 141–154. Bloomington, Indiana (2003)
21. Van Rijsbergen, K.: The Geometry of Information Retrieval. Cambridge University Press, Cambridge (2004)
22. Aerts, D., Czachor, M., D'Hooghe, B.: Towards a quantum evolutionary scheme: violating bell's inequalities in language. In: Gontier, N., Van Bendegem, J.P., Aerts, D. (eds.) Evolutionary Epistemology, Language and Culture - A Non Adaptationist Systems Theoretical Approach. Springer, Heidelberg (2006)
23. Widdows, D.: Geometry and Meaning. CSLI Publications, University of Chicago Press (2006)
24. Osherson, D.N., Smith, E.E.: On the adequacy of prototype theory as a theory of concepts. Cognition 9, 35–58 (1981)
25. Hampton, J.A.: Inheritance of attributes in natural concept conjunctions. Memory & Cognition 15, 55–71 (1987)
26. Hampton, J.A.: Overextension of conjunctive concepts: Evidence for a unitary model for concept typicality and class inclusion. Journal of Experimental Psychology: Learning, Memory, and Cognition 14, 12–32 (1988)
27. Hampton, J.A.: Disjunction of natural concepts. Memory & Cognition 16, 579–591 (1988)
28. Hampton, J.A.: The combination of prototype concepts. In: Schwanenflugel, P. (ed.) The Psychology of Word Meanings. Erlbaum, Hillsdale (1991)
29. Hampton, J.A.: Conceptual combination: Conjunction and negation of natural concepts. Memory & Cognition 25, 888–909 (1997)
30. Hampton, J.A.: Conceptual combination. In: Lamberts, K., Shanks, D. (eds.) Knowledge, Concepts, and Categories, pp. 133–159. Psychology Press, Hove (1997)
31. Osherson, D.N., Smith, E.E.: Gradedness and conceptual combination. Cognition 12, 299–318 (1982)

32. Rips, L.J.: The current status of research on concept combination. Mind & Language 10, 72–104 (1995)
33. Smith, E.E., Osherson, D.N.: Conceptual combination with prototype concepts. Cognitive Science 8, 357–361 (1988)
34. Springer, K., Murphy, G.L.: Feature availability in conceptual combination. Psychological Science 3, 111–117 (1992)
35. Storms, G., De Boeck, P., Van Mechelen, I., Ruts, W.: Not guppies, nor goldfish, but tumble dryers, Noriega, Jesse Jackson, panties, car crashes, bird books, and Stevie Wonder. Memory & Cognition 26, 143–145 (1998)
36. Pitowsky, I.: Quantum Probability, Quantum Logic. Lecture Notes in Physics, vol. 321. Springer, Heidelberg (1989)
37. Aerts, S., Aerts, D.: When can a data set be described by quantum theory? In: Bruza, P., Lawless, W., van Rijsbergen, K., Sofge, D., Coecke, B., Clark, S. (eds.) Proceedings of the Second Quantum Interaction Symposium, Oxford 2008, pp. 27–33. College Publications, London (2008)
38. Aerts, D., D'Hooghe, B.: Classical logical versus quantum conceptual thought: examples in economics, decision theory and concept theory. In: Bruza, P.D., Sofge, D., Lawless, W., Van Rijsbergen, C.J., Klusch, M. (eds.) QI 2009. LNCS (LNAI), vol. 5494, pp. 128–142. Springer, Heidelberg (2009)
39. Tversky, A., Shafir, E.: The disjunction effect in choice under uncertainty. Psychological Science 3(5), 305–309 (1992)
40. Tversky, A., Kahneman, D.: Judgments of and by representativeness. In: Kahneman, D., Slovic, P., Tversky, A. (eds.) Judgment under uncertainty: Heuristics and biases. Cambridge University Press, Cambridge (1982)
41. Tversky, A., Kahneman, D.: Extension versus intuitive reasoning: The conjunction fallacy in probability judgment. Psychological Review 90(4), 293–315 (1983)
42. Bar-Hillel, M., Neter, E.: How alike is it versus how likely is it: A disjunction fallacy in probability judgments. Journal of Personality and Social Psychology 65, 1119–1131 (1993)
43. Gavanski, I., Roskos-Ewoldsen, D.R.: Representativeness and conjoint probability. Journal of Personality and Social Psychology 61, 181–194 (1991)

Extracting Spooky-Activation-at-a-Distance from Considerations of Entanglement

Peter Bruza[1], Kirsty Kitto[1], Douglas Nelson[2], and Cathy McEvoy[3]

[1] Faculty of Science and Technology, Queensland University of Technology
p.bruza@qut.edu.au, kirsty.kitto@qut.edu.au
[2] Department of Psychology, University of South Florida
dneslon@cas.usf.au
[3] School of Ageing Studies, University of South Florida
cmcevoy@cas.usf.edu

Abstract. Following an early claim by Nelson & McEvoy [19] suggesting that word associations can display 'spooky action at a distance behaviour', a serious investigation of the potentially quantum nature of such associations is currently underway. This paper presents a simple quantum model of a word association system. It is shown that a quantum model of word entanglement can recover aspects of both the Spreading Activation model and the Spooky model of word association experiments.

1 Modelling Words and Meaning

Human beings are adept and drawing context-sensitive associations and inferences across a broad range of situations ranging from the mundane to the creative inferences that lead to scientific discovery. Such reasoning has a strong pragmatic character and is transacted with comparatively scarce cognitive assets. However, despite our apparent proficiency at drawing inferences, and our ability to express words in such a manner that other people can (usually) understand the meaning that we are trying to convey, our theoretical understanding of how this process occurs has been slow to develop.

The field of cognitive science has recently produced an ensemble of semantic models which have an encouraging, and at times impressive track record of replicating human information processing, such as human word associations norms [16, 4, 14, 15, 11, 12, 24, 13, 25]. The term "semantic" derives from the intuition that words seen in the context of a given word contribute to its meaning, or, more colloquially expressed, the meaning of a word is derived from the "company it keeps" [8]. In order to progress in our understanding of how meaning is generated from sets of words in a language we must understand the way in which the mental lexicon of that language is generated during language acquisition, and how it works once created in the mind of a specific individual.

1.1 The Mental Lexicon

The mental lexicon of a language refers to the words of a language, but its structure is represented by the associative links that bind this vocabulary together.

P. Bruza et al. (Eds.): QI 2009, LNAI 5494, pp. 71–83, 2009.

Such links are acquired through experience and the vast and semi-random nature of this experience ensures that words within this vocabulary are highly interconnected, both directly and indirectly through other words. For example, during childhood development and the associated acquisition of English, the word *planet* becomes associated with *earth, space, moon,* and so on. Even within this set, *moon* can itself become linked to *earth* and *star* etc. Words are so associatively interconnected with each other that they meet the qualifications of a 'small world' network wherein it takes only a few associative steps to move from any one word to any other in the lexicon [26]. Because of such connectivity individual words are not represented in long-term memory as isolated entities but as part of a network of related words. However, depending upon the context in which they are used, words can take on a variety of different meanings and this is very difficult to model [7].

Much evidence shows that for any individual, seeing or hearing a word activates words related to it through prior learning. Understanding how such activation affects memory requires a map of links among known words, and free association provides one reliable means for constructing such a map [21]. In free association experiments, words are presented to large samples of participants who produce the first associated word to come to mind. The probability or strength of a pre-existing link between words is computed by dividing the production frequency of a response word by its sample size. For example, the probabilities that *planet* produces *earth* and *mars* are 0.61 and 0.10, respectively, and we say that *earth* is a more likely or a stronger associate of *planet* than *mars*.

Just like the nonlocality experiments of quantum theory, human memory experiments require very careful preparation of the state to be tested. For example, in extralist cuing, participants typically study a list of to-be-recalled target words shown on a monitor for 3 seconds each (e.g., *planet*). The study instructions ask them to read each word aloud when shown and to remember as many as possible, but participants are not told how they will be tested until the last word is shown. The test instructions indicate that new words, the test cues, will be shown and that each test cue is related to one of the target words just studied (e.g., *universe*). These cues are not present during study (hence, the name extralist cuing). As each cue is shown, participants attempt to recall its associatively related word from the study list. In contrast, during intralist cuing the word serving as the test cue is presented with its target during study (e.g., *universe planet*). Participants are asked to learn the pairing, but otherwise the two tasks are the same. It appears that more associate-to-associate links benefit recall, and two competing explanations for this phenomenon have been proposed: Spreading Activation, and Spooky Activation At a Distance.

This paper will demonstrate an intriguing connection between these two explanations, obtained by making the assumption that words can become entangled in the human mental lexicon.

1.2   Isn't Entanglement Correlation?

Entanglement is a phenomenon unique to quantum behaviour. If a system consisting of two components becomes entangled then it cannot be thought of as separate anymore; a description of one component without reference to the other will, in some cases, fail. Indeed, an entangled quantum system will generally exhibit an intercomponent agreement with reference to any combination of measurement settings. This is of particular importance for a system that becomes spatially extended, as in the case where the two components are taken a long way from each other. Here we find that quantum systems display correlation instantaneously in response to what might even be a delayed choice of measurement setting [1], and yet cannot be used to transmit information between two observers, and thus does not actually violate Special Relativity [17]. This is in contrast with classical scenarios of correlation. In a classical situation a system is in a pre-existing state, and this is discovered through the process of measurement. Not so with a quantum system, where the process of measurement can actively influence the outcome itself. This fundamental difference between the two types of system was first alluded to in the by now famous EPR[1] debate, but was only inescapably highlighted with the more subtle (and recent) results surrounding the contextuality of quantum systems (see [10] or [2] for a good introduction to these ideas). An entangled quantum system is very different from a correlated classical system; no pre-existing elements of reality [6] have been found that can explain the agreement that is obtained between distant measuring devices that are set to determine the state of a quantum system.

To make these ideas more concrete, let us consider a specific example of classical correlation. If the same number is written on two pieces of paper, enclosed in two envelopes, and sent to Alice and Bob at two distant ends of the Universe, the information obtained upon opening of one of the envelopes will instantly correlate with the state of the other envelope at the other end of the universe. However, these correlated pieces of paper are not entangled. The number on the two pieces of paper can be regarded as a hidden variable, or element of reality; even before we open the envelope it exists, in both envelopes. Upon opening the envelope at one end of the Universe we find out what that number is, and hence know what number is already inscribed upon the other piece of paper. The quantum analogue of this scenario would be far stranger. The situation most similar to the nonlocal effects exhibited by entangled quantum systems would involve Alice, at one end of the Universe choosing to write a number upon her blank piece of paper when she opens her envelope, and then finding that Bob, upon opening his envelope found exactly the same number upon his piece of paper at the other end of the Universe. Obviously this does not happen.

However, we might ask if similar cases of intercomponent dependency, or spooky-activation-at-a-distance, exist for systems beyond the field of physics.

We shall now look at the problem of modelling associate-to-associate links in the human mental lexicon, before showing how the assumption that associates

---

[1] Einstein–Podolsky–Rosen.

might be entangled in a subject's cognitive state can lead to a new model of word associations.

2   Modelling Associate-to-Associate Links

Figure 1 shows a hypothetical target word having two target-to-associate links in a subject's cognitive state. There is also an associate-to-associate link between Associates 1 and 2, and an associate-to-target link from Associate 2 to the Target t. The values on the links indicate relative strengths estimated via free associa-tion. Nelson et al. have investigated reasons for the more likely recall of words having more associate-to-associate links [20]. Two competing explanations for why associate-to-associate links benefit recall have been proposed.

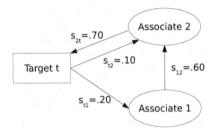

Fig. 1. A hypothetical target with two associates and single associate-to-target and associate-to-associate links. From Nelson, McEvoy, and Pointer ([20]).

The first is the Spreading Activation equation, which is based on the classic idea that activation spreads through a fixed associative network, weakening with conceptual distance (e.g., [5]):

$$S(t) = \sum_{i=1}^{n} S_{ti} S_{it} + \sum_{i=1}^{n} \sum_{j=1}^{n} S_{ti} S_{ij} S_{jt} \tag{1}$$

$$= (.10 \times .70) + (0.20 \times 0.60 \times 0.70) \tag{2}$$

$$= 0.154 \tag{3}$$

where n is the number of associates and $i \neq j$. $S(t)$ denotes the strength of implicit activation of target t due to study, S_{ti} target-to-associate activa-tion strength, S_{it} associate-to-target activation strength (resonance), and S_{ij} associate-to-associate activation strength (connectivity). Multiplying link strengths produces the weakening effect. Activation ostensibly travels from the target to and among its associates and back to the target in a continuous chain, and the target is strengthened by activation that returns to it from pre-existing connections involving two and three-step loops. More associate-to-associate links create more three-step loops and theoretically benefit target recall by increasing its activation strength in long-term memory. Importantly, note that the effects of associate-to-associate links are contingent on the number and strength of associate-to-target links because they allow activation to return to the target. If associate-to-target links were absent, even the maximum number of associate-to-associate links would have no effect on recall because activation could not return to the target.

In contrast, in the 'Spooky Activation at a Distance' equation, the target activates its associative structure in synchrony:

$$S(t) = \sum_{i=1}^{n} S_{ti} + \sum_{i=1}^{n} S_{it} + \sum_{i=1}^{n}\sum_{j=1}^{n} S_{ij} \tag{4}$$

$$= 0.20 + 0.10 + 0.70 + 0.60 \tag{5}$$

$$= 1.60 \tag{6}$$

where $i \neq j$; S_{ti}, target-to-associate i strength; S_{it}, associate i-to-target strength (resonance); S_{ij}, associate i-to-associate j strength (connectivity).

This equation assumes that each link in the associative set contributes additively to the target's net strength. The beneficial effects of associate-to-associate links are not contingent on associate-to-target links. Stronger target activation is predicted when there are many associate-to-associate links even when associate-to-target links are absent. In fact, associate-to-target links are not special in any way. Target activation strength is solely determined by the sum of the link strengths within the target's associative set, regardless of origin or direction.

3   Entanglement of Words

How should we represent the combination of words in the human mental lexicon? QT uses the tensor product, $\otimes$, to denote composite systems. Consider the case of $m = 2$ study words: u and v presented to a group of subjects. Let us assume that, when cued, the subjects recall neither target word. In this case we could write:

$$|u\rangle \otimes |v\rangle = |0\rangle \otimes |0\rangle = |00\rangle \tag{7}$$

where the notation $|00\rangle$ is just shorthand for the tensor product state $|0\rangle \otimes |0\rangle$ describing the composite system of two negative outcomes. If word u alone was recalled then we would write $|u\rangle \otimes |v\rangle = |10\rangle$, whereas in the converse case we would write $|01\rangle$ and finally, if both words were recalled then the tensor product would yield the state $|11\rangle$.

However, this straightforward scenario is not the only form of situation possible in the quantum formalism. Superposition states can also occur, and these are important as they can represent the situation where the words u and v may be more likely to be recalled in one context than another. Assume that we can represent one subject's cognitive state with reference to the combined targets u and v as a 2 q-bit register that refers to their states of 'recalled' and 'not recalled' in combination. Thus, if we represent the target words using the standard superpositions $|u\rangle = a_0|0\rangle + a_1|1\rangle$ and $|v\rangle = b_0|0\rangle + b_1|1\rangle$, (where $a_0^2 + a_1^2 = 1$

and $b_0^2 + b_1^2 = 1$), then it is possible to denote the state of the combined system by writing the tensor product

$$|u\rangle \otimes |v\rangle = (a_0|0\rangle + a_1|1\rangle) \otimes (b_0|0\rangle + b_1|1\rangle) \qquad (8)$$
$$= a_0b_0|00\rangle + a_1b_0|10\rangle + a_0b_1|01\rangle + a_1b_1|11\rangle, \qquad (9)$$

where $|a_0b_0|^2 + |a_1b_0|^2 + |a_0b_1|^2 + |a_1b_1|^2 = 1$. This is the most general state possible. It represents a quantum combination of the above four possibilities, obtained using a tensor multiplication between the states $|u\rangle$ and $|v\rangle$. In contrast to the simple cases discussed above, here no state of recall is 'the' state, rather, we must cue the subject and elicit a response from them before we can talk about a word being 'recalled' or 'not recalled'. Indeed, a different cue might elicit a very different response, and the quantum formalism could deal with this via a change of basis.

It is important to realise however, that (9) is *not* the only form of state that can be obtained from combination of $|u\rangle$ and $|v\rangle$ in the quantum formalism.

The other form of state, an *entangled state* is one that it is impossible to write as a product. As an example of an entangled state, we might consider the the state ψ where the words u and v are either *both* recalled, or both *not* recalled in relation to a cue q. One representation of this scenario is given by the following state:

$$\psi = \frac{1}{\sqrt{2}}(|00\rangle + |11\rangle). \qquad (10)$$

This seemingly innocuous state is one of the so-called *Bell states* in QT. It is impossible to write as a product state, thus it differs markedly from (9). The fact that entangled systems cannot be expressed as a product of the component states makes them non-separable. More specifically, there are no coefficients which can decompose equation (10) into a product state exemplified by equation (9) which represents the two components of the system, u and v, as independent of one another. For this reason ψ is not written as $|u\rangle \otimes |v\rangle$ as it can't be represented in terms of the component states $|u\rangle$ and $|v\rangle$.

4   An Analysis of Spooky-Activation-at-a-Distance in Terms of Entanglement

Nelson and McEvoy have recently begun to consider the Spooky-activation-at-a-distance formula in terms of quantum entanglement, claiming that "The activation-at-a-distance rule assumes that the target is, in quantum terms, entangled with its associates because of learning and practicing language in the world. Associative entanglement causes the studied target word to simultaneously activate its associate structure" [19, p3]. The goal of this section is to formalise this intuition. At the outset, it is important that the quantum formalism be able to cater for the set size and connectivity effects described elsewhere [3]. Recall both set size and associative connectivity have demonstrated

Table 1. Matrix corresponding to hypothetical target shown in Figure 1

	t	a_1	a_2
t		0.2	0.1
a_1			0.6
a_2	0.7		
	$p_t = 0.7$	$p_{a_1} = 0.2$	$p_{a_2} = 0.7$

time and again robust effects on the probability of recall. Because the Spooky-activation-at-a-distance formula sums link strengths irrespective of direction, it encapsulates the idea that a target with a large number of highly interconnected associates will generate a high activation level during study.

Table 1 is a matrix representation of the associative network of the hypothetical target t shown in Figure 1. The last line of the matrix represents the summation of free association probabilities for a given word. For example,

$$p_{a_2} = \Pr(a_2|t) + \Pr(a_2|a_1) \tag{11}$$
$$= 0.1 + 0.6 \tag{12}$$
$$= 0.7 \tag{13}$$

These free association probabilities may be added as it is assumed that each free association experiment is independent, that is associate a_2 being recalled in relation to cue t is assumed independent of it being recalled in relation to the cue a_1. Since free association experiments require a subject who has not been primed in any way beyond that provided by the cue, this appears to be a reasonable simplifying assumption.

Both Spreading Activation and Spooky-activation-at-a-distance approaches assume that free association probabilities determine the strength of activation of a word during study; they only differ in the way this activation strength is computed. Viewing free association probabilities in this way allows the matrix to be considered as a many bodied quantum system modelled by three qubits. Figure 2 depicts the system, here, each word is in a superposed state of being activated $|1\rangle$, or not $|0\rangle$. Note how each summed column with a non-zero probability leads to a qubit.

For ease of exposition in the following analysis, we shall change variables. The probabilities depicted in table 1 are related to the probability densities of figure 2 by taking their square root: e.g. $\pi_t^2 = p_t$. Using such a change of variables, the state of the target word t would be written as:

$$|t\rangle = \sqrt{\bar{p}_t}|0\rangle + \sqrt{p_t}|1\rangle = \bar{\pi}_t|0\rangle + \pi_t|1\rangle, \tag{14}$$

where the probability of recall due to free association is $p_t = \pi_t^2$, and $\bar{p}_t = 1 - p_t = \bar{\pi}_t^2$ represents the probability of a word not being recalled. Thus, the states of the individual words are represented as follows in order to avoid cluttering the analysis with square root signs:

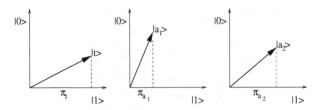

Fig. 2. Three bodied quantum system of words

$$|t\rangle = \bar{\pi}_t|0\rangle + \pi_t|1\rangle \tag{15}$$
$$|a_1\rangle = \bar{\pi}_{a_1}|0\rangle + \pi_{a_1}|1\rangle \tag{16}$$
$$|a_2\rangle = \bar{\pi}_{a_2}|0\rangle + \pi_{a_2}|1\rangle \tag{17}$$

where $\bar{\pi}_t = 1 - \pi_t, \bar{\pi}_{a_1} = 1 - \pi_{a_1}$ and $\bar{\pi}_{a_2} = 1 - \pi_{a_2}$.

As detailed in the previous section, tensor products are used to model many bodied quantum systems. The state ψ_t of the most general combined quantum system is given by the tensor product of the individual states:

$$\psi_t = |t\rangle \otimes |a_1\rangle \otimes |a_2\rangle \tag{18}$$
$$= (\bar{\pi}_t|0\rangle + \pi_t|1\rangle) \otimes (\bar{\pi}_{a_1}|0\rangle + \pi_{a_1}|1\rangle) \otimes (\bar{\pi}_{a_2}|0\rangle + \pi_{a_2}|1\rangle) \tag{19}$$
$$\begin{aligned} = & \bar{\pi}_t\bar{\pi}_{a_1}\bar{\pi}_{a_2}|000\rangle + \pi_t\bar{\pi}_{a_1}\bar{\pi}_{a_2}|100\rangle + \bar{\pi}_t\pi_{a_1}\bar{\pi}_{a_2}|010\rangle + \pi_t\pi_{a_1}\bar{\pi}_{a_2}|110\rangle \\ & + \bar{\pi}_t\bar{\pi}_{a_1}\pi_{a_2}|001\rangle + \pi_t\bar{\pi}_{a_1}\pi_{a_2}|101\rangle + \bar{\pi}_t\pi_{a_1}\pi_{a_2}|011\rangle + \pi_t\pi_{a_1}\pi_{a_2}|111\rangle \end{aligned} \tag{20}$$

The intuition behind this expression is an enumeration of the possibilities of the states of the qubits. So, $|111\rangle$ represents the state in which all respective qubits collapse onto their state $|1\rangle$. In other words, $|111\rangle$ denotes the state of the system in which words t, a_1 and a_2 have all been *activated* due to study of target t. The probability of observing this is given by the taking the square of the product $\pi_t\pi_{a_1}\pi_{a_2}$. Conversely, the state $|000\rangle$ corresponds to the situation in which none of the words have been activated.

The state ψ_t of the three-bodied system does not capture Nelson & McEvoy's intuition that the studied target word t simultaneously activates its associative structure. Recall their suggestion that target t activates its associates in synchrony: When target t is studied it activates *all* of its associates, or none at all. Interestingly, this idea is quite easily captured by the assumption that the state ψ_t evolves into a Bell entangled state ψ_t' of the form:

$$\psi_t' = \sqrt{p_0}|000\rangle + \sqrt{p_1}|111\rangle. \tag{21}$$

This formula represents a superposed state in which the entire associative structure of a subject is in a state of potential activation ($|111\rangle$) or potential non-activation ($|000\rangle$), with the probabilities of these states occurring given by p_0 and p_1 respectively.

How should we ascribe values to the probabilities p_0 and p_1? In QT these values would be determined by the unitary dynamics evolving ψ_t into ψ_t'. As such dynamics are yet to be worked out for cognitive states, we are forced to speculate. One approach is to assume the lack of activation of the target is determined solely in terms of lack of recall of any of the associates. That is,

$$p_0 = \bar{p}_t \bar{p}_{a_1} \bar{p}_{a_2} \tag{22}$$

Consequently, the remaining probability mass contributes to the activation of the associative structure as a whole. Thus, starting from the assumption of a Bell entangled state, we find that the probability p_1 expresses Nelson & McEvoy's intuition:

$$p_1 = 1 - \bar{p}_t \bar{p}_{a_1} \bar{p}_{a_2} \tag{23}$$
$$= 1 - (1 - p_t)(1 - p_{a_1})(1 - p_{a_2}) \tag{24}$$
$$= 1 - (1 - p_t - p_{a_1} + p_t p_{a_1} - p_{a_2} + p_t p_{a_2} + p_{a_1} p_{a_2} - p_t p_{a_1} p_{a_2}) \tag{25}$$
$$= \underbrace{p_t + p_{a_1} + p_{a_2}}_{A} + \underbrace{p_t p_{a_1} p_{a_2}}_{B} - \underbrace{(p_t p_{a_1} + p_t p_{a_2} + p_{a_1} p_{a_2})}_{C} \tag{26}$$

Term A corresponds to the summation of the free association probabilities in the above matrix. In other words, term A corresponds exactly to the Spooky-activation-at-a-distance formula (See equation 4). At such, the assumption of a Bell entangled state provides partial support for the summation of free association probabilities which is embodied by the Spooky-activation-at-a-distance equation. Ironically perhaps, term B corresponds to free association probabilities multiplied according to the directional links in the associative structure. This is expressed in the second term of the spreading activation formula (See equation 1). In other words, departing from an assumption of entanglement leads to an expression of activation strength which combines aspects of both Spooky-activation-at-a-distance and Spreading Activation.

The third term C is more challenging to interpret. In a more complete model capable of generating some form of evolutionary dynamics we might expect that it would arise from the underlying structure of the Hilbert space used. Here, it has arisen from apparently sensible assumptions about how probabilities should be amassed in the the Bell entangled state. What significance can be drawn from this term? The beginnings of an answer can be formulated by returning to figure 1. When seen in the context of actual values, the term C has a significant compensating effect, subtracting 0.77 from the summation of the Spooky term with the spreading activation term:

$$p_1 = A + B - C \tag{27}$$
$$= (0.7 + 0.2 + 0.7) + (0.7 \times 0.2 \times 0.7)$$
$$\quad - (0.7 \times 0.2 + 0.7 \times 0.7 + 0.2 \times 0.7) \tag{28}$$
$$= 0.928 \tag{29}$$

Thus, according to this analysis, the strength of activation p_1 lies somewhere between Spreading Activation and Spooky-activation-at-a-distance. Based on a

substantial body of empirical evidence, Nelson and McEvoy have argued persua-
sively spreading activation *underestimates* strength of activation [23]. Here we
have seen that when departing from an assumption that the associative structure
is Bell entangled, a cautious preliminary conclusion is that Spooky-activation-
at-a-distance overestimates the strength of activation and it is term C which
compensates for this.

We conclude this section with some remarks how the entanglement model
above assists in the explanation of some experimental results that have not been
well accounted for in current models of the mental lexicon. Attempts to map the
associative lexicon soon made it clear that some words produce more associates
than others. This feature is called 'set size' and it indexes a word's associative
dimensionality [18, 22]. It was also revealed that the associates of some words
are more interconnected than others. Some words have many such connections
(e.g., *moon-space*, *earth-planet*), whereas some have none, and this feature is
called "connectivity" [20]. Thus, experiments have shown that in addition to
link strengths between words, the set size and connectivity of individual words
have powerful effects on word recall, which existing theories cannot generally
explain.

This is an interesting result for the model suggested here, as according to
the above analysis we might surmise that the more associates a target has,
the more qubits are needed to model it. When these are tensored the resulting
space will have a higher dimensionality. Therefore a large set size is catered
for by a tensor space of higher dimensionality. Conversely, interconnectivity is
catered for by larger probabilities in the initial superposed states of the respective
qubits. Consider once again the matrix in table 1. In the general case, when the
associative structure is highly interconnected the sums of probabilities in the last
row will tend to be higher. These will contribute to higher activation strength as
they are summed (as defined by probability p_1) in the same way as in Spooky-
activation-at-a-distance. So it is possible to have a high activation strength even
though the tensor space has high dimensionality. When the associative structure
in not interconnected these probabilities will be low and hence lead to lower
strength of activation.

5   Conclusions and Future Directions

Obviously the model presented here is overly simple. The human mental lexicon
is a vast and highly interconnected network consisting of thousands of words, and
its associative structure is far more complex than the simple toy model in figure 1
could ever hope to represent. However, some promising initial results have been
obtained from this very simplified analysis. We have seen some evidence that a
quantum approach can model set size and connectivity effects, and a prediction
has been made that Spooky-activation-at-a-distance overestimates associative
behaviour by some factor. While equation (26) makes a very concrete suggestion
about how large this overestimation might be, it is unlikely that the situation
will prove this simple.

Firstly, this is a very basic toy model, addressing the behaviour of only one target and two associates. A fully developed theory would have to be stated in terms of all the targets and all of their associates. It is very hard to extrapolate from this model to a more realistic one. One possibility would involve applying the apparatus of Statistical Mechanics in a density matrix approach to model sets of targets and their associates, but no results have been obtained here to date. A related issue surrounds the meaning of the term C in equation (26) as more associative structure is added. Here, we would see C gaining more and more terms (since A and B essentially correspond to the two extreme values in the expansion of the bracketed term in (24)). This does not seem plausible, and the way in which this term would act as it became larger is yet to be established.

In addition to these immediate issues, much work remains to be done in this area. Given the large amount of data collected about word association norms [21] we might expect that experiments can be performed that might distinguish between the varying predictions of the different models, and work is underway to generate results here. A physically and cognitively motivated time evolution equation capable of generating the Bell-type state (10) is essential before we can consider this model to be truly quantum(like) and initial ruminations about how this might be achieved are presented in [9].

Ending on a slightly more positive note, we consider the most important result of this article to be the indication that a straight tensorial combination of associate words in the mental lexicon is not particularly representative of the intuition that words and their associates are activated in synchrony. Given that a Bell-type entangled state provides a far more likely candidate for the behaviour of word association networks this result thus provides some first steps towards establishing evidence that human cognitive structures have some quantum(like) behaviour.

Acknowledgements. This project was supported in part by the Australian Research Council Discovery grant DP0773341 to P. Bruza and K. Kitto, and by grants from the National Institute of Mental Health to D. Nelson and the National Institute on Ageing to C. McEvoy.

References

[1] Aspect, A., Dalibard, J., Roger, G.: Experimental test of Bell's inequalities using time varying analyzers. Physical Review Letters 49(25), 1804–1807 (1982)

[2] Ballentine, L.: Quantum Mechanics: A modern development. World Scientific, Singapore (1998)

[3] Bruza, P.D., Kitto, K., Nelson, D., McEvoy, C.L.: Entangling words and meaning. In: Proceedings of the Second Quantum Interaction Symposium (QI 2008). College Publications (2008)

[4] Burgess, C., Livesay, K., Lund, K.: Explorations in context space: words, sentences, discourse. Discourse Processes 25(2&3), 211–257 (1998)

[5] Collins, A.M., Loftus, E.F.: A spreading-activation theory of semantic processing. Psychological Review 82(6), 407–428 (1975)
[6] Einstein, A., Podolsky, B., Rosen, N.: Can quantum-mechanical description of physical reality be considered complete? Phyical Review 47, 777–780 (1935)
[7] Gabora, L., Rosch, E., Aerts, D.: Toward an ecological theory of concepts. Ecological Psychology 20, 84–116 (2008)
[8] Kintsch, W.: Predication. Cognitive Science 25, 173–202 (2001)
[9] Kitto, K., Bruza, P.D., Sitbon, L.: Generalising Unitary Time Evolution. In: Bruza, P., Sofge, D., Lawless, W., van Rijsbergen, K., Klusch, M. (eds.) QI 2009. LNCS (LNAI), vol. 5494, pp. 17–28. Springer, Heidelberg (2009)
[10] Laloë, F.: Do we really understand quantum mechanics? Strange correlations, paradoxes, and theorems. American Journal of Physics 69(6), 655–701 (2001)
[11] Landauer, T.K., Dumais, S.T.: A solution to Plato's problem: The latent semantic analysis theory of acquisition, induction and representation of knowledge. Psychological Review 104, 211–240 (1997)
[12] Landauer, T.K., Foltz, P.W., Laham, D.: An introduction to latent semantic analysis. Discourse Processes 25(2&3), 259–284 (1998)
[13] Levy, J.P., Bullinaria, J.A.: Learning lexical properties from word usage patterns: Which context words should be used? In: French, R.F., Sounge, J.P. (eds.) Connectionist Models of Learning, development and Evolution: Proceedings of the Sixth Neural Computation and psychology Workshop, pp. 273–282. Springer, Heidelberg (1999)
[14] Lowe, W.: What is the dimensionality of human semantic space? In: Proceedings of the 6th Neural Computation and Psychology workshop, pp. 303–311. Springer, Heidelberg (2000)
[15] Lowe, W.: Towards a theory of semantic space. In: Moore, J.D., Stenning, K. (eds.) Proceedings of the Twenty-Third Annual Conference of the Cognitive Science Society, pp. 576–581. Lawrence Erlbaum Associates, Mahwah (2001)
[16] Lund, K., Burgess, C.: Producing high-dimensional semantic spaces from lexical co-occurrence. Behaviour Research Methods, Instruments & Computers 28(2), 203–208 (1996)
[17] Maudlin, T.: Quantum non-locality and relativity: metaphysical intimations of modern physics. Aristotelian society series, vol. 13. Blackwell publishers limited, Oxford (1994)
[18] Nelson, D., McEvoy, C.L.: Encoding context and set size. Journal of Experimental Psychology: Human Learning and Memory 5(3), 292–314 (1979)
[19] Nelson, D., McEvoy, C.L.: Entangled associative structures and context. In: Bruza, P.D., Lawless, W., van Rijsbergen, C.J., Sofge, D. (eds.) Proceedings of the AAAI Spring Symposium on Quantum Interaction. AAAI Press, Menlo Park (2007)
[20] Nelson, D., McEvoy, C.L., Pointer, L.: Spreading activation or spooky action at a distance? Journal of Experimental Psychology: Learning, Memory and Cognition 29(1), 42–52 (2003)
[21] Nelson, D., McEvoy, C.L., Schreiber, T.: The university of South Florida, word association, rhyme and word fragment norms. Behavior Research Methods, Instruments & Computers 36, 408–420 (2004)
[22] Nelson, D., Schreiber, T.A., McEvoy, C.L.: Processing implicit and explicit representations. Psychological Review 99(2), 322–348 (1992)
[23] Nelson, D.L., McEvoy, C.L.: Entangled Associative Structures and Context, Stanford University. AAAI Press, Menlo Park (2007)

[24] Patel, M., Bullinaria, J.A., Levy, J.P.: Extracting semantic representations from large text corpora. In: French, R.F., Sounge, J.P. (eds.) Connectionist Models of Learning, Development and Evolution: Proceedings of the Fourth Neural Computation and Psychology Workshop, pp. 199–212. Springer, Heidelberg (1997)

[25] Sahlgren, M.: Towards a Flexible Model of Word Meaning. In: The AAAI Spring Symposium 2002, Stanford University, Palo Alto, California, USA, March 25-27 (2002)

[26] Steyvers, M., Tenenbaum, J.B.: Graph theoretic analysis of semantic networks: Small worlds in semantic networks. Cognitive Science 29(1), 41–78 (2005)

Quantum Amplitude Amplification Algorithm: An Explanation of Availability Bias

Riccardo Franco

Abstract. In this article, I show that a recent family of quantum algorithms, based on the quantum amplitude amplification algorithm, can be used to describe a cognitive heuristic called availability bias. The amplitude amplification algorithm is used to define quantitatively the ease of a memory task, while the quantum amplitude estimation and the quantum counting algorithms to describe cognitive tasks such as estimating probability or approximate counting.

1 Introduction

The idea that human judgements and decision-making can evidence quantum mechanics behaviour has a great deal of intuitive appeal, and it is at the basis of a recent research topic, which can be called quantum cognition. A number of authors have explored such idea, like [1] for decision making, or [2] and [3] for human judgements. The quantum-like models there proposed seem to adequately describe the experimental results: however, the potentialities of the quantum formalism have not been fully explored in cognitive science, mainly for what concerns the quantum parallelism and a characterization of quantum algorithms in terms of human tasks.

In the present article, I propose to describe the experimental results concerning the availability heuristic with the *quantum amplitude amplification, quantum amplitude estimation* and *quantum counting* algorithms [8]: the first is a recent generalization of Grover's algorithm [9], while the other two algorithms are applications of the amplitude amplification, followed by a quantum Fourier transform. I show that these algorithms are able to model some important experimental results of cognitive science relevant to availability heuristic: in particular, the amplitude amplification algorithm gives a mathematical characterization of the ease to recall items or concepts, while the amplitude estimation/counting algorithms allow to introduce a formal connection between such ease and judgements of probability/frequency about facts. Here I do not discuss about the physical possibility for human mind to perform quantum algorithms: I only consider from a formal point of view the problem of defining mathematically the ease to remember.

Grover's algorithm is an important quantum algorithm based on quantum parallelism which allows to search in an unsorted database with high number of items faster than any classical algorithm (quadratic speedup). One of the first attempts to use such algorithm in cognitive science (more precisely a generalization [10]) has been done by Franco [11] to describe the influence of emotions

P. Bruza et al. (Eds.): QI 2009, LNAI 5494, pp. 84–96, 2009.

on the ease to remember. The same quadratic speedup is provided by amplitude amplification algorithm, based on quantum parallelism.

This article attempts to model within such quantum framework the *availability heuristic*, a human cognitive bias that causes people to estimate frequency or probability on the basis of *how easily* they can recall or imagine instances of whatever it is they are trying to estimate. The article is structured as follows: in section 2 I describe the main features of availability heuristic, while in sections 3, 4 and 5 I introduce the main features of the quantum algorithms based on the amplitude amplification, and in section 6 I compare the model's predictions with the experimental data.

2   Availability Bias

Availability is a human cognitive bias that causes people to estimate the probability or the size of particular categories of items on the basis of *how easily* they can recall or imagine them. Such heuristic, discovered in 1973 by psychologists Amos Tversky and Daniel Kahneman (2002 Nobel Prize in Economics) [4], is at the root of many other human biases and culture-level effects. Two important examples of availability heuristic introduced by Tversky and Kanneman [4], which I will discuss in a detailed way in section (6), are the word frequency experiment, and the risk frequency experiment. In the word frequency experiment, replicated and discussed later by Seldmeier et al. [5], subjects have to estimate the likelihood of letter R in the first or in the third position of English words. Even if the letter R in the English language appears more frequently in the third than in the first position, the most part of participants judged the first position to be more likely. In the risk frequency experiment, replicated and discussed later by Lichtenstein et al. [6] and Hertwig et al. [7], it is evidenced that subjects assess the risk of heart attack among middle-aged people by recalling such occurrences among ones acquaintances. Both experiments thus evidence that judged frequencies are higher for categories which provide a better cue for recalling instances of them.

Several researchers have pointed out that the notion of the availability heuristic presented in [4] has been only vaguely sketched and is consistent with several different mechanisms. In fact, within the availability heuristic two major mechanisms have been identified, which seem to coexist and to play an important role: 1) the *ease to remember*, due for example to items' vividness in memory, and 2) the *overestimation/underestimation* of low/high frequencies respectively.

In the definition of the availability heuristic it is important to operationalize the ease with which the memory processes are performed. In particular, two different definitions have been used widely in availability experiments, giving an experimental measure of the ease of the memory task: 1) the *availability-by-number*, the produced number of good items in a fixed time, and 2) the *availability-by-speed*, the retrieval time for a fixed number of good items. In general, the availability experiments involve two (sometimes different) groups of subjects: one which performs the memory task, and one that performs the judgements about the probability/number of items. Thus the availability experiments verify a positive correlation between the measure of ease in the memory task (by using the availability-by-number or the availability-by-speed) and the quantitative judgements performed by the subjects. I will focus the attention on the following two categories of experiments: *judgements of probability*, where subjects judge the probability of events, like in the word frequency experiment of [4], and *judgements of number*, where subjects judge the number of instances/occurrences of an event, like in the risk frequency experiment [4] (where subjects estimate the incidence of a disease in terms of number of ill persons in a year).

In the next sections, I will explore the correlation between availability-by-number and the judgements of probability/number. The main idea is that subjects, when perform their judgements, do not perform complete recalling tasks, which seem to be required by the concept of availability-by-number. I will show that that the amplitude estimation algorithm is based only on partial recalling tasks, and thus can capture simultaneously the concepts of availability-by-number and the overestimation/underestimation for low/high probabilities.

3   Amplitude Amplification Algorithm

The amplitude amplification algorithm, invented by Brassard et al. [8], is a generalization of Grover's algorithm, and it can be used for solving the following problem: let us consider N items and a boolean function $f : \{0, 1, ..., N - 1\} \rightarrow \{0, 1\}$, which partitions the items into t *good items* (those for which f is equal to 1), and $N - t$ *bad items* (those for which f is equal to 0). It is evident that such algorithm can be used to model the retrieval tasks in cognitive science. For example, the experiment of Tversky and Kahneman [4] relevant to words with letter R in first or third position can be represented as a partitioning of English words in two categories: the good items (words with R at first position) and the bad items (words with R at third position). Even if the mathematical details of the algorithm are described in next subsection, I now present the main features, reducing to the minimum the formalism. The intuitions here presented are similar to those preliminarly exposed in Franco [11].

The quantum amplification algorithm is also called **QSearch**(A, f), where the parameter f is the boolean function previously defined, while A is an operator, whose meaning is to define the initial weight a relevant to good items. Like Grover's algorithm, the algorithm is composed by three main parts:

1) The *initial state*, in which the N items are encoded into the elements of a basis of a N-dimensional vector space. An important feature of the amplitude amplification algorithm, which differences it from Grover's algorithm, is that the items within such initial state can have different weights: in particular, the parameter a is the probability to measure a good item in such initial state. In Grover's algorithm we always have $a = t/N$. The initial state can be interpreted, in the context of cognitive processes, as a *guessing state*, representing the initial mental weights relevant to the items. If $a > t/N$, this means that the good items have initially more relevance than the bad items. If the guessing state is a flat

distribution over all the items ($a = t/N$), this means that the subjects have no preliminar idea about good/bad items.

2) The *amplification engine*, which is an iterative process allowing to enhance the weights of the good items: at each step the boolean function f is evaluated simultaneously over all the items, and the weights of the good items are enhanced through interference effects. Differently from Grover's algorithm, the efficiency of the amplification engine depends on the *guessing state*: the algorithm succeeds after a number of iterations proportional to $1/\sqrt{a}$. If $a = t/N$ the algorithm is equal to Grover's algorithm, and the required number of steps is proportional to $\sqrt{N/t}$. It is important to note that a classic algorithm would imply a number of steps proportional to N/t, while Grover's algorithm allows for a quadratic speedup, that is a number of steps proportional to $\sqrt{N/t}$. The amplitude amplification algorithm allows for a further speedup when the guessing state is such that $a > t/N$, because the number of required steps is proportional to $1/\sqrt{a} < \sqrt{N/t}$: the initial guessing state gives higher weight to the good items than to the bad items, making faster the retrieval process.

The interpretation of such amplification engine in the context of cognitive tasks is in terms of subconscious processes: they allow for parallelism in the evaluation of the boolean function over all the items, but they need a number of iterations proportional to $1/\sqrt{a}$ to amplify the probability of good items. In other words, the subjects are able to apply $f(x)$ on each item x (thus deciding if each item is good or bad). The algorithm suggests that such decision procedure is performed in a parallel and subconscious way, thus faster than in a serial way.

3) A *measure* on the final state. The algorithm modifies the initial guess state, producing a final state which contains almost only good states. Thus a final measure produces one of the good items, and the recall task is finished. This fact represents in my description the conscious act of remembering.

The amplitude amplification algorithm gives a simple mathematical definition of the *ease* to retrieve in terms of the availability-by-speed: the time required to find a good item is proportional to $1/\sqrt{a}$, where a is the initial guessing parameter: a high value of a gives a short time to retrieve a good item. The parameter a represents how vivid are the good items in memory before retrieving them.

We finally note that the *availability-by-number* (the number of good items that subjects can remember in a fixed time) is proportional, in our model, to $\sqrt{a}$. In the word frequency experiment, where subjects have to write English words with letter R in first or third position [4], the time to produce the word is lower with R as first letter than as the third letter. Thus I assume that the guess state contains a set of N items (the most common English words), and the weight for the words beginning with R is higher than for those with R at third position.

3.1 Mathematical Details for the Amplitude Amplification Algorithm

In the quantum formalism, the partition of N items into good and bad items leads to consider a N−dimensional Hilbert space, whose computational basis is $\{|0\rangle, |1\rangle, ..., |N-1\rangle\}$: each vector corresponds to a particular item. Thus the function f introduces a partition of H into a *good subspace* (spanned by the vectors $|x\rangle$ for which $f(x) = 1$) and a *bad subspace* (spanned by the vectors $|x\rangle$ for which $f(x) = 0$). Thus any superposition $|s\rangle = \sum_x \psi(x)|x\rangle$ can be written as $|s\rangle = |\psi_0\rangle + |\psi_1\rangle$, where $|\psi_1\rangle$ is the superposition of good vectors ($f(x) = 1$) and $|\psi_0\rangle$ is the superposition of bad vectors ($f(x) = 1$).

The algorithm presents the following steps:

1) *Initial state*: prepare the vector $A|0\rangle = |\psi_0\rangle + |\psi_1\rangle$, where A is a quantum algorithm which uses no measurement, and $a = \langle\psi_1|\psi_1\rangle$ is the probability to measure a good state. If A is the quantum Fourier transform $F_N : |x\rangle \rightarrow N^{-1/2}\sum_{y=0}^{N-1} e^{2\pi i x y}|y\rangle$, we have a uniform superposition of vector states with amplitude $N^{-1/2}$, and $a = t/N$ (as in standard Grover's algorithm).

2) *Amplification engine*: apply the operator $Q = -AS_0A^{-1}S_f$, where S_0 and S_f are conditional phase inversion operators (S_0 changes the sign of the amplitude if and only if the state is the zero state $|0\rangle$, while S_f conditionally changes the sign of the amplitudes of the good states).

3) *Measure* the final state: obtain one of the search results, measuring the resulting state in the computational basis.

It can be shown that after $\lfloor\pi/4arcsin(\sqrt{a})\rfloor$ iterations (where $\lfloor x\rfloor$ is the rounding of x) the measured outcome is good with probability at least $max(a, 1-a)$. If we have a high number of items N and $a \ll N$, then the optimal number of iterations is proportional to $1/\sqrt{a}$. If A is the quantum Fourier transform the optimal number of iterations is proportional to $\sqrt{N/t}$, which corresponds to the speedup of Grover's algorithm. If $a > t/N$, the number of iterations is lower than $\sqrt{N/t}$.

4   The Quantum Amplitude Estimation Algorithm

The quantum amplitude estimation algorithm [8] allows to estimate the amplitude of a quantum state by applying at different steps the amplitude amplification algorithm. From a cognitive point of view, it allows to estimate with a good precision the probability a to find a good item (according to the partitioning introduced by function f) when the opinion state about the N items is the initial guessing state. Even if the mathematical details of the algorithm are described in next subsection, I now present its main features, reducing to the minimum the formalism. The algorithm can be decomposed in three parts:

1) *Initial state*: it is composed by the guessing state, as described before.

2) *Parallel amplifications*: different instances of the amplification engine are applied in a parallel way, with different numbers of iterations. Thus we have a double level of parallelism: in each step of the amplification engine the function $f(x)$ is applied simultaneously to the items, and this works simultaneously for each instance of the amplification engine.

3) *Analysis* of the different amplifications: since the efficiency of each amplification engine depends on the parameter a, the analysis of different instances of the amplification process with different number of iterations allows to estimate a, with a few standard deviations, after a number of evaluations of f proportional to $1/\sqrt{a}$.

This algorithm is particularly important for the study of cognitive processes, because it allows to describe the tasks where subjects produce subjective probabilities relevant to events. In this sense, it provides the formal link between a quantum-like approach describing choices (for example, [1]) and a quantum-like approach describing subjective probabilities (for example, [2]): choices are the effect of simple measurements on quantum states, while the subjective probabilities are the result of a quantum amplitude estimation algorithm applied on the same state. In the context of availability heuristic, the present algorithm can be used to describe the experiment of [4] presented in the introduction about the likelihood of letter R in the first or in the third position of English words (word frequency experiment). The retrieve process for words with R in first or third position involves two different partitioning of English words and two different amplification processes with parameters a and a'. In other words, we assume that subjects' mental state (the guess state) involves N words, and that the weight in such state relevant to words with R in first and third position is a and a' respectively. According to our model, the ease to recall words with R in first position can be described by the availability-by-number and is proportional to $\sqrt{a}$, and the estimated probability to recall words with R in first position is near to a. Thus if subjects recall more words with R in first position than in third ($\sqrt{a} > \sqrt{a'}$), then the estimated probability to find a word with letter R in first position is higher than the estimated probability to find words with R in third position ($a > a'$).

Like for the amplitude amplification algorithm, also in this case the produced estimated probability can be described as the result of subconscious amplification processes (with evaluations of function f) and a final analysis and measure.

4.1   Mathematical Description of Amplitude Estimation Algorithm

The amplitude estimation algorithm, called **Est_Amp**(A, f, M), is able to estimate the amplitude of $|\psi_1\rangle$ (good states superposition) in $A|0\rangle$. It is based on the amplitude amplification algorithm. In particular:

1) *Initial state*: prepare the vector $F_M|0\rangle A|0\rangle$, formed by two distinct registers: the first has dimension M, while the second has dimension N. We recall that F_M is the quantum Fourier transform $F_M : |x\rangle \to M^{-1/2} \sum_{y=0}^{M-1} e^{2\pi i x y} |y\rangle$.

2) *Parallel amplifications*: apply the operator $\Lambda_M(Q)$, defined by $|j\rangle|z\rangle \rightarrow |j\rangle Q^j|z\rangle$ with $0 \leq j \leq M$, where $Q = -AS_0A^{-1}S_f$ is the standard amplitude amplification engine. In other words, operator $\Lambda_M(Q)$ applies in a parallel way different degrees of amplification, from 0 to M, to the guess state $A|0\rangle$.

3) *Find the period of the wave function*: apply F_M^{-1} to the first register and measure it, obtaining an integer y. The estimated amplitude is then $\tilde{a} = sin^2(\pi y/M)$: the accuracy of such estimate is given in Theorem 12 in [8], where it is shown that the difference between the real and the estimated probabilities is at least

$$|\tilde{a} - a| = \Delta(a, M) \leq 2\pi \frac{\sqrt{a(1-a)}}{M} + \frac{\pi^2}{M^2} \qquad (1)$$

In particular, to obtain a probability estimate with a few standard deviations, we have to choose $M = \lfloor 1/\sqrt{a} \rfloor$. As we will study in section 6, in case of $a \simeq 1$, the error can only reduce the estimate from a to $a - \Delta(a, M)$. Analogously, when $a \simeq 0$, the error can only increase the estimate from a to $a + \Delta(a, M)$. We will call these effects overestimation and underestimation of low/high probabilities respectively.

5   The Quantum Counting Algorithm

The quantum counting algorithm [8] allows, given a boolean function f defined on a set X of N items, to estimate the number of elements of X for which the function f is true $t = |\{x \in X | f(x) = 1\}|$. In other words, the algorithm allows to estimate the size of the subset of *good items* (those for which $f(x) = 1$). The best classical strategy is to evaluate f on random elements of X: thus the number of evaluations in order to have a good estimate of t is proportional to N. On the contrary, the quantum counting algorithm produces good estimates for such number in approximatively $\sqrt{N}$ steps (quadratic speedup).

   The quantum counting algorithm can be considered as an application of the previous amplitude estimation algorithm. In fact, if the guessing state assigns the same weight to all the items, then the estimated probability relevant to the good items is near to t/N: the approximate number of good items can be obtained by multiplying such estimated probability by N. I propose here a simple generalization of the quantum counting algorithm, which I will discuss in mathematical details in the next subsection: if the guessing state assigns non-uniform weights to the items, the probability relevant to good items is $a \neq t/N$. If for example $a > t/N$, then the estimated number of items is near to $aN > t$: we have an overestimation of the number of items, due to the guessing state in the amplification process.

   Such simple generalization allows to describe the risk frequency experiment in [4] , where subjects were asked to judge the annual mortality (in the United States) associated with a wide range of risks, including motor vehicle accidents, poisoning by vitamins, and lung cancer. The experimental results show a positive correlation between the annual estimated mortality rate of a disease and the number of recalled occurrences of the same disease in subjects' social circle (i.e., family, friends, and acquaintances). In fact the same parameter a is involved both in the recalling process and in the approximate counting process: thus the ease to recall (proportional to $\sqrt{a}$) entails a higher estimated rate for the same disease(aN).

5.1 Mathematical Description of Quantum Counting Algorithm

Given a boolean function f over a discrete set X with N elements, the quantum counting algorithm **Count**(F_N, f, M) can be written as a special case of the amplitude estimation: $\tilde{t} = N \times$ **Est_Amp**(F_N, f, M). If we use, instead of the Fourier transform F_N, a generic operator A, the quantum counting algorithm **Count**(A, f, M) does not produce a correct estimate of t, the number of good items. However, if $a > t/N$, the modified counting algorithm produces an estimate $\tilde{t} > t$, while if $a < t/N$, it produces an estimate $\tilde{t} < t$.

6 Predictions of the Model and Experimental Results

Given a memory task, the *availability-by-number* can be mathematically defined as the mean number R_I of recalled items (in a fixed time) in the good class. The main predictions of the quantum model based on the previously defined algorithms are:

1) The availability-by-number R_I (the mean number of recalled good items) is approximatively proportional to the square root of the estimated number of good items $\tilde{t} = N\tilde{a}$:

$$R_I \simeq g\sqrt{\tilde{t}} = g\sqrt{N\tilde{a}}. \tag{2}$$

The proportionality constant g depends on the experimental variables (time to recall/estimate, nature of items). In fact, since $1/\sqrt{a}$ is the number of steps required to recall a good item, the number of recalled elements R_I (in a fixed time) is proportional to $\sqrt{a}$, and thus approximatively proportional to $\tilde{t}$ or to $\tilde{a}$ (the output of the amplitude estimation algorithm **Est_Amp**(A, f, M)).

2) The regression mechanism: the estimation algorithm **Est_Amp**(A, f, M) is subjected to precision errors, which change the estimate from probability a to probability $\tilde{a}$. The error $|\tilde{a} - a| = \Delta(a, M)$ is defined as in equation (1) and it is a descending function of precision M. However, for high probabilities ($a \simeq 1$) the final judgment is underestimated, since $\tilde{a} = a - \Delta(a, M)$. On the other side, for low probabilities ($a \simeq 0$) the final judgment is overestimated, since $\tilde{a} = a + \Delta(a, M)$. In other words, the algorithm seems to be affected by problems of underestimation/overestimation in the extreme cases. Even if this could seem a weakness of the algorithm, in our case this is a nice feature, since it takes into account in a natural way a well known and widely observed bias.

In the next subsections, I compare such predictions with the main experimental results relevant to the availability heuristic.

6.1   Word Frequency Experiment

The first class of experiments I consider here is the "judgement of word frequency", invented by Tversky and Kahneman (1973) [4] and based on the following question: *Consider the letter R: is R more likely to appear in the first position or in the third position?*. This experiment has been done with the consonants (K, L, N, R, V), which in the English language appear more frequently in the third than in the first position. The most part of participants (105 among 152) judged the first position to be more likely. The explanation given by Tversky and Kahnemann is that people estimate the number of words based on the ease with which they can recall them, which is the availability heuristic: the first letter provides a better cue for recalling instances of words than does the third letter.

Seldmeier et al. (1998) [5] repeated the experiment with German words and with a more complete set of letters in first and second position. In particular, the authors compared the actual and the judged proportions (estimated ratio of words with the letter in first/second position) with some mathematical models like the availability-by-number (defined in equation 2) and the regressed frequencies. The *regressed-frequencies hypothesis* assumes that the mind keeps track of the frequencies of individual letters in different positions. It further assumes that low frequencies are overestimated and high frequencies are underestimated. The amount of this "regression" (toward the mean of all letter frequencies) is assumed to be 70%, following Attneave's (1953) results. Thus, given a letter, if $A(1)$ is the actual relative frequency of words in first position, the judged relative frequency of words in first position is $A(1) - 0.7[A(1) - 0.5]$. For example, for letter C we have an actual value of 13.5%, which produces a proportion of 39%.

Figure 2 in [5] represents the main results: similarly to Kahneman's results, for some consonants that are actually less frequent in the first position (C, R, N, and L), subjects overestimated the proportion of words which present such letter in first position. However, for other consonants that are actually more frequent in the first position (S, F and G), subjects underestimated the proportion of words which present such letter in first position. By analyzing figure 2, the authors conclude that the regressed-frequencies mechanism is the model that best fits the subjects judgements.

However, I present here two observations: 1) it is not clear how the subjects know the actual proportions in order to compute the regressed frequencies, and 2) the authors have regressed respect to 50% also the availability-by-number/speed, which is not fully justified (since this is not the result of a judgement). On the contrary, in the present work we do not regress the availability-by-number. Moreover, we note that the estimated proportion of words with the letter in first position relevant to those with the same letter in first or second position can be written in our formalism as $\tilde{a}$. Thus, equation (2) states that the square root of the judged probability $\tilde{a}$ is approximatively proportional to the availability-by-number R_I. In table 1 we show that the normalized judged probability $\tilde{a}$ and the normalized availability-by-number R_I are approximately equal, mostly when the actual proportions are far from 1. For the judged proportions which are near to 1,

Table 1. Comparison of actual proportion, availability-by-number, estimated proportion and square root of estimated proportion

Letter	Actual proportion	Av.-by-number R_I	Estimated prop. $\tilde{a}$	Square root of est. prop. $\sqrt{\tilde{a}}$
C	0.1	**0.53**	0.28	**0.53**
R	0.21	**0.6**	0.47	**0.69**
N	0.32	**0.68**	0.42	**0.65**
L	0.43	**0.65**	0.48	**0.69**
B	0.73	**0.62**	0.62	**0.79**
S	0.85	**0.65**	0.6	**0.77**
F	0.91	**0.55**	0.68	**0.82**
G	0.99	**0.77**	0.7	**0.84**

instead, the prediction 1 is less well verified: this could be explained with higher statistical errors, due to the very low number of words produced with the letter in the third position. Moreover, according to prediction 2, the judged proportions which are near to 1 should be corrected in order to take into account to the regression mechanism: in particular, such proportions would be incremented, producing an even bigger difference from the availability-by-number.

6.2   Risk Frequency Experiment

Tversky and Kahneman [4] exemplified the availability heuristic with the fact that one may assess the risk of heart attack among middle-aged people by recalling such occurrences among ones acquaintances. Lichtenstein et al. [6] explored the judgments of risk frequency, by asking participants to judge the mortality rate (in the United States) associated with a wide range of risks, including motor vehicle accidents, poisoning by vitamins, and lung cancer. The experiment evidenced a confirm of the availability heuristic and of the regression mechanism for high and low frequencies.

Hertwig et al. [7] replicated such experiments, in order to compare the results with different theoric models: subjects were asked to estimate the annual mortality rate or the incidence rate of a particular disease, or to recall its occurrences in their social circle (i.e., family, friends, and acquaintances) and to write down the number of instances they could retrieve. Two mechanisms, the availability-by-number and the regressed frequency, conformed best to people's estimates of absolute risk frequencies. First of all we note that figure 1 in [7] evidences clearly an underestimation or overestimation of estimated frequencies relative to the real frequencies in the extreme cases, which is consistent with formula (1). In order to compare the experimental data with prevision of my model, I consider the data kindly sent by the authors of [7], calling $\tilde{t}$ the median estimated frequencies relevant to a particular disease, and R_I the number of retrieved occurrences of that disease (availability-by-number). While the overestimation/underestimation of low/high frequencies are quite evident, the direct

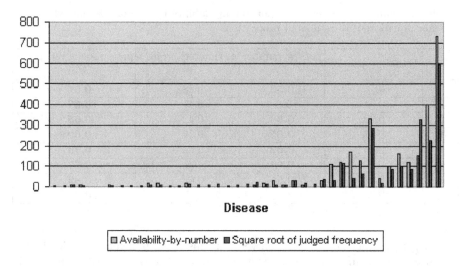

Fig. 1. Comparison of the rescaled availability-by-number $R_I g$ versus the square root of the judged frequency $\sqrt{\tilde{t}}$ for different diseases

comparison of $\tilde{t}$ and R_I evidences a positive correlation, but not a linear relation. However, according to prevision 1 of my model, the availability-by-number R_I should be proportional to $\sqrt{\tilde{t}}$, the square root of the non-regressed judged annual incidence. Figure 1 evidences, for different diseases, that $R_I g$ is quite similar to $\sqrt{\tilde{t}}$, with the constant coefficient $g = 10$, giving a preliminar evidence of prevision 1 of my model. For high judged frequencies, we can also see an underestimation effect, which can be probably be imputed to the regression mechanism of prediction 2.

7   Conclusions

In this article I show how three important quantum algorithms can model the experimental results of availability heuristic. I introduce the amplitude amplification algorithm to give a mathematical characterization of the ease to recall items or concepts. Then I present the amplitude estimation/counting algorithm, establishing a connection between the ease to retrieve and the judgements of probability/frequency about facts. The quantum description of availability heuristic, and in particular the use of quantum algorithms, has some advantages: 1) the economy of a quantum description, which seems to be consistent with a large number of cognitive heuristics (see for example [2], and [1]), while the classic alternatives are ad-hoc models with a very weak mathematical apparatus. 2) The ease of recall and the regression mechanism are naturally taken into account by the quantum model. 3) The quantum amplitude estimation and the quantum counting algorithms involve partial searches, which is consistent with the availability-by-number measure. 4) The availability experiments presented in this article describe judgements which involve in theory a great number of computations. In the example of word frequency, the English language contains about 500000 words, and the task previously described could reasonably involve computations over a subset of about 25000 more common words. This considerations make stronger the quantum-like point of view, since the quantum algorithms here proposed manifest a quadratic speed up, and thus are faster than any classic algorithm.

As noted by Manin [12], some human tasks, such as playing chess or speech generation and perception, require a great number of computations per second, as is evidenced by efficient chess playing software (based on classical algorithms). Since the characteristic time of neuronal processing is about 10^{-3} seconds, it seems difficult that a classical model could describe such tasks: in the word frequency experiments [4], the set of words on which perform the computation is in theory of 500000 elements, thus making a classic algorithm modelling the cognitive processes more difficult to apply than fast quantum algorithms. Moreover, the parallel use of quantum chips controlled by a classical computer, as well as the possibility to perform quantum search by using parallel queries [13] makes the quantum approach competitive also respect to the classic parallel computing. However, in this article I do not explore in a detailed way such point of view: in fact, this is a preliminar article showing a possible use of quantum algorithms to provide a formal description of memory and estimation tasks. Perhaps, in future we could find that other quantum algorithms are more suitable for such description.

There are some questions which need further investigations: 1) How the amplitude estimation processes can be influenced by a change in the partitioning? For example, we can choose to partition the items into two different ways. 2) Availability heuristic is relevant not only with estimated probabilities or approximate counting, but also with generic evaluations, like described in [14]. It should be investigated if the algorithm used in the present algorithm can be generalized also to generic evaluations (like for example course ratings). 3) Recent availability experiments show that the judged probabilities can be modified by previous tasks: thus a belief updating quantum model should be developed.

References

1. Busemeyer, J.R., Matthew, M., Wang, Z.: An information processing explanation of disjunction effects. In: Sun, R., Miyake, N. (eds.) The 28th Annual Conference of the Cognitive Science Society and the 5th International Conference of Cognitive Science, pp. 131–135. Erlbaum, Mahwah (2006)
2. Franco, R.: Quantum mechanics, Bayes' theorem and the conjunction fallacy, arXiv:quant-ph/0703222v2
3. Franco, R., Busemeyer, J.R.: A quantum probability explanation for the inverse fallacy. Psychonomic Review & Bulletin (2008) (to appear)
4. Tversky, A., Kahneman, D.: Availability: A heuristic for judging frequency and probability. Cognitive Psychology 5, 207–232 (1973)

5. Seldmeier, P., Hertwig, R., Gigerenzer, G.: Are Judgments of the Positional Frequencies of Letters Systematically Biased Due to Availability? Journal of Experimental Psychology: learning, Memory, and Cognition 24(3), 754–770 (1998)
6. Lichtenstein, S., Slovic, P., Fischhoff, B., Layman, M., Combs, B.: Judged frequency of lethal events. Journal of Experimental Psychology: Human Learning and Memory 4, 551–578 (1978)
7. Hertwig, R., Pachur, T., Kurzenhauser, S.: Judgments of Risk Frequencies: Tests of Possible Cognitive Mechanisms. Journal of Experimental Psychology: Learning, Memory, and Cognition 2005 31(4), 621–642 (2005)
8. Brassard, G., Hoyer, P., Mosca, M., Tapp, A. (eds.): Quantum amplitude amplification and estimation (2000), http://arxiv.org/abs/quant-ph/0005055
9. Grover, L.: Quantum mechanics helps in searching for a needle in a haystack. Phys. Rev. Lett. 79, 325 (1997)
10. Long, G.L.: Grover algorithm with zero theoretical failure rate. Phys. Rev. A64, 022307 (2001)
11. Franco, R.: Grover's algorithm and human memory (2008), http://xxx.lanl.gov/pdf/0804.3294
12. Manin, Y.I.: Why quantum computing (1999), quant-ph/9903008v1
13. Grover, L.K., Radhakrishnan, J.: Quantum search for multiple items using parallel queries (2004), arXiv:quant-ph/0407217v1
14. Fox, C.R.: The availability heuristic in classroom: How soliciting more criticism can boost your course ratings. Judgement and Decision Making 1, 8690 (2006)

Quantum Coherence without Quantum Mechanics in Modeling the Unity of Consciousness

Marcin Jan Schroeder

Akita International University
Akita, Japan
mjs@aiu.ac.jp

Abstract. The primary objective of the paper is to demonstrate how the phenomenal unity of consciousness, when interpreted as a result of integrative functions of the brain can be modeled in terms of algebraic properties of the quantum-mechanical formalism detached from its physical interpretation. The model proposed here is going one step beyond extracting the formal property of quantum coherence by making the transition from the lattice theoretic formalism to more suitable and slightly more general framework of closure spaces, but this generalization is independent from the basic idea of using quantum-mechanical formalism in the explanation of possibly non-quantum-mechanical processes in the brain.

Keywords: Quantum coherence, Quantum logic, Information integration processing, Consciousness, Closure space.

1 Introduction

The paper presents a general outline of the model of information integration in the brain using a formalism derived from the algebraic (lattice theoretic) formulation of quantum mechanics. Although the formal, algebraic characteristic of the quantum coherence plays a fundamental role in the model, there is no assumption made that the brain mechanisms considered as physical objects are actually quantum-mechanical systems. The model has a purely theoretical character and at this point no attempt has been made to identify specific functional elements in the brain responsible for its implementation in the cognitive functions. Naturally, the simplest candidates for such functional units would be neurons, but the model does not depend on such a choice.

The primary objective of the paper is to demonstrate how the phenomenal unity of consciousness, when interpreted as a result of integrative functions of the brain can be modeled in terms of algebraic properties of the quantum-mechanical formalism detached from its physical interpretation. In fact, the model proposed here is going one step beyond extracting the formal property of quantum coherence by making the transition from the lattice theoretic formalism to more suitable and slightly more general framework of closure spaces, but this generalization is independent from the basic idea of using quantum-mechanical formalism in the explanation of possibly non-quantum-mechanical processes.

P. Bruza et al. (Eds.): QI 2009, LNAI 5494, pp. 97–112, 2009.
© Springer-Verlag Berlin Heidelberg 2009

There is a natural question why quantum coherence has been selected as the fundamental characteristic for the model of consciousness. There is no doubt, that the challenge of going beyond what others (e.g. K. H. Pribram, S. R. Hameroff, R. Penrose) [1,2,3] have managed to achieve in explaining the unity of consciousness by quantum-mechanical coherence of brain structures, without reaching the ultimate goal, makes the subject attractive. But there are more important reasons.

The idea of the need for a new holistic conceptual and methodological framework for the study of phenomena occurring in complex systems (for instance biological or social) exhibiting properties different from those derived from its constituents, has never produced any formalism which could be used effectively in the description of the work of brain. There was a period of hope and excitement generated by Ludwig von Bertalanffy's vision which declared the arrival of a new age of inquiry: "General system theory, therefore, is a general science of 'wholeness' which up till now was considered a vague, hazy, and semi-metaphysical concept. In elaborate form it would be a logico-mathematical discipline, in itself purely formal but applicable to various empirical sciences."[4] But what was vague, hazy and semi-metaphysical has remained such in the next forty years. The only instance, in the decades of attempts, for a successful formalization of the compounds which manifested unity going beyond association of mechanically interacting independent units has been achieved independently in quantum mechanics. Thus, first argument could be simply the lack of alternatives. If not quantum-mechanical superposition of states, what else can be considered as a candidate for the integrative mechanisms in the brain?

There are also other compelling arguments for looking in quantum-mechanical formalisms for mathematical models of brain mechanisms. The formalism of quantum mechanics, especially its lattice theoretic version of the so called quantum logic, involves many concepts fundamental in several disciplines of mathematics which historically have developed from the human inquiry of the physical reality. Actually, the concept of quantum logic is a direct generalization of the lattice theoretic formalism for projective geometry. There are similar links to topology, probability, and logic (which has been the reason for the name "quantum logic".) Thus, the model of cognitive functions in such terms offers a promising point of departure for more specific study of the conscious experience of geometric or topological relations on one hand, and of cognitive processes related to language or logical reasoning on the other.

Finally, there is a good reason to consider in the context of consciousness studies the formalism of an empirical discipline such as quantum mechanics. Our conscious experience is an empirical process in which we are exploring our environment. Unless we believe that the consciousness is an accidental, purposeless side-product of the evolution, it is quite obvious that our perception of the surrounding world is a form of experimental procedure, and the most thorough analysis of such procedures has been made in the foundations of quantum mechanics. It is worth to notice, that the quantum mechanical description of the physical systems is inseparable from the consideration of the measuring apparatus, the characteristic which seems highly relevant for the analysis of brain's exploration of the physical reality.

As declared above, there is no intention in this paper to identify the functional elements in the brain which could be involved in the process of information integration. There is a legitimate question in what way such a purely theoretical model can be useful in the study of consciousness. The answer requires a short exposition of the problem to be solved, which in more elaborate format has been presented elsewhere [5].

Phenomenal unity has been recognized as a central feature of consciousness already by William James. The attempts to identify the mechanism producing this unity out of the variety of sensory inputs have been always employing some form of the infamous homunculus haunting all studies of consciousness. The most recent empirical research, although directed by the conviction that "subjective experience is integrated information" [6] which for the author of the present paper is the most promising approach to the study of consciousness, did not manage to escape the homunculus fallacy. Experimental studies of information integration, such as research reported by Tononi and Edelman [7], assume that the integration must be manifested by simultaneity of neuronal excitation. Of course, this assumption is re-introducing a homunculus whose watch plays the role of an integrator. There is no reason to believe that the subjective unity is any way related to external time or space measurement.

The brain is integrating sensory information not necessarily arriving simultaneously. For instance, a sequence of sounds is being integrated into a melody. Thus, the central question is what exactly information integration is. Only after we have a model of integration or integrated information, we can try to identify the functional units which implement the mechanism in the brain. The present paper is an attempt to use an algebraic property of quantum formalism corresponding to coherence to build such a model.

2 Information and Information Integration

Since the main objective of the paper is to develop a model of consciousness, we have to start from the clarification of its status. Consciousness is here considered a result of information integration, or simply it is integrated information. The reference to information requires in turn a definition of this commonly used (and abused) term. The definition used here, based on the one-many relationship, has been presented in more elaborate form elsewhere [8]. Thus, in short information is understood as an identification of a variety. Whenever there is a variety ("many") and there is anything that gives unity to this variety ("one",) it is information. There are two inseparable aspects of information, which determine the dominant characteristics of its manifestation. When the one is a result of a selection out of the many, the information can be characterized as selective. When the many are unified into the one, for instance through internal interactions or an imposed structure, the information has structural manifestation. However, it is always possible to link one of the manifestations with the other.

It is easy to recognize in the selective aspect of information the subject of Shannon's information theory, in which a probability distribution of the selection is utilized to define the measure of information in terms of entropy. Kolmogorov-Chaitin algorithmic information approach is concerned more with the structural manifestation of information. Since either of the two aspects can be easily reinterpreted in terms of the other, information defined above has a well defined measure, although it is a matter of continuing discussion which of the multiple measures introduced in literature is most adequate [9].

For a physical system, information carried by it is closely associated with the concept of a physical state considered in the context of the variety of all potential states, although typically the state is given epistemic status, while in this approach information has ontological status.

Information may be a subject of transformation, either in the sense of change of its manifestation, for instance from a selective form to structural. Also, information can be transferred from one variety to another. It is the latter case which in the case of the selective information was the subject of the classical Shannon's analysis of communication. Finally, we can consider a qualitative transformation of information integration. The concept of information, as defined above has already integrative character, as information may be understood as that which unifies a variety, or which reduces it to unity. The measures of information reflect the range of the variety and the degree to which information is making it one. However the mutual relationships of the elements of the united variety escape the measurement. This aspect of information may in the future find quantitative expression, but at the moment it has been considered only as a structural characteristic.

Now, we can proceed to the more specific presentation of the idea of information processing and integration. We will start from considering a very simple example of information processing in which the level or character of integration is not changed, but the manifestation of information is changing from the selective to structural. It is the Young-Helmholtz simplified mechanism of color vision. It has to be emphasized that this mechanism of color vision serves in this paper only as a familiar example of a model of information processing. The goal of this paper is very far from the analysis of what actually happens in the eye, as the present author is convinced that there are no processes occurring in the retina which could explain unity of consciousness.

Recent research shows that the activation of receptors in the retina may involve quantum processes which could make the actual process not suitable for our purpose [10]. But, even without quantum effects, the process of color recognition is much more complicated by the involvement of relative intensities of the three basic colors, which in this paper is ignored to provide the simplest example of the process transforming selective manifestation of information into structural one without any change in the level of information integration used as a point of departure for the analysis of more integrative transformations. Considering the more accurate model of vision would not have provided more insight into the process of information transformation, but would have obscured the underlying structures.

We can think about this three-color mechanism as a gate with the eight input channels distinguished by the eight colors of incoming light, from black, through so called pure colors of violet, green and red, to yellow, pink and blue, and to white. The input light characterized by one of these colors (selective information) is activating an appropriate combination of the three types of receptors (the pattern of activation is the information in the structural form.) If the incoming light is monochromatic, there is a bijective correspondence between the colors and the patterns of excited receptors.

However, we can consider mixed light coming. Then, when the mix consists of red and green light, the resulting pattern is different from either of patterns corresponding to red or green. Instead, the excited receptors correspond to yellow color. But when we have a mix of red and yellow light, the pattern belongs to yellow color, too. It is easy to recognize that the mechanism of color detection is defining on the set of colors a partial order. If the mix of two lights is coming and the resulting pattern of receptor excitation belongs to one of them, this element is greater. In the Young-Helmholtz model this partial order is a Boolean algebra with three atoms of "pure colors" (violet, green, red) three of mixed colors (blue, yellow, pink,) black color as the least element and white as the greatest. Now we can generalize this form of a gate as a system consisting of an atomic Boolean algebra whose all elements are input channels (or connected to input channels) and whose atoms are output channels. Activation of each of input channels produces a pattern of responding activation of a set of atoms. This type of a gate has been called a Venn gate, as Venn diagrams provide a good illustration of its functioning [11].

Similar gate (or rather module) is well known in computer architecture as a priority encoder. [12] Here too, we have four (or more) input channels and two (or more) output channels, but the module is structured not by a Boolean algebra, but by a linear order of priority. Each input channel is encoded in a unique way as a combination of the output channels, but whenever two input channels are activated instead of one, the output corresponds to that of higher priority in the linear ordering imposed by the encoder. Thus, when two input channels are activated, the output configuration is always identical with the output configuration for one of incoming signals.

There are of course many other possible gates of this type, each inducing different partial order on the incoming information while encoding the selection of the input channels in a unique way in form of a pattern of activated output channels. Thus, there is a natural question about the possible generalization (which has been called by the author a generalized Venn gate) in which the underlying structure does not have to be a Boolean algebra or linear order, but some partially ordered set. For instance we could substitute for the Boolean algebra its generalization in the form of quantum logic. This substitution is not as bizarre as it may seem at first moment. The explanation of the Young-Helmholtz mechanism of color vision is not much different from the way how quantum logics are introduced in terms of yes-no experiments [13].

Summarizing, the central role in the model presented here is given to what has been called a generalized Venn gate in which the input selective information coming in multiple channels is transformed into the output structural information, with or without change in the level of information integration.

In purely classical (disintegrated) case of simple Venn gate, the processing unit is modeled by an atomic Boolean algebra. Each of input channels is activating an element of the Boolean algebra, which produces an activation of an appropriate combination of the atoms corresponding to the input element. This pattern of activation forms the structural output. The process can be understood as a form of "logarithmic set operation." The reference to logarithm is justified by the fact that in the finite case to the 2^n input channels (elements of the Boolean algebra) correspond n output channels (atoms of the algebra.) Thus, the process is exactly opposite to the formation of the "power set."

In the case of a purely quantum Venn gate, the processing unit is modeled by an irreducible, atomistic, orthomodular, complete lattice (so called "quantum logic") which is not distributive, and therefore not a Boolean algebra, but the process otherwise is similar. Each of possible inputs is activating some combination of atoms of the lattice.

It is easy to identify underlying mathematical concepts. Generalized Venn gates at the first step of generalization can be defined as injective functions which map the elements of a complete atomistic lattice into subsets of atoms in such a way that the image of each element of the lattice is the set of atoms whose join is this element.

It is worth to mention that no non-trivial quantum gate can be finite, so if we want to have a model of the brain with the finite number of functional units (e.g. neurons,) each with the finite number of states (e.g. on/off), we have to look for a different irreducible generalized Venn gate. This is a consequence of the fact that every quantum logic of dimension higher than 2 (each element of the lattice is a join of at least three atoms) cannot be finite [14].

While other conditions defining quantum logic are simply generalizing the concept of a Boolean algebra, the condition of irreducibility is of the highest importance for us. In quantum mechanics, it is equivalent to the assumption that the Superposition Principle is completely unrestricted for all states of the system. In the description of actual physical systems, we have always the so called superselection rules which are limiting superposition. To the presence of superselection rules corresponds decomposition of the quantum logic into purely quantum-theoretic indecomposable factors. In the case of a Boolean algebra (purely classical case) the decomposition is complete into the direct product of factors consisting of the trivial two-element (indecomposable) Boolean algebras.

Thus, direct product irreducibility (indecomposability into a direct product) is the essence of the quantum character of the system, and at the same time a reflection of unlimited application of the Superposition Principle. This is the reason why in the model of the process of information integration in the brain presented here this property has been considered fundamental. Now, direct product irreducibility is a basic property of algebraic systems which does not require specific concepts defining quantum logics or any other association with quantum mechanics as a physical theory. It is applied here to the structures which model information processing to achieve information integration in purely formal analogy to quantum mechanical superposition.

Now, we can see that the particular choice of the quantum logic structure for the generalized Venn gate is not necessary in order to achieve some form of integration. The author proposed in the earlier papers to look for different structures to be used for generalized Venn gates among the lattices of closed subsets for a wider class of closure operators than the lattice of closed subspaces of a Hilbert space which appears in the standard quantum theory as quantum logic [11]. However, in such a case it may be necessary to consider for the output set, whose pattern of activation encodes the structural information, the elements which are not necessarily atoms of the lattice of closed subsets, as such a lattice may not even have atoms, or the set of atoms may not be sufficient to encode all input.

There is quite extensive literature on the lattices in which join- and meet-irreducible elements play the analogue role to that of the atoms in Boolean algebras or quantum logics as considered in the context of the lattice representation theory [15]. It turns out that modeling of information processing by Venn gates utilizing only partial order is essentially identical with the use of closure spaces in the simplest cases involving finite distributive lattices, not interesting for us due to direct product reducibility. In general, there is a correspondence between partially ordered sets and closure spaces defined on the same set. However, there are many such spaces which correspond to the same partial ordering. In the case of quantum Venn gates for instance, it is the choice of orthocomplementation on the lattice that fixes one specific closure space out of many associated with the partial order [16]. We have to remember at this point that the traditional terminology may result in confusion. To be precise, we should say "orthomodular ortho-lattice" instead of "orthomodular lattice." The same lattice can admit two different orthocomplementations producing non-isomorphic structures, one orthomodular and the other not. Thus, it should not be a surprise that different orthocomplementations correspond to essentially different closure spaces. It is a little bit less obvious that once orthocomplementation is introduced, the corresponding closure operation is unique.

The ortho-lattice formalism has many formal advantages over that based on partial order only. But, there are instances in which the conceptual framework of lattices and orthocomplementations is still too narrow for the purpose of constructing generalized Venn gates. For instance, if we want to consider a generalized Venn gates processing information related to geometric relations within synthetic geometry (going beyond projective geometry) or to topological properties, the more general language of closure spaces has clear advantage [11]. This is the main (but not only) reason for developing the model of information integration in the framework of closure spaces, in which of course quantum mechanics can be, and actually is formulated as a special case.

3 Closure Space Formulation of the Model

This section presents a short outline of the mathematical formulation of the model of information integration going beyond the formulation in terms of partial order as described above. Someone familiar with the subject of closure spaces may find short introductory portion of this section redundant, but its inclusion makes the paper self-contained.

A closure operator on set S is a function f from the power set of S to itself such that for all $A, B \subseteq S$: $A \subseteq f(A)$, $A \subseteq B \Rightarrow f(A) \subseteq f(B)$, and $f(A) = f(f(A))$.

The set S with a closure operator f form a closure space $<S, f>$. We will use letter symbols (consisting of a capital letter or a sequence of small letters preceding capital letter) for the properties of closure operators, and when an operator f satisfies a condition indicated with for instance letters xY, we will write $f \in xY(S)$ or simply $f \in xY$. The exceptions are symbols T_0 and T_1 with the long tradition in topology.

Every closure operator on a set S is uniquely defined by the Moore family of its closed subsets $f\text{-}Cl = \{A \subseteq S: A = f(A)\}$, and every Moore family $\mathfrak{I}$ of subsets of S, i.e. family of sets which includes S and is closed with respect to arbitrary intersections, is a family of closed sets for the unique closure operator defined by $f(A) = \cap\{B \in \mathfrak{I}: A \subseteq B\}$. It is easy to see that for every closure operator its family of closed sets forms a complete lattice L_f with respect to the set inclusion.

If a closure operator f is defined on a set S, it defines a closure operation g on every subset B of S called a restriction of f to B: $\forall A \subseteq B: g(A) = f(A) \cap B$. If $B \in f\text{-}Cl$, then $\forall A \subseteq B: g(A) = f(A)$ and the closure space $<B, g>$ is called a subspace of $<S, f>$.

As mentioned above, there is a bijective correspondence between finite partially ordered sets and finite closure spaces with closure operator f satisfying the following three additional conditions, which illustrates the point of overlap of the concepts of gates in terms of partial ordering and in terms of closure spaces:

(fA) $\forall A, B \subseteq S, f(A) \cup f(B) = f(A \cup B)$.
(N) $f(\varnothing) = \varnothing$ $(f \in N(S))$.
(T_0) $\forall x, y \in X, x \in f(\{y\}) \Rightarrow y \in f(\{x\})$.

The correspondence is based on the relationship between the partial order and the topological closure on singleton sets: $x \leq y$ iff $x \in f(\{y\})$ [15].
There is a stronger condition than T_0 known in topology as T_1:

(T_1) $\forall x \in S, f(\{x\}) = \{x\}$.

Of course, for closure operators satisfying this condition the partial ordering generated as above is trivial, and usually the closure operation cannot be reconstructed from the ordering, i.e. from the closures of singleton sets. However, many closure spaces of special interest for our purpose, such as geometric or topological closure spaces satisfy this condition.

In a closure space a subset A is called generating if f(A)=S, it is said to be generating for a set B, if f(A)=B. A subset A of S which is a minimal subset generating its own closure is called independent. The latter condition is usually formulated in an equivalent way: $\forall x \in A: x \notin f(A \setminus \{x\})$. A generating and independent subset is called a base. The term "base" has been originally introduced in closure spaces as a generalization of the concept of a base in the vector spaces. When we consider the closure operator defined by the Moore family of vector subspaces, the two concepts coincide. However, in general bases have very different properties. For instance, they do not have to be equicardinal, as the following example shows.

Example 1. Let S be a set with two disjoint proper, nonempty subsets T and U, and let the closure operator f be defined by f(A)=S, if $T \subseteq A$ or $U \subseteq A$, f(A)=A otherwise. Then, both T and U are bases. If they are not equicardinal, they provide example of bases of different cardinality in the same closure space.

For the purpose of defining a generalized Venn gate in terms of closure spaces it is necessary to introduce a new concept of a frame for a closure space. Since the study of the properties of frames has not been published yet, the following text is presented in a more formal way.

Definition 1. Let $<S,f>$ be a closure space and B its subset (not necessarily proper). B is a frame for $<S,f>$, if $\forall A \subseteq S \ \exists B_A \subseteq B: f(A) = f(B_A)$. A frame is proper, if B is a proper subset of S; it is a minimal frame, if there is no proper subset of B which is a frame. The closure space is simple if it does not have proper frame.

Proposition 1. The condition defining a frame is equivalent to the following: $\forall A \subseteq S: f(A) = f(B \cap f(A))$, where the equality can be replaced by the inclusion $\subseteq$.

Proof: Obviously B∩f(A) can be selected as B_A, which shows that the condition is necessary. To show that it is sufficient, observe that $f(A)=f(B_A))\subseteq f(B\cap f(A))\subseteq f(A)$.

Since all set S is always a (trivial) improper frame, each closure space has at least one frame. For simple spaces it is the only frame. Naturally, we are interested in the least possible subsets of closure spaces which are frames. It is obvious that B\f(∅) is always a frame, whenever B is a frame in <S,f>.

Although, obviously in every finite closure space there exists a minimal frame, in infinite spaces there may be no minimal frames at all.

We already have identified a convenient equivalent for the definition of a frame in Proposition 1. However, there are several other equivalent conditions for a subset B of the closure space <S,f> to be a frame.

Proposition 2. *The following conditions for a subset B of S are equivalent:*
- a) *B is a frame for <S,f>,*
- b) $\forall A \subseteq S: A \subseteq f(B\cap f(A))$,
- c) $\forall x \in S: x \in f(B\cap f(\{x\}))$,
- d) $\forall x \in S: f(\{x\})=f(B\cap f(\{x\}))$,
- e) $\forall C,D \subseteq S: f(C)\cap B=f(D)\cap B \Rightarrow f(C)=f(D)$,
- f) $\forall C \subseteq S: C \in f\text{-}Cl \Rightarrow f(B\cap C)=C$.

Proof: The equivalence of first two conditions is the subject of the preceding proposition.

For the equivalence of the second and third condition, we have to show only one direction of the implication. Suppose that ∀x∈ S: x∈ f(B∩f({x})), but some subset A of S is not a subset of f(B∩f(A)). This means that there exists y in A, such that y∉ f(B∩f(A)), but f(B∩f({y}))⊆f(B∩f(A)), hence y∉ f(B∩f({y})), which contradicts the assumption.

The equivalence of the third and fourth condition is straightforward. The necessity of the fifth condition for the subset B to be a frame is obvious.

We will show the sufficiency of the transposition of the implication in the fifth condition for B to be a frame ∀C,D⊆S: f(C)≠f(D)⇒f(C)∩ B≠f(D)∩B. Suppose there exists a subset A of S, such that f(A)≠f(B∩f(A)). Then by the assumption f(A)∩B≠f(B∩f(A))∩B, which in this case must be a strict inclusion of f(B∩f(A)) ∩B in f(A)∩B. But we have always the inclusions f(A)∩B⊆f(B∩f(A)) and f(A)∩B⊆B, therefore f(A)∩B⊆f(B∩f(A))∩B, a contradiction concluding the proof.

The equivalence of the sixth condition with the condition from the preceding proposition for B to be a frame is obvious.

Remark. As a consequence of the fourth condition, every T_1 closure space <S,f> (in which every one-element subset, or singleton, is closed, and therefore the empty set is also closed) must be simple, as the only set intersecting with all singletons is all set S.

Also, it is worth to mention that in the fifth condition the arbitrary subsets C and D cannot be replaced by singletons, as the following example shows.

Example 2. Let <S,f> be a closure space, T be a proper, nonempty subset of S, and f be defined for any subset A of S by: $f(A) = A$ if $A \subseteq T$, and $f(A)=S$ otherwise. Then, the set T satisfies the condition $\forall x,y \in S$: $f(\{x\}) \cap T=f(\{y\}) \cap T \Rightarrow f(\{x\})=f(\{y\})$, but T is not a frame for <S,f>.

Although the correspondence between finite partially ordered sets and T_0 closure spaces provides many examples where the concept of a frame can be considered purely in terms of the order relation, the need for more general concept of a frame in terms of closure spaces becomes clear in the structures build over infinite sets and for closure operators which are not bound by the T_0 condition.

Let's return to the issue of the relationship between frames and bases in closure spaces. The concept of a base in a closure space has been studied thoroughly, but actually its usefulness is very limited. The main reason for the interest in bases results from the importance of the concept of a base in vector spaces. However, the construction of arbitrary element of the vector space from the elements of the base does not have any counterpart in the general case of closure spaces. We cannot even generate all subspaces of the vector space using closures of subsets of the base. The main motivation for the introduction of the concept of a frame was the fact that the structure of closed subsets can be recovered from the closure operation acting on a base. For this reason it is interesting to observe the relationship between the two concepts.

Obviously, every frame B is a generating set ($f(B)=S$). However, it does not have to be an independent set, and therefore it does not have to be a base. Since independent sets are minimal sets generating their closure it follows that if a frame is a base, it has to be a minimal frame. However, Example 1 shows a case of a simple closure space (with only trivial frame of all set S) in which there exist proper subsets T and U which are bases. Thus the concepts of frames and of bases are essentially different, although not mutually exclusive. The following is a simple example of a frame which is a base.

Example 3. Let $\{S_i: i \in I\}$ be a partition of S, and the closure operator f on S be defined by $f(A)= \cup\{S_i: i \in I$ and $S_i \cap A \neq \varnothing\}$. Then every subset B of S, such that $\forall i \in I$: $|B \cap S_i|=1$ is a minimal frame which also is a base.

It is possible to characterize the subsets which are both frames and bases.

Proposition 3. *A frame B in a closure space <S,f> is a base iff the restriction g of the closure operation f to set B satisfies $\forall A \subseteq B: g(A)=A$.*

Proof: We need to show only that the condition for independence of a subset B is equivalent to $\forall A \subseteq B$: $f(A) \cap B=A$. B is f-independent if $\forall x \in B$: $x \notin f(B \setminus \{x\})$, or equivalently, if it is a minimal set generating its closure $f(B)$. Suppose B is independent, but there exists a subset of A of B which is a proper subset of $f(A) \cap B$ (obviously it must be a subset). Then, A is not generating its closure, and therefore A is not independent, which contradicts the fact that every subset of an independent set B must be independent. This proves that our condition is necessary for the independence of B. It is also sufficient, as the negation of the defining condition $\exists x \in B$: $x \in f(B \setminus \{x\})$ contradicts our condition applied to the set $A= B \setminus \{x\}$.

We will finish the exposition of the concept of frames by considering the relationship between the lattices of closed subsets for $<X,f>$ and $<B,g>$, where B is a frame and g is a restriction of the closure operator f to B .

Proposition 4. *Let L_f and L_g be the complete lattices of closed subsets in $<S,f>$ and $<B,g>$ respectively, and g be the restriction of f to a frame B in $<S,f>$. Then L_f and L_g are isomorphic.*

Proof: Let φ be a function from L_f and L_g defined for every f-closed subset C by $\varphi(C)=B\cap C$. Since B is a frame and C is f-closed, $g(B\cap C)= f(B\cap f(C))\cap B= f(C)\cap B= B\cap C$, i.e. $B\cap C$ is g-closed. The function is surjective, as if $C=g(C)\subseteq B$, then $f(C)\cap B=C$, and therefore $\varphi(f(C))=f(C)\cap B=C$. That it is also injective follows directly from the fifth condition for frames in Proposition 2. Finally it is obviously isotone, and its inverse defined for every g-closed subset A by $\varphi^{-1}(A)=f(A)$ is also isotone. Thus, φ is a lattice isomorphism.

Corollary. *Let f_1 and f_2 be closure operators defined on the same set S, g_1 and g_2 be their respective restrictions to a subset B of S, which is a frame for both closure spaces $<S,f_1>$ and $<S,f_2>$. Then, from the equality of the restrictions $g_1=g_2$ follows isomorphism of the lattice of f_1-closed subsets of S and the lattice of f_2-closed subsets of S.*

It is easy to see that whenever lattice L_f is atomistic, its atom space is isomorphic with a minimal frame for the closure space. For the closure space with algebraic lattice L_f, the set of join-irreducible elements assumes the role of minimal frame. However, in more general cases the choice of the frame may be arbitrary, as not always minimal frames exist. Also, for a large class of closure spaces there exists only trivial frame equal to all set S.

Now we are ready to describe a generalized Venn gate in terms of a closure space $<S,f>$. Its input channels are elements of the set on which is defined the lattice L_f of all f-closed subsets, its output channels are elements of a frame B of $<S,f>$.

Definition 2. *Let f be a closure operator on set S and B be a frame of $<S,f>$ A generalized Venn gate is a function φ which is mapping all elements of L_f, the lattice of all f-closed subsets of S, to the set of subsets of B, such that $\forall A\in L_f$: $f(\varphi(A))= A$. The gate is integrating if the lattice L_f is (direct –product) irreducible.*

Thus, when one of the input channels represented by the elements of the lattice of closed subsets is activated, the elements of the frame which generate this closed subset are activated. The lattice L_f can be considered the logic of the gate.

Simple Venn gate corresponds to the trivial closure operator f on S, which to every subset A of S assigns as its closure itself, i.e. $f(A)=A$. In this case the lattice of closed subsets is the Boolean algebra of all subsets of S.

Quantum Venn gate can be realized using as the closure space a complex Hilbert space with the Moore family of all its closed subspaces. The resulting lattice of closed subspaces is of course a quantum logic, and the frame consists of all elements of norm one (quantum mechanical states), or the isomorphic space of one-dimensional sub-spaces (atoms of the lattice of closed subspaces).

As mentioned above, the generalized Venn gates defined by closure spaces give opportunity to link the model with mathematical formalisms of several disciplines relevant for the study of cognition. A few examples will be given to conclude the presentation of the outline of the model. In each case the closure space will be identified by its defining properties.

Logic is the discipline of natural special interest for the study of cognition, although our model of information integration is intended for the study of the unity of consciousness, well below the level of logical inference. However, information integration may play an important role at all levels of cognition, as will be explained below. In logic, Tarski's logical consequence closure operator requires only one defining condition called "finite character":

$$(fC) \ \forall A \subseteq S \forall x \in S: x \in f(A) \Rightarrow \exists \ B \in Fin(A): x \in f(B),$$

where $Fin(A)$ indicates the set of all finite subsets of A.

The gates for processing information related to topological properties of perceived environment are associated with the best known instance of the topological closure which satisfies conditions (fA) and (N) above and appropriate separation condition such as for instance T_0, or T_1.

More complicated is the situation with constructing a gate processing geometric relations, as the closure spaces studied in the context of geometry are related to specific aspects of geometric configurations. The oldest and the best known are closure spaces related to projective and affine geometries.

The starting point to the study of geometric closure spaces is the selection of basic properties. It is always assumed that every geometry is defined as a closure space $<S,f>$ in which $f \in NT_1(S)$, which means that the closure of the empty set is the empty set, and that the closure of point sets (sets with one element) is the same point set. In addition to that two additional conditions, more geometric are assumed. The first is the finite character property of Tarski, introduced above to define logical consequence operator and the "exchange property" (of Steinitz):

$$(wE) \ \forall A \subseteq S \forall x,y \in S: x \notin f(A) \ \& \ x \in f(A \cup \{y\}) \Rightarrow y \in f(A \cup \{x\})$$

The agreement regarding the choice of axioms ends here. The formulation of projective or affine geometries in terms of closure operators splits into a wide range of different, sometimes non-equivalent theories.

A projective geometry is frequently defined by only one additional condition for a geometry called the "projective law":

$$(pL) \ \forall A,B \subseteq S \ \& \ A,B \neq \emptyset \Rightarrow f(A \cup B) = \{f(\{x,y\}): x \in f(A) \ \& \ y \in f(B)\}.$$

However, such geometry may have very strange properties contradicting our spatial intuition (e.g. different lines intersecting in more than one point,) so other conditions are sometimes added.

In geometries defined as closure spaces $(f \in NT_1 fCwE(S))$ the additional condition making such a structure consistent with our intuition of spatial relations gives a special role to the closures of pairs of points (lines):

$$\forall A \subseteq S: [A=f(A) \text{ iff } \forall x,y \in A: f(\{x,y\}) \subseteq A].$$

Thus, projective geometries are sometimes defined by the projective law and the condition of linearity (above).

To maintain the usual relationship between projective and affine geometries, the definition of the latter includes the usual condition of Euclid's "fifth postulate":

(fP) $\forall x,y,z,p,q,r \in S: f(\{p,q\}) \subseteq f(\{x,y,x\}) \ \& \ r \notin f(\{p,q\}) \Rightarrow$

$\exists t,u \in f(\{x,y,x\}): t \neq u \ \& \ r \in f(\{t,u\} \ \& \ f(\{p,q\} \cap f(\{t,u\})=\varnothing$

and every other closure of two points satisfying this conditions is identical with $f(\{t,u\})$.

Also the condition called "strong planarity," which is satisfied automatically by projective geometries, has to be assumed so in order to maintain the relationship between the two forms of geometry. Strong planarity adds to the planarity:

$$\forall A \subseteq S: [A=f(A) \text{ iff } \forall x,y,z \in A: f(\{x,y,x\}) \subseteq A] \text{ the condition:}$$

(sP) $\forall A \subseteq S \forall p,q \in S \forall r \in f(A): p \in f(A \cup \{q\}) \Rightarrow \exists s \in f(A): p \in f(\{q,r,s\}).$

This conceptual framework gives complete translation of projective and affine geometries into the language of closure spaces, but does not allow recovery of all geometry without going outside of it. All earlier or recent attempts to recover either Hilbert's Axioms of Order or the concept of convexity are referring to external concepts such as for instance concept of orientation.

Considered as separate structures, convex geometries are defined as closure spaces $<S,f>$ such that $f \in NT_1 fC$ and that f satisfies the "anti-exchange" condition:

(awE) $\forall A \subseteq S \forall x,y \in S: x \neq y$ and $x \notin f(A)$ and $x \in f(A \cup \{y\}) \Rightarrow y \notin f(A \cup \{x\}).$

It is easy to see that the anti-exchange condition is a generalization of the basic property of Hilbert's "betweenness," which also is related to exchange property. However, the connection of such convex geometries with projective and affine geometries on one hand, and synthetic geometry on the other is not as simple as could be expected, unless we assume some additional strong conditions.

There is a natural question about properties common for both types of geometries. Of course, in both cases we have $f \in NT_1 fC(S)$. Also, it is obvious that in both cases we have:

(linearity) $\forall A \subseteq S: [A=f(A) \text{ iff } \forall x,y \in A: f(\{x,y\}) \subseteq A]$, or at least

(planarity) $\forall A \subseteq S: [A=f(A) \text{ iff } \forall x,y,z \in A: f(\{x,y,x\}) \subseteq A].$

Notice that Hilbert's Axioms of Connection are related to the first of the conditions when $f(\{x,y\})$ is interpreted as a line, and at the same time his Axioms of Order are used to define convexity by using the same condition when $f(\{x,y\})$ is interpreted as a segment.

Thus, we have two basic forms of closure spaces describing two aspects of spatial relations. All former attempts to construct a closure space combining abstract form of direction and abstract form of convexity have been going beyond closure space formalism making them not suitable for our purpose of looking for the characteristics of the mechanisms integrating information about spatial relations. Also, in either case topological properties of point configurations are not considered. The problem of closure space formulation of synthetic geometry is still open. Any further work on constructing a generalized Venn gate for processing geometric information depends on its solution.

4 Processing of Symbolic Information

Quantum logics have been used extensively to analyze concept formation and concept structures [17]. The relationship between the present work and such an approach to cognitive studies is not obvious. In this paper, the study of information integration is focused on the issues related to information at the very basic level of the states of the environment, their correlates in the brain and the mechanisms which produce phenomenal consciousness. However, it is possible to build the bridge between the two levels. In the following a brief outline of a speculative construction is provided.

Concept formation is quite obviously related to the symbolic character of information. The concept of a symbol has been a subject of long philosophical and linguistic disputes. From the point of view on information presented above, the symbolic representation can be considered a process in which big volume of information is replaced in information processing by small one, as a result of the limited capacity of the brain mechanisms processing information. It is a well known fact that the brain is "chunking" information when the number of perceived items exceeds the "magical number seven."

We can expect that the concept formation is a process in which relatively small volume of information (symbol) is used to represent a bigger volume of integrated information (meaning). The fact that in our conscious experience objects identified by concepts have definite identity suggests that the process of information integration similar to that of total conscious experience has place. Further study is necessary to find out whether conceptual thinking requires introduction of some form of "superselection rules" disintegrating perceived reality into conceptual components. Another possibility would be the parallel existence of two independent levels of integration, one of total conscious experience, the other of objects identified by concepts. In any case, the quantum properties of concepts identified in the literature of the subject may have explanation in the integration of information within concepts.

5 Conclusion

The attempts to identify brain mechanisms responsible for integration of information carried by neuro-psychologists have been thus far based on the assumption that integration is reflected by temporal or spatial coincidence of neural activity. However, the assumption that such simultaneity or spatial proximity is an indication of integration leads to a homunculus fallacy. Therefore, the fundamental question in the study of integrative functions of the brain is how to understand integration of information. Only after we have a model of information integration or integrated information, we can try to identify the implementation of the mechanism. At present, we simply do not know what we are looking for.

One of promising directions in the study of information integration was directed by the assumption that information integration is a result of quantum coherence of some regions of the brain. With time, the hope for identification of quantum mechanical cognitive processes faded away. However, there is possibility that the unity of consciousness is not achieved by quantum mechanical coherence *per se*, but can be the outcome of a mechanism of similar characteristics as quantum coherence. Since quantum coherence can be identified with irreducibility of quantum logic, this algebraic property can be extracted from the quantum mechanical formalism and applied to the structure modeling information integration. The structure presented in this paper is based on the formalism of closure spaces which in a special case can be identified with quantum logic. However, the formalism is much more general and can be used to define several structures of special interest in cognitive studies.

References

1. Pribram, K.H., Nuwer, M., Baron, R.: The holographic hypothesis of memory structure in brain function and perception. In: Atkinson, R.C., Krantz, D.H., Luce, R.C., Suppes, P. (eds.) Contemporary Developments in Mathematical Psychology. W. H. Freeman, San Francisco (1974)
2. Hameroff, S.R., Penrose, R.: Orchestrated reduction of quantum coherence in brain microtubules: a model for consciousness. Journal of Consciousness Studies 3, 36–53 (1996)
3. Penrose, R.: Shadows of the Mind: A Search for the Missing Science of Consciousness. Oxford University Press, Oxford (1994)
4. von Bertalanffy, L.: General System Theory: Foundations, Development, Applications. George Braziller, New York (1968)
5. Schroeder, M.J.: Philosophical Reflection and Scientific Inquiry of Information. In: Conference on Philosophy's Relevance in Information Science, Paderborn, Germany (submitted, 2008)
6. Tononi, G.: The Information Integration Theory of Consciousness. In: Velmans, M., Schneider, S. (eds.) The Blackwell Companion to Consciousness. Blackwell, Malden (2007)
7. Tononi, G., Edelman, G.M.: Consciousness and Complexity. Science 282, 1846–1851 (1998)
8. Schroeder, M.J.: Philosophical Foundations for the Concept of Information: Selective and Structural Information. In: Proceedings of the Third International Conference on the Foundations of Information Science (2005), http://www.mdpi.org/fis2005
9. Schroeder, M.J.: An Alternative to Entropy in the Measurement of Information. Entropy 6, 388–412 (2004)
10. Ramakrishna, C., Rajagopal, A.K., Usha Devi, A.R.: Quantum Mechanical Basis of Vision, http://lanl.arxiv.oreg/pdf/0804.0190
11. Schroeder, M.J.: Model of structural information based on the lattice of closed subsets. In: Kobayashi, Y., Adachi, T. (eds.) Proceedings of The Tenth Symposium on Algebra, Languages, and Computation, Toho University, pp. 32–47 (2006)
12. Morris, M.M., Kime, C.R.: Logic and Computer Design Fundamentals. Prentice-Hall, London (1997)
13. Jauch, J.M.: Foundations of Quantum Mechanics. Addison-Wesley, Reading (1968)

14. Ivert, P.-A., Sjödin, T.: On the impossibility of a finite propositional lattice for quantum mechanics. Helvetica Physica Acta 51, 635–636 (1978)
15. Davey, B. A., Pristley, H. A.: Introduction to Lattices and Order. Cambridge University Press, Cambridge (1992)
16. Schroeder, M.J.: Logico-algebraic structures for information integration in the brain. In: Proceedings of RIMS 2007 Symposium on Algebra, Languages, and Computation, Kyoto University, pp. 61–72 (2007)
17. Aerts, D., D'Hooghe, B.: Classical Logical versus Quantum Conceptual Thought: Examples in Economy, Decision Theory and Concept Theory,
http://xxx.lanl.gov/pdf/0810.5332

An Exploration of Type Indeterminacy in Strategic Decision-Making

Jerome R. Busemeyer[1] and Ariane Lambert-Mogiliansky[2]

[1] Indiana University
jbusemeyer@indiana.edu
[2] Paris School of Economics
alambert@pse.ens.fr

Abstract. In this paper we explore an extension of the Type Indeterminacy model of decision-making to strategic decision-making. A 2×2 game is investigated. We first show that in a one-shot simultaneous move setting the TI-model is equivalent to the standard Bayes-Harsanyi model. We then let the game be preceded by a cheap-talk promise game. We show in an example that in the TI-model the promise stage can have an impact on the next following behavior when the standard Bayes-Harsanyi model predicts no impact whatsoever. The TI approach differs from other behavioral approaches in identifying the source of the effect of cheap-talk promises in the intrinsic indeterminacy of the players' type.

Keywords: quantum indeterminacy, type, strategic decision-making, game.

1 Introduction

This paper belongs to a very recent and rapidly growing literature where formal tools of Quantum Mechanics are proposed to explain a variety of behavioral anomalies in social sciences and in psychology (see e.g. [1,2,4,5,7,9,10,14,17,18]).

The use of quantum formalism in game theory was initiated by Eisert et al. [8] who propose that models of quantum games can be used to study how the extension of classical moves to quantum ones can affect the analysis of a game.[1] Another example is La Mura [16] who investigates correlated equilibria with quantum signals in classical games. In this paper we introduce some features of an extension of the Type Indeterminacy (TI) model of decision-making [15] from simple decisions to strategic decisions. We study, in two different settings, a 2x2 game with options, to cooperate and to defect and we refer to it as a Prisoner Dilemma, PD[2]. In the first setting, the players move simultaneously and the game is played once. In the second setting, the simultaneous move PD game is preceded by a promise exchange

---

[1] From a game-theoretical point of view the approach consists in changing the strategy spaces, and thus the interest of the results lies in the appeal of these changes.

[2] This is for convenience, as we shall see that the game is not perceived as a true PD by all possible types of a player.

P. Bruza et al. (Eds.): QI 2009, LNAI 5494, pp. 113–127, 2009.

game. Our aim is to illustrate how the TI approach can provide an explanation as to why cheap talk promises matter.[3] There exists a substantial literature on cheap talk communication games. The approach in this paper does not belong to the literature on communication games. The cheap talk promise exchange stage is used to illustrate the possible impact of pre-play interaction. Various behavioral theories have also been proposed to explain the impact of cheap talk promises when standard theory predicts that there is none. They most often rely on very specific assumptions amounting to adding ad-hoc elements to the utility function (a moral cost for breaking promises) or emotional communication [11]. Our approach provides an explanation relying on a fundamental structure of the model i.e., the quantum indeterminacy of players' type. An advantage of our approach is that the type indeterminacy hypothesis also explains a variety of other so called behavioral anomalies such as framing effects, cognitive dissonance [15], the disjunction effect [3] or the inverse fallacy [10].

A main interest with TI-game is that the Type Indeterminacy hypothesis can modify quite significantly the way we think about games. Indeed, a major implication of the TI-hypothesis is to extend the field of strategic interactions. This is because actions impact not only on the payoffs but also on the profile of types, i.e., on who the players *are*. In a TI-model, players do not have a deterministic (exogenously given) type. The types change along the game together with the chosen actions (which are modelled as measurements of the type). We provide an example showing that an initially non-cooperative player can be (on average) turned into a rather cooperative one by confronting him with a tough player in a cheap talk promise exchange game.

Not surprisingly we find that there exists no distinction in terms of predictions between the standard Bayesian and Type Indeterminacy approaches in a simultaneous move context. The two models yield distinct predictions under the following conditions: i. at least one player makes more than one move; ii. those moves correspond to non-commuting Game Situations[4]; iii. a first-coming move separates between "potential" types that would otherwise interfere in the determination of the outcome of a next-coming interaction. We show that under those conditions a move with no informational content or payoff relevance still impacts on the outcome of the game.

2   A TI-Model of Strategic Decision-Making

2.1   Generals

In the TI-model a simple decision situation is represented by an *observable*[5] called a *DS*. A decision-maker is represented by his state or *type*. A type is a vector $|t_i\rangle$ in a Hilbert space. The measurement of the observable corresponds

---

[3] Cheap talk promises are promises that can be broken at no cost.

[4] A Game Situation is an operator that measures the type of a player, see below.

[5] An observable is a linear operator.

to the act of choosing. Its outcome, the chosen item, actualizes an *eigentype*[6] of the observable (or a *superposition*[7] of eigentypes if the measurement is coarse). It is information about the preferences (type) of the agent. For instance consider a model where the agent has preferences over sets of three items, i.e. he can rank any 3 items from the most preferred to the least preferred. Any choice experiment involving three items is associated with six eigentypes corresponding to the six possible rankings of the items. If the agent chooses a out of $\{a, b, c\}$ his type is projected onto some superposition of the rankings $[a > b > c]$ and $[a > c > b]$. The act of choosing is modelled as a measurement of the (preference) type of the agent and it impacts on the type i.e., it changes it (for a detailed exposition of the TI-model see [15]). How does this simple scheme change when we are dealing with strategic decision-making?

We denote by *GS* (for Game Situation) an observable that measures the type of an agent in a strategic situation, i.e. in a situation where the outcome of the choice, in terms of the agent's utility, depends on the choice of other agents as well. The interpretation of the outcome of the measurement is that the chosen action is a *best reply* against the opponents' expected action. This interpretation parallels the one in the simple decision context. There, we interpret the chosen item as the *preferred one* in accordance with an underlying assumption of (basic) rationality i.e., the agent maximizes his utility (i.e., chooses what he prefers). The notion of revealed preferences and a fortiori of revealed best-reply is problematic however. A main issue here is that a best reply is a response to an *expected* play. When the expected play involves subjective beliefs there may be a problem as to the measurability of the preferences. This is in particular so if subjective beliefs are quantum properties.[8] But in the context of maximal information games (which means that the initial types are pure types)we are dealing objective probabilities so it is warranted to talk about revealed best-reply.

TI-games are game with type indeterminate players, i.e., games characterized by uncertainty. In particular, players do not know the payoff of other players. The standard (classical) approach to incomplete information in games is due to Harsanyi. It amounts to transforming the game into a game of imperfect information where Nature moves at the beginning of the game and selects, for each player, one among a multiplicity of possible types (payoff functions). A player's own type is his private information. But in a TI-game the players may not even know their *own payoff*. This is true even in TI-game of *maximal information* where all players are represented by pure types.[9] In this paper we focus on TI-games of maximal information. Can the Harsanyi approach be extended to TI-games? We shall argue that the TI-paradigm gives even more content to Harsanyi's approach. What is a fictitious Nature's move in Harsanyi's

---

[6] The eigentypes are the types associate with the eigenvalues of the observable i.e., the possible outcomes of the measurement of the *DS*.

[7] A superposition is a linear combination of the form $\sum \lambda_i |t_i\rangle$; $\sum \lambda_i^2 = 1$.

[8] If subjective beliefs and preferences are quantum properties that do not commute then they cannot be measured them simultaneously.

[9] For a discussion about pure and mixed types (states) see Section 3.2 in [5].

setting becomes a real move (a measurement) with substantial implications. And the theoretical multiplicity of types of a player becomes a real multiplicity of "selves".

Types and eigentypes

We use the term *type* to refer to a *quantum pure state* of a player. A pure type is maximal information about the player i.e., about his payoff function.[10] But because of (intrinsic) indeterminacy, the type is *not* complete information about the payoff function in all games simultaneously not even to the player himself.

In a TI-game we also speak about the *eigentypes* of any specific game M, these are *complete information* about the payoff functions *in a specific static game* M. Any eigentype of a player knows his own M-game payoff function but he may not know that of the other players. The eigentypes of a TI-game M are identified with their payoff function in that game.

So we see that while the Harsanyi approach only uses a single concept, i.e., that of type and it is identified both with the payoff function and with the player. In any specific TI-game M we must distinguish between the type which is identified with the player and the eigentype which is identified with the payoff function in game M. A helpful analogy is with multiple-selves models (see e.g., [19] and [12]). In multiple-selves models, we are most often dealing with two "levels of identity". These two levels are identified with short-run impulsive selves on the one side and a long-run "rational self" on the other side. In our context we have two levels as well: the level of the player (the type) and the level of the selves (the eigentypes) which are to be viewed as potential incarnations of the player *in a specific game.*[11]

A central assumption that we make is that the reasoning leading to the determination of the best-reply is performed at the level of the eigentypes of the game. This key assumption deserves some discussion. What we have in mind is very much in line with quantum computing. What is happening in the head of a player is some form of parallel reasoning, all the active (with non-zero coefficient of superposition) eigentypes perform their own strategic thinking. Another way to put it is that we assume that the player is able to reason from different perspectives. Note that this is not as demanding as it may at first appear. Indeed we are used in standard game theory to the assumption that players are able to put themselves "in the skin" of other players to think out how those will play in order to be able to best-respond to that.

As in the basic TI-model, the outcome of the act of choosing, here a *move*, is information about the (actualized) type of the player. The act of choosing changes the type from some initial type to the actualized one. We call GO or Game Operator, a complete collection of (commuting) GS (each defined for a specific opponent). The outcome of a GO is an eigentype of the game, it gives information about how a player plays against any possible opponent in a

---

[10] The payoff of a player is a function of all the players' actions.

[11] A superposition is a linear combination of the form $\sum \lambda_i |t_i\rangle$; $\sum \lambda_i^2 = 1$.

specific game. Each player is an independent system i.e., there is no entanglement between players.[12]

We next investigate an example of a maximal information (see below for precise definition) two-person game. The objective is to introduce some basic features of TI-games in a simple context and to illustrate an equivalence and some distinctions between the Bayesian approach and the TI-approach.

2.2   A Single Interaction

Consider a 2X2 symmetric game, M, and for concreteness we call the two possible actions cooperate (C) and defect (D) (as in a Prisoner's Dilemma game but as we shall see below for certain types, it is a coordination game) and we define the preference types of game M also called the M-eigentypes as follows:

θ_1 : prefers to cooperate whatever he expects the opponent to do;

θ_2 : prefers to cooperate if he expects the opponent to cooperate with probability $p > q$ (for some $q \leq 1$) otherwise he prefers to defect;

θ_3 : prefers to defect whatever he expects the opponent to do.

An example of these types is in the payoff matrices below where we depict the row player's payoff:

$$\theta_1 : \begin{pmatrix} & C\ D \\ C & 10\ 5 \\ D & 0\ 0 \end{pmatrix}, \quad \theta_2 : \begin{pmatrix} & C\ D \\ C & 10\ 0 \\ D & 6\ 8 \end{pmatrix}, \quad \theta_2 : \begin{pmatrix} & C\ D \\ C & 0\ 0 \\ D & 10\ 5 \end{pmatrix}$$

Note that these types are complete characterization in the sense that they give the player's payoff for any action of the opponent.

We shall now proceed to investigate this simultaneous move TI-game. We note immediately that θ_1 and θ_3 are non-strategic while θ_2 is, i.e., his best-reply will depend on what he expects the opponent to do. The initial types are generally not eigentypes of the game under consideration. Let player 1 be described by the superposition:

$$|t_1\rangle = \lambda_1 |\theta_1\rangle + \lambda_2 |\theta_2\rangle + \lambda_3 |\theta_3\rangle, \sum \lambda_i^2 = 1.^{13} \tag{1}$$

We shall first be interested in the optimal play of player 1 when he interacts with a player 2 of different eigentypes. Suppose he interacts with a player 2 of eigentype θ_1. Using the definitions of the eigentypes θ_i above and (1), we know by Born's rule[14] that with probability $\lambda_1^2 + \lambda_2^2$ player 1 plays C (because θ_2's best-reply to θ_1 is C) and he collapses on the (superposed) type $|t_1'\rangle = \frac{\lambda_1}{\sqrt{\lambda_1^2+\lambda_2^2}} |\theta_1\rangle + \frac{\lambda_2}{\sqrt{\lambda_1^2+\lambda_2^2}} |\theta_2\rangle$. With probability λ_3^2 he plays D and collapses on the

---

[12] In future research we intend to investigate the possibility of entenglement between players.

[13] As in the original TI-model we use real numbers.

[14] The calculus of probability in Quantum Mechanics is done according to Born's rule which defines the probability for the different eigentypes is given by the square of the coefficients of superposition.

eigentype $|\theta_3\rangle$. If instead player 1 interacts with a player 2 of type $|\theta_3\rangle$ then with probability λ_1^2 he plays C and collapses on the eigentype $|\theta_1\rangle$ and since $|\theta_2\rangle's$ best-reply to $|\theta_3\rangle$ is D, with probability $\lambda_2^2 + \lambda_3^2$ he plays D and collapses on type $|t"_1\rangle = \frac{\lambda_2}{\sqrt{\lambda_3^2 + \lambda_2^2}} |\theta_2\rangle + \frac{\lambda_3}{\sqrt{\lambda_3^2 + \lambda_2^2}} |\theta_3\rangle$.

We note that the probabilities for player 1's moves depends on the opponent's type and corresponding expected play - as usual. More interesting is that, as a consequence, the *resulting type* of player 1 also depends on the type of the opponent. This is because in a TI-model the act of choice is a measurement of the own type that changes it. We interpret the resulting type as the initial type modified by the measurement. In a one-shot context, this is just an interpretation since formally it cannot be distinguished from a classical informational interpretation where the resulting type captures our revised beliefs about player 1.

We now consider a case when player 2 is indeterminate as well, it is given by

$$|t_2\rangle = \gamma_1 |\theta_1\rangle + \gamma_2 |\theta_2\rangle + \gamma_3 |\theta_3\rangle, . \sum \gamma_i^2 = 1. \tag{2}$$

From the point of view of the eigentypes of a player (the θ_i), the situation can be analyzed as a standard situation of incomplete information. We consider two examples:

Example 1. Let $\lambda_1^2 \geq q$, implying that the eigentype type θ_2 of player 2 cooperates and let $\gamma_1^2 + \gamma_2^2 \geq q$ so the eigentype θ_2 of player 1 cooperates as well.

Example 2. Let $\lambda_1^2 \geq q$ so the eigentype θ_2 of player 2 cooperates but now let $\gamma_1^2 + \gamma_2^2 < q$ so here the eigentype θ_2 of player 1 prefers to defect.

In Example 1 the types θ_1 and θ_2 of both players pool to cooperate. So in particular player 1's resulting type is a superposition of $|\theta_1\rangle$ and $|\theta_2\rangle$ with probability $(\lambda_1^2 + \lambda_2^2)$ and it is the eigentype $|\theta_3\rangle$ with probability λ_3^2. In Example 2, player 1's type θ_2 and θ_3 pool to defect so player 1's resulting type is a superposition of $|\theta_2\rangle$ and $|\theta_3\rangle$ with probability $\lambda_2^2 + \lambda_3^2$ and $|\theta_1\rangle$ with probability λ_1^2. So we see again how the resulting type of player 1 varies with the initial (here superposed) type of his opponent.

When both players play a best reply to each other we have an equilibrium more precisely:

Definition
A static TI-equilibrium of a game M is

  i. A profile of strategies that form a Nash equilibrium[15] i.e., such that each one of the M−eigentypes of each player maximizes his expected utility given the (superposed) type of his opponent and the strategies played by the opponent's eigentypes.

  ii. A corresponding profile of expected resulting types, one for each player.

---

[15] A Nash equilibrium is a strategy profile such that each player's strategy maximizes his utility given the strategies of the other players.

For concreteness we shall now solve for the TI-equilibrium of this game in a numerical example. Suppose the initial types are

$$|t_1\rangle = \sqrt{.7}\,|\theta_1\rangle + \sqrt{.2}\,|\theta_2\rangle + \sqrt{.1}\,|\theta_3\rangle\,, \tag{3}$$

$$|t_2\rangle = \sqrt{.2}\,|\theta_1\rangle + \sqrt{.6}\,|\theta_2\rangle + \sqrt{.2}\,|\theta_3\rangle\,. \tag{4}$$

Given the payoff matrices above, the threshold probability q that rationalizes the play of C for the eigentype θ_2 is $q = .666$. For the ease of presentation, we let $q = .7$. We know that the θ_2 of player 2 cooperates since $\lambda_1^2 = .7 \geq q$ and so does the θ_2 of player 1 since $\gamma_1^2 + \gamma_2^2 = .8 > q$.

In the TI-equilibrium of this game player 1 plays C with probability .9 and collapses on $|t_1'\rangle = \frac{\sqrt{.7}}{\sqrt{.7+.2}}\,|\theta_1\rangle + \frac{\sqrt{.2}}{\sqrt{.7+.2}}\,|\theta_2\rangle$ and with probability .1 player 1 plays D and collapses on $|\theta_3\rangle$. Player 2 plays C with probability .8 and collapses on $|t_2'\rangle = \frac{\sqrt{.4}}{\sqrt{.4+.4}}\,|\theta_1\rangle + \frac{\sqrt{.4}}{\sqrt{.4+.4}}\,|\theta_2\rangle$ and with probability .2, he plays D and collapses on $|\theta_3\rangle$.

We note that the *mixture actually played* by player 1 (.9C, .1D) is *not* the best reply of any of his eigentypes. The same holds for player 2. The eigentypes are the "real players" and they play pure strategies.

We end this section with a comparison of the TI-game approach with the standard incomplete information treatment of this game where the square of the coefficients of superposition in (1) and (2) are interpreted as players' beliefs about each other. The sole substantial distinction is that in the Bayes-Harsanyi setting the players privately learn their own type *before* playing while in the TI-model they learn it in the process of playing. A player is thus in the same informational situation as his opponent with respect to his own play. However under our assumption that all the reasoning is done by the eigentypes, the classical approach and the TI-approach are indistinguishable. They yield the same equilibrium outcome. The distinctions are merely interpretational.

Statement 1
The TI-model of a simultaneous one-move game is equivalent to a Bayes-Harsanyi model.

A formal proof of Statement 1 can be found in our companion paper "TI-game 2".

This central equivalence result should be seen as an achievement which provides support for the hypotheses that we make to extend the basic TI-model to strategic decision-making. Indeed, we do want the non-classical model to deliver the same outcome in a simultaneous one-move context.[16] We next move to a setting where one of the players is involved in a sequence of moves. This is the simplest setting in which to introduce the novelty brought about by the type indeterminacy hypothesis.

---

[16] Indeed we know that quantum indeterminacy cannot be distinguished from incomplete information in the case of a single measurement. A simultaneous one-move game corresponds to two single measurements performed on two non-entangled systems.

2.3   A Multi-stage TI-game

In this section we introduce a new interaction involving player 1 and a third player, a promise game. We assume that the GS representing the promise game do not commute with the GS representing the game M (described in the previous section).[17] Player 1 and 3 play a promise game where they choose between either making a non-binding promise to cooperate with each other in game M or withholding from making such a promise. Our objective is to show that playing a promise exchange game - with a third player - can increase the probability for cooperation (decrease the probability for defection) between the player 1 and 2 in a next following game M. Such an impact of cheap-talk promises is related to experimental evidence reported in Frank (1988).

We shall compare two situations called respectively protocol I and II. In protocol 1 player 1 and 2 play game M. In protocol II we add a third player, 3, and we have the following sequence of events:

step 1 Player player 1 and 3 play a promises exchange game N, described below.

step 2 Player 1 and 2 play M.

step 3 Player 1 and 3 play M.[18]

The promise exchange game

At *step 1*, player 1 and 3 have to select one of the two announcements: "I promise to play cooperate", denoted, P, and "I do not promise to play cooperate" denoted $no - P$. The promises are cheap-talk i.e., breaking them in the next following games has no implications for the payoffs i.e., at step 2 or step 3.

There exists three eigentypes in the promise exchange game:

τ_1 : prefers to never make cheap-talk promises - let him be called the "honest type";

τ_2 : prefers to make a promise to cooperate if he believes the opponent cooperates with probability $p \geq q$ (in which case he cooperates whenever he is of type θ_2 or θ_1 or any superposition of the 2). Otherwise he makes no promises - let him be called the "sincere type";

τ_3 : prefers to promise that he will cooperate whatever he intends to do - he can be viewed as the "opportunistic type".

Information assumptions

Before moving further to the analysis of the behavior in protocol II we have to make clear the information that the players have at the different stages of the game. Specifically we assume that:

i. All players know the statistical correlations (conditional probabilities) between the eigentypes of the two (non-commuting) games.[19]

---

[17] To each game we associate a collection of GS each of which measures the best reply a possible type of the opponent.

[18] The reason why we have the interaction at *step 3* is essentially to motivate the promise exchange game. Our main interest will focus on the interaction at *step 2*.

[19] So in particular they can compute the correlation between the *plays* in the different games.

ii. At *step 2*, player 2 knows that player 1 has interacted with player 3 but he does not know the outcome of the interaction.

The classical model. We first establish that in the classical setting we have the same outcome in protocol I and at *step 2* of protocol II. We already know from Statement 1 that the analysis of a TI model of game M is fully equivalent with the classical Bayes-Harsanyi analysis of the corresponding incomplete information game.

We investigate in turn how the interaction between player 1 and 3 at *step 1* affects the incentives and/or the information of player 1 and 2 at *step 2*. Let us first consider the case of player 1. In a classical setting, player 1 knows his own type, so he learns nothing from the promise exchange stage. Moreover the announcement he makes is not payoff relevant to his interaction with player 2. So the promise game has no direct implication for his play with player 2. As to player 2, the question is whether he has reason to update his beliefs about player 1. Initially he knows $|t_1\rangle$ from which he derives his beliefs about player 1's equilibrium play in game M. By our informational assumption (i) he also knows the statistical correlations between the eigentypes of the two games from which he can derive the expected play conditional on the choice at the promise stage. He can write the probability of e.g., the play of D using the conditional probability formula:

$$p(D) = p(P)\,p(D|\,P) + p(no-P)\,p(D|\,no-P).\tag{5}$$

He knows that player 1 interacted with 3 but he does not know the outcome of the interaction. Therefore he has no new element from which to update his information about player 1. We conclude that the introduction of the interaction with player 3 at *step 1* leaves the payoffs and the information in the game M unchanged. Hence, expected behavior at *step 2* of protocol II is the same as in protocol I.

The TI-model. Recall that the GS representing the promise game do not commute with the GS representing the game M. We now write eq. (1) and (2) in terms of the eigentypes of game N, i.e., of the promise stage eigentypes:

$$|t_1\rangle = \lambda_1'\,|\tau_1\rangle + \lambda_2'\,|\tau_2\rangle + \lambda_3'\,|\tau_3\rangle \text{ and } |t_3\rangle = \gamma_1'\,|\tau_1\rangle + \gamma_2'\,|\tau_2\rangle + \gamma_3'\,|\tau_3\rangle.$$

Each one of the $N-$eigentype can in turn be expressed in terms of the eigentypes of game M:

$$\begin{aligned}
|\tau_1\rangle &= \delta_{11}\,|\theta_1\rangle + \delta_{12}\,|\theta_2\rangle + \delta_{13}\,|\theta_3\rangle \tag{6}\\
|\tau_2\rangle &= \delta_{21}\,|\theta_1\rangle + \delta_{22}\,|\theta_2\rangle + \delta_{23}\,|\theta_3\rangle\\
|\tau_3\rangle &= \delta_{31}\,|\theta_1\rangle + \delta_{32}\,|\theta_2\rangle + \delta_{33}\,|\theta_3\rangle
\end{aligned}$$

where the δ_{ij} are the elements of the basis transformation matrix (see the last subsection below). Assume that player 3 is (initially) of type θ_3 with probability close to 1, we say he is a "tough" type. We shall investigate the choice of between P and *no-P* of player 1 i.e., the best response of the eigentypes τ_i of player 1.

By definition of the τ_i type, we have that τ_1 always plays no-P and τ_3 always play P. Now by assumption, player 3 is of type θ_3 who never cooperates. Therefore, by the definition of τ_2, player 1 of type τ_2 chooses not to promise to cooperate, he plays $no- P$.

This means that at *step 1* with probability $\lambda_1'^2 + \lambda_2'^2$ player 1 plays $no- P$ and collapses on $|\widehat{t_1}\rangle = \frac{\lambda_1'}{\sqrt{(\lambda_1'^2 + \lambda_2'^2)}} |\tau_1\rangle + \frac{\lambda_2'}{\sqrt{(\lambda_1'^2 + \lambda_2'^2)}} |\tau_2\rangle$. With probability $\lambda_3'^2$ he collapses on $|\tau_3\rangle$.

We shall next compare player 1's propensity to defect in protocol I with that propensity in protocol II. For simplicity we shall assume the following correlations: $\delta_{13} = \delta_{31} = 0$, meaning that the honest type τ_1, never systematically defects and that the opportunistic guy τ_3 never systematically cooperate.

Player 1's propensity to defect in protocol I
We shall consider the same numerical example as before i.e., given by (3) and (4) so in particular we know that θ_2 of player 1 cooperates so $p(D||t_1\rangle) = \lambda_3^2$. But our objective in this section is to account for the indeterminacy due to the fact that in protocol I the promise game is *not* played. We have

$$|t_1\rangle = \lambda_1' |\tau_1\rangle + \lambda_2' |\tau_2\rangle + \lambda_3' |\tau_3\rangle$$

and using the formulas in (6) we substitute for the $|\tau_i\rangle$

$$|t_1\rangle = \lambda_1' (\delta_{11} |\theta_1\rangle + \delta_{12} |\theta_2\rangle + \delta_{13} |\theta_3\rangle) + \lambda_2' (\delta_{21} |\theta_1\rangle + \delta_{22} |\theta_2\rangle + \delta_{23} |\theta_3\rangle)$$
$$+ \lambda_3' (\delta_{31} |\theta_1\rangle + \delta_{32} |\theta_2\rangle + \delta_{33} |\theta_3\rangle).$$

Collecting the terms we obtain

$$|t_1\rangle = (\lambda_1' \delta_{11} + \lambda_2' \delta_{21} + \lambda_3' \delta_{31}) |\theta_1\rangle + (\lambda_1' \delta_{12} + \lambda_2' \delta_{22} + \lambda_3' \delta_{32}) |\theta_2\rangle +$$
$$(\lambda_{13}' \delta + \lambda_2' \delta_{23} + \lambda_3' \delta_{33}) |\theta_3\rangle.$$

We know from the preceding section that both $|\theta_1\rangle$ and $|\theta_2\rangle$ choose to cooperate so

$$p(D||t_1\rangle) = p(|\theta_3\rangle ||t_1\rangle).$$

Using $\delta_{13} = 0$, we obtain the probability for player 1's defection in protocol I:

$$p(D||t_1\rangle)_M = (\lambda_2' \delta_{23} + \lambda_3' \delta_{33})^2 = \lambda_2'^2 \delta_{23}^2 + \lambda_3'^2 \delta_{33}^2 + 2\lambda_2' \delta_{23} \lambda_3' \delta_{33}. \qquad (7)$$

Player 1's propensity to defect in protocol II
When the promise game is being played, i.e. the measurement N is performed, we can (as in the classical setting) use the conditional probability formula to compute the probability for the play of D

$$p(D||t_1\rangle)_{MN} = p(P)p(D|P) + p(no - P)p(D|no - P). \qquad (8)$$

Let us consider the first term: $p(P)p(D|P)$. We know that $p(P) = p(|\tau_3\rangle) = \lambda_3'^2$. We are now interested in $p(D|P)$ or $p(D|\tau_3\rangle)$. $|\tau_3\rangle$ writes as a superposition of the θ_i with θ_1 who never defects, θ_3 who always defect while θ_2's propensity to defect depends on what he expects player 2 to do. We cannot take for granted that player 2 will play in protocol II as he plays in protocol I. Instead we assume for now that eigentype θ_2 of player 2 chooses to cooperate (as in protocol I) because he expects player 1's propensity to cooperate to be no less than in protocol I. We below characterize the case when this expectation is correct. Now if θ_2 of player 2 chooses to cooperate so does θ_2 of player 1 and $p\left(D\,|\tau_3\right)) = \delta_{33}^2$ so

$$p\left(P\right)p\left(D|\,P\right) = \lambda_3^{2\prime}\delta_{33}^2$$

We next consider the second term of (8). The probability for $p\left(no-P\right)$ is $\left(\lambda_1^{2\prime}+\lambda_2^{2\prime}\right)$ and the type of player 1 changes, he collapses on $|\widehat{t_1}\rangle = \dfrac{\lambda_1^\prime}{\sqrt{\left(\lambda_1^{2\prime}+\lambda_2^{2\prime}\right)}}\,|\tau_1\rangle + \dfrac{\lambda_2^\prime}{\sqrt{\left(\lambda_1^{2\prime}+\lambda_2^{2\prime}\right)}}\,|\tau_2\rangle$. Since we consider a case when θ_2 of player 1 cooperates, the probability for defection of type $|\widehat{t_1}\rangle$ is $\left(\dfrac{\lambda_1^\prime}{\sqrt{\left(\lambda_1^{2\prime}+\lambda_2^{2\prime}\right)}}\right)^2\delta_{13}^2 + \left(\dfrac{\lambda_2^\prime}{\sqrt{\left(\lambda_1^{2\prime}+\lambda_2^{2\prime}\right)}}\right)^2\delta_{23}^2$.

Recalling that $\delta_{13} = 0$, we obtain that $p\left(no-P\right)p\left(D|\,no-P\right)$ is equal to

$$\left(\lambda_1^{2\prime}+\lambda_2^{2\prime}\right)\left(\dfrac{\lambda_2^\prime}{\sqrt{\left(\lambda_1^{2\prime}+\lambda_2^{2\prime}\right)}}\right)^2\delta_{23}^2 = \lambda_2^{2\prime}\delta_{23}^2$$

which gives

$$p\left(D\,||t_1\rangle\right)_{MN} = \lambda_2^{2\prime}\delta_{23}^2 + \lambda_3^{2\prime}\delta_{33}^2 \tag{9}$$

Comparing formulas in (7) and (9) :

$$p\left(D\,||t_1\rangle\right)_{MN} - p\left(D\,||t_1\rangle\right)_M = -2\lambda_2^\prime\delta_{23}\lambda_3^\prime\delta_{33} \tag{10}$$

which can be negative or positive because the interference terms only involves amplitudes of probability i.e., the square roots of probabilities. The probability to play defect decreases (and thus the probability for cooperation increases) when player 1 plays a promise stage whenever $2\lambda_2^\prime\delta_{23}\lambda_3^\prime\delta_{33} < 0$. In that case the expectations of player 2 are correct and we have that the θ_2 type of both players cooperate which we assumed in our calculation above.[20]

Result 1. *When player 1 meets a tough player 3 at step 1, the probability for playing defect in the next following M game is not the same as in the M game alone,* $p\left(D\,||t_1\rangle\right)_M - p\left(D\,||t_1\rangle\right)_{MN} \neq 0$.

It is interesting to note that $p\left(D\,||t_1\rangle\right)_{MN}$ is the same as in the classical case, it can be obtained from the same conditional probability formula.

In order to better understand our *Result 1*, we now consider a case when player 1 meets with a "soft" player 3, i.e., a θ_1 type, at *step 1*.

---

[20] For the case the best reply of the θ_2 types changes with the performance of the promise game, the comparison between the two protocols is less straightforward.

The soft player 3 case

In this section we show that if the promise stage is an interaction with a soft player 3 there is no effect of the promise stage on player 1's propensity to defect and thus no effect on the interaction at *step 2*.

Assume that player 3 is (initially) of type θ_1 with probability close to 1. What is the best reply of the N-eigentypes of player 1, i.e., how do they choose between P and *no-P*? By definition we have that τ_1 always plays *no-P* and τ_3 always play P. Now by the assumption we just made player 3 is of type θ_1 who always cooperates so player 1 of type τ_2 chooses to promise to cooperate, he plays P.

This means that at *t=1* with probability $\lambda_1'^2$ he collapses on $|\tau_1\rangle$ and with probability $\lambda_2'^2 + \lambda_3'^2$ player 1 plays P and collapses on $|\hat{t}_1\rangle = \dfrac{\lambda_2'}{\sqrt{\lambda_2'^2+\lambda_3'^2}}|\tau_2\rangle + \dfrac{\lambda_3'}{\sqrt{\lambda_2'^2+\lambda_3'^2}}|\tau_3\rangle$. We shall compute the probability to defect of that type.[21] We first the type vector $|\hat{t}_1\rangle$ in terms of the M-eigentypes,

$$|\hat{t}_1\rangle = \left(\frac{\lambda_2'}{\sqrt{\lambda_2'^2 + \lambda_3'^2}}\right)(\delta_{21}|\theta_1\rangle + \delta_{22}|\theta_2\rangle + \delta_{23}|\theta_3\rangle)$$

$$+ \left(\frac{\lambda_3'}{\sqrt{\lambda_2'^2 + \lambda_3'^2}}\right)(\delta_{31}|\theta_1\rangle + \delta_{32}|\theta_2\rangle + \delta_{33}|\theta_3\rangle)$$

As we investigate player 1's M-eigentypes' best reply, we again have to make an assumption about player 2's expectation. And the assumption we make is that he believes that player 1's propensity to defect is unchanged, so as in protocol I the θ_2 of both players cooperate and only θ_3 defects. We have

$$p\left(D\,||\hat{t}_1\rangle\right)_{MN} = \left[\frac{\lambda_2'}{\sqrt{\lambda_2'^2 + \lambda_3'^2}}\delta_{23} + \frac{\lambda_3'}{\sqrt{\lambda_2'^2 + \lambda_3'^2}}\delta_{33}\right]^2$$

$$p\left(D\,||\hat{t}_1\rangle\right)_{MN} = \frac{1}{\lambda_2'^2 + \lambda_3'^2}\left[\lambda_2'^2\delta_{23}^2 + \lambda_3'^2\delta_{33}^2 + 2\lambda_2'\lambda_3'\delta_{23}\delta_{33}\right]$$

The probability for defection is thus

$$p\left(D\,||t_1\rangle\right)_{MN} = P(\tau_1)\,p\left(D\,||\tau_1\rangle\right) + P\left(\hat{t}_1\right)p\left(D\,||\hat{t}_1\rangle\right) =$$

$$0 + \left(\lambda_2'^2 + \lambda_3'^2\right)\frac{1}{\lambda_2'^2 + \lambda_3'^2}\left[\lambda_2'^2\delta_{23}^2 + \lambda_3'^2\delta_{33}^2 + 2\lambda_2'\lambda_3'\delta_{23}\delta_{33}\right] = \lambda_2'^2\delta_{23}^2 + \lambda_3'^2\delta_{33}^2 + 2\lambda_2'\lambda_3'\delta_{23}\delta_{33}.$$

Comparing with eq. (7) of protocol I we see that here

$$p\left(D\,||t_1\rangle\right)_M = p\left(D\,||t_1\rangle\right)_{MN}$$

There is NO effect of the promise stage. This is because the interference effects are still present. We note also that player 2 was correct in his expectation about player 1's propensity to defect.

---

[21] Recall that τ_1 never defects.

Result 2. *If player 1's move at step 1 does not separate between the N-eigentypes that would otherwise interfere in the determination of his play of D at step 2 then* $p(D||t_1))_M = p(D||t_1))_{MN}$.

Let us try to provide an intuition for our two results. In the absence of a promise stage (protocol I) both the sincere and opportunistic type coexist in the mind of player 1. Both these two types have a positive propensity to defect. When they coexist they interfere positively(negatively) to reinforce(weaken) player 1's propensity to defect. When playing the promise exchange game the two types may either separate or not. They separate in the case of a tough player 3. Player 1 collapses either on a superposition of the honest and sincere type (and chooses *no-P*) or on the opportunistic type (and chooses *P*). Since the sincere and the opportunistic types are separated (by the first measurement, game N) there is no more interference. In the case of a soft player 3 case, the play of the promise game does not separate the sincere from the opportunistic guy, they both prefer P. As a consequence the two Neigentypes interfere in the determination of outcome of the next following M game as they do in protocol I.

In this example we demonstrated that in a TI-model of strategic interaction, a promise stage does make a difference for players' behavior in the next following performance of game M. The promise stage makes a difference because it may destroy interference effects that are present in protocol I.

Quite remarkably the distinction between the predictions of the classical and the TI-game only appears in the *absence* of the play of a promise stage (with a tough player). Indeed the probability formula that applies in the TI-model for the case the agent undergoes the promise stage (9) is the same as the conditional probability formula that applies in the standard classical setting.

A few words about the structure of the example. In the example above we are dealing with a type space Θ which has six elements. These elements go in two families corresponding to the two games i.e., M: $\{\theta_1, \theta_2, \theta_3\}$ and the promise game, N: $\{\tau_1, \tau_2, \tau_3\}$. So for instance the strategic type θ_1 is defined as a mapping from the simplex of the opponent possible types into actions θ_1 : $\Delta(\{\theta_1, \theta_2, \theta_3\}) \rightarrow A$ where A is the set of actions, $A = \{C, D\}$. It is interpreted as the best reply of player 1 against player 2 in the M game. Similarly τ_1 is defined by a mapping τ_1 : $\Delta(\{\tau_1, \tau_2, \tau_3\}) \rightarrow A'$, where $A' = \{P, no - P\}$ is interpreted as the best reply of player 1 to player 3 in the promise exchange game. The corresponding GS are indexed by the type of the opponent.

Our type space is a three dimensional Hilbert space where, $\{|\theta_1\rangle, |\theta_2\rangle, |\theta_3\rangle\}$ and $\{|\tau_1\rangle, |\tau_2\rangle, |\tau_3\rangle\}$ are two alternative basis. So in contrast with a standard Harsanyi type space where all types are alternatives (orthogonal) to each other, here $|\theta_1\rangle \perp |\theta_2\rangle$ and $|\tau_1\rangle \perp |\tau_2\rangle$ but $|\theta_i\rangle$ is not orthogonal to $|\tau_i\rangle$, $i = 1, 2, 3$. The two games are incompatible measurements of the type of a player. A basis transformation matrix links the eigentypes of the two GO M and N :

$$\begin{pmatrix} \langle\tau_1|\theta_1\rangle = \delta_{11} & \langle\tau_1|\theta_2\rangle = \delta_{12} & \langle\tau_1|\theta_3\rangle = \delta_{13} \\ \langle\tau_2|\theta_1\rangle = \delta_{21} & \langle\tau_2|\theta_2\rangle = \delta_{22} & \langle\tau_2|\theta_3\rangle = \delta_{23} \\ \langle\tau_3|\theta_1\rangle = \delta_{31} & \langle\tau_3|\theta_2\rangle = \delta_{32} & \langle\tau_3|\theta_3\rangle = \delta_{33} \end{pmatrix}.$$

Since there are three eigentypes and only two actions, two of the eigentypes must pool in their choice. The corresponding *GS* are coarse measurements of the type.

3   Concluding Remarks

In this paper we have explored an extension of the Type Indeterminacy model of decision-making to strategic decision-making in a maximal information context. We did that by means of an example of a 2X2 game that we investigate in two different settings. In the first setting the game is played directly. In the second setting the game is preceded by a promise exchange game. We first find that in a one-shot setting the TI-model is equivalent to the standard Bayes-Harsanyi approach to games of incomplete information. This is no longer true in the sequential move setting. We give an example of circumstances under which the predictions of the two models are not the same. We show that the TI-model can provide an explanation for why a cheap-talk promises matter. The promise game can separates between types and destroys interference effects that otherwise contribute to the determination of the propensity to defect in the next following game.

Last we want to emphasize the very explorative character of this paper. A companion paper TI-game 2 develops the basic concepts and solutions of TI-games. We believe that this avenue of research has a rich potential to explain a variety of puzzles in (sequential) interactive situations and to give new impulses to game theory.

References

1. Busemeyer, J.R., Wang, Z., Townsend, J.T.: Quantum Dynamics of Human Decision-Making. Journal of Mathematical Psychology 50, 220–241 (2006)
2. Busemeyer, J.R.: Quantum Information Processing Explanation for Interaction between Inferences and Decisions. In: Proceedings of the Quantum Interaction Symposium. AAAI Press, Menlo Park (2007)
3. Busemeyer, J.R., Matthew, M., Wang, Z.: An information processing explanation of the Disjuction effect. Sun, Miyake (eds.), pp. 131–135 (2006)
4. Busemeyer, J.R., Santuy, E., Lambert-Mogiliansky, A.: Distinguishing quantum and markov models of human decision making. In: Proceedings of the the second interaction symposium (QI 2008), pp. 68–75 (2008)
5. Danilov, V.I., Lambert-Mogiliansky, A.: Measurable Systems and Behavioral Sciences. Mathematical Social Sciences 55, 315–340 (2008)
6. Danilov, V.I., Lambert-Mogiliansky, A.: Decision-making under non-classical uncertainty. In: Proceedings of the the second interaction symposium (QI 2008), pp. 83–87 (2008)
7. Deutsch, D.: Quantum Theory of Propability and Decisions. Proc. R. Soc. Lond. A 455, 3129–3137 (1999)
8. Eisert, J., Wilkens, M., Lewenstein, M.: Quantum Games and Quantum Strategies. Phys. Rev. Lett. 83, 3077 (1999)

9. Franco, R.: The conjunction Fallacy and Interference Effects (2007), arXiv:0708.3948v1
10. Franco, R.: The inverse fallacy and quantum formalism. In: Proceedings of the the second interaction symposium (QI 2008), pp. 94–98 (2008)
11. Frank, H.R.: Passion within Reason. W.W. Norton & company, New York (1988)
12. Fudenberg, D., Levine, D.: A Dual Self Model of 1446-1476 (2006)
13. Fudenberg, D., Tirole, J.: Game Theory. MIT Press, Cambridge (1991)
14. Yu, K.A.: A Model of Quantum-like decision-making with application to psychology and Cognitive Sciences (2007), http://arhiv.org/abs/0711.1366
15. Lambert-Mogiliansky, A., Zamir, S., Zwirn, H.: Type-indeterminacy - A Model for the KT-(Kahneman and Tversky)-man (2007); Journal of Mathematical Psychology (forthcoming, 2009), ArXiv:physics/0604166
16. La Mura, P.: Correlated Equilibria of Classical Strategic Games with Quantum Signals. In: Game Theory and Information 0309001 EconWPA (2003)
17. La Mura, P.: Decision Theory in the Presence of Risk and Uncertainty. In: Mimeo. Leipzig Graduate School of Business (2005)
18. La Mura, P.: Prospective expected utility. In: Proceedings of the the second interaction symposium (QI 2008), pp. 87–94 (2008)
19. Strotz, R.H.: Myopya and Time Inconsistency in Dynamic Utility Maximization. Review of Economic Studies 23(3), 165–180 (1956)

Classical Logical Versus Quantum Conceptual Thought: Examples in Economics, Decision Theory and Concept Theory

Diederik Aerts and Bart D'Hooghe

Leo Apostel Center, Brussels Free University, Brussels, Belgium
diraerts@vub.ac.be, bdhooghe@vub.ac.be
http://www.vub.ac.be/CLEA/aerts/

Abstract. Inspired by a quantum mechanical formalism to model concepts and their disjunctions and conjunctions, we put forward in this paper a specific hypothesis. Namely that within human thought two superposed layers can be distinguished: (i) a layer given form by an underlying classical deterministic process, incorporating essentially logical thought and its indeterministic version modeled by classical probability theory; (ii) a layer given form under influence of the totality of the surrounding conceptual landscape, where the different concepts figure as individual entities rather than (logical) combinations of others, with measurable quantities such as 'typicality', 'membership', 'representativeness', 'similarity', 'applicability', 'preference' or 'utility' carrying the influences. We call the process in this second layer 'quantum conceptual thought', which is indeterministic in essence, and contains holistic aspects, but is equally well, although very differently, organized than logical thought. A substantial part of the 'quantum conceptual thought process' can be modeled by quantum mechanical probabilistic and mathematical structures. We consider examples of three specific domains of research where the effects of the presence of quantum conceptual thought and its deviations from classical logical thought have been noticed and studied, i.e. economics, decision theory, and concept theories and which provide experimental evidence for our hypothesis.

1 Introduction

We put forward in this paper a specific hypothesis. Namely that in human thought as a process two specifically structured and superposed layers can be identified:

(i) A first layer that we call the 'classical logical' layer. The thought process within this layer is given form by an underlying classical logical conceptual process [1]. The manifest process itself may be, and generally will be, indeterministic, but the indeterminism is due to a lack of knowledge about the underlying deterministic classical process. For this reason the process within the classical logical layer can be modeled by using a classical Kolmogorovian probability description [2], and eventually, in an idealized form, it could be modeled as a stochastic process.

P. Bruza et al. (Eds.): QI 2009, LNAI 5494, pp. 128–142, 2009.

(ii) A second layer that we call the 'quantum conceptual' layer. The thought process within this layer is given form under influence of the totality of the surrounding conceptual landscape, where the different concepts figure as individual entities, also when they are combinations of other concepts, contrary to how this is the case in the classical logical layer where combinations of concepts figure as classical combinations of entities and not as individual entities. In this sense one can speak of a phenomenon of 'conceptual emergence' taking place in this quantum conceptual layer, certainly so for combinations of concepts. Quantum conceptual thought has been identified in different domains of knowledge and science related to different, often as paradoxically conceived, problems in these domains. The sorts of measurable quantities being able to experimentally identify quantum conceptual thought have been different in these different domains, depending on which aspect of the conceptual landscape was most obvious or most important for the identification of the deviation of classically expected values of these quantities. For example, in a domain of cognitive science where representations of concepts are studied, and hence where concepts and combinations of concepts, and relations of items, exemplars, instances or features with concepts are considered, measurable quantities such as 'typicality', 'membership', 'similarity' and 'applicability' have been studied and used to experimentally put into evidence the deviation of what classically would be expected for the values of these quantities [3,4,5,6,7,8,9,10,11,12,13]. In decision theory measurable quantities such as 'representativeness', 'qualitative likelihood' 'similarity' and 'resemblance' have played this role [14,15,16,17,18,19,20,21,22]. In a domain such as economics one has considered measurable quantities such as 'preference', 'utility' and 'presence of ambiguity' to put into evidence the deviation of classical values of these quantities [23,24]. The quantum conceptual thought process is indeterministic in essence, i.e. there is not necessarily an underlying deterministic process independent of the context. Hence, if analyzed deeper with the aim of finding more deterministic sub processes, unavoidably effects of context will come into play. Since all concepts of the interconnected web that forms the landscape of concepts and combinations of them attribute as individual entities to the influences reigning in this landscape, and more so since this happens dynamically in an environment where they are all quantum entangled structurally speaking, the nature of quantum conceptual thought contains aspects that we strongly identify as holistic and synthetic. However, the quantum conceptual thought process is not unorganized or irrational. Quantum conceptual thought is as firmly structured as classical logical thought but in a very different way. We believe that the reason why science has hardly uncovered the structure of quantum conceptual thought is because it has been believed to be intuitive, associative, irrational, etc... meaning 'rather unstructured'. As a consequence its structure has not been sought for consistently since believed to be hardly existent. An important second hypothesis that we put forward in this paper is that an idealized version of this quantum conceptual thought process, or a substantial part of it, can be modeled as a quantum mechanical process. To indicate this idealization or this part we have called it 'quantum conceptual thought'. Hence we believe that important aspects of the basic structure of quantum conceptual thought can be uncovered as a consequence of this quantum structure modeling.

The distinction of two modes of thought has been proposed by many and in many different ways. Already Sigmund Freud in his seminal work 'The interpretation of dreams' made the proposal of considering thought as consisting of two processes, which he called primary and secondary [25], a distinction that became popularized as conscious and subconscious. Somewhat later William James introduced the idea of 'two legs of thought' where he specified one as 'conceptual', being exclusive, static, classical and following the rules of logic, and the other one as 'perceptual', being intuitive and penetrating. He expressed the opinion that 'just as we need two legs to walk, we also need both conceptual and perceptual modes to think' [26]. Remark that James used the connotation 'conceptual' to indicate the classical logical mode, contrary to us using 'conceptual' mainly with respect to the quantum structured mode. Jean Piaget, in his study of child thought, introduced 'directed or intelligent thought' which is conscious and follows the rules of logic and 'autistic thought' which is subconscious and not adapted to reality in the sense that it creates a dream world [27]. More recently, Jerome Bruner introduced the 'paradigmatic mode of thought', transcending particularities to achieve systematic categorical cognition where propositions are linked by logical operators, and the 'narrative mode thought', engaging in sequential, action-oriented, detail-driven thought, where thinking takes the form of stories and 'gripping drama' [28]. One of the authors of the present article has investigated aspects of different modes of thought and the influence of their presence on human cognitive evolution [29].

There are some fundamental differences between earlier versions in the literature of different modes of thought and the hypothesis about different layers of thought put forward in the present article. First of all, it is the specific mathematical structure of the quantum model, elaborated by one of the authors for the description of the combination of concepts in [30,31,32], that defines the structural aspects of the two layers of thought that we put forward in this article and how they are intertwined. This means that the nature of this double layered structure follows from a mathematical model for experimental data on the non classical aspects of thought identified in these different domains. Secondly, and directly related to the first difference, the double structure that we mention is a complex quantum entanglement, technically meaning a 'superposition of two modes of thought' rather than an individual or separated or eventually parallel existence. In [30,31,32] this entangled structure is analyzed in great detail, and it is shown that the presence of the two layers and the specific way they are entangled follows from the quantum field nature of the model developed in [31,32]. Also in the following of the present article we put forward some of these details. We believe that the superposed layers of thought have connections with the approaches of 'two modes of thought' that have been considered in history [25,26,27,28,29], and are planning to find out more about the details of such correspondences in future research. Actually we have worked mostly on the explanatory power for the specific examples in the different domains that we mentioned, and the rest of this article will focus on this.

The effects of the presence of quantum conceptual thought are observed in situations where deviations of classical logical thought are apparent in a systematic and intersubjective way, i.e. such that the effect can be measured and proven to be not due to chance and be repeated quantitatively. In sections 2, 3 and 4 we give examples of three specific domains of research where the effects of the presence of quantum conceptual thought and its deviations from classical logical thought have been noticed and studied, i.e. economics, decision theory, and concept theories. In section 5 we illustrate in detail the functioning of the two modes of thought on the specific example of the 'disjunction of concepts', since it was indeed the quantum modeling of the disjunction of concepts in [30,31,32] that made us propose the basic hypothesis of the presence of the two superposed layers of classical logical thought and of quantum conceptual thought within the human thought process as analyzed in the present article. There is a growing research activity in applying quantum structures to domains of science such as economics [33,34,35,36] and psychology and cognition [37,38,39,40,41,30,31,32] and language and artificial intelligence [42,43,44,45,46,47], and the study of the two layers of thought put forward in the present article is a contribution to this.

2   Quantum Conceptual Versus Classical Logical Thought in Economics

Almost seven decades ago, John von Neumann and Oskar Morgenstern founded a new branch of interdisciplinary research by applying game theory to economics [48,49]. Expected utility functions are used to model the preferences of rational agents over different ventures with random prospects, i.e. 'betting preferences' over what can be called *lotteries*. One of the basic principles of the von Neumann-Morgenstern theory is Savage's Sure-Thing Principle (STP) [50]. Savage introduced this principle in the following story: *A businessman contemplates buying a certain piece of property. He considers the outcome of the next presidential election relevant. So, to clarify the matter to himself, he asks whether he would buy if he knew that the Democratic candidate were going to win, and decides that he would. Similarly, he considers whether he would buy if he knew that the Republican candidate were going to win, and again finds that he would. Seeing that he would buy in either event, he decides that he should buy, even though he does not know which event obtains, or will obtain, as we would ordinarily say.* This assumption is the independence axiom of expected utility theory: independence means that if a person is indifferent between simple lotteries L_1 and L_2, the agent is also indifferent between L_1 mixed with an arbitrary simple lottery L_3 with probability p and L_2 mixed with L_3 with the same probability p.

Problems in economics such as Allais paradox [23] and Ellsberg paradox [24] show an inconsistency with the predictions of the expected utility hypothesis, namely a violation of the STP. As an illustration we consider the situation put forward by Daniel Ellsberg [24] which was mostly meant to illustrate the so called 'ambiguity aversion': persons prefer 'sure choices' over 'choices that contain ambiguity'. Consider the following hypothetical experiment. Imagine an urn known to contain 30 red balls and 60 black and yellow balls, the latter in unknown proportion. One ball is to be drawn at random from the urn. To 'bet on Red' will mean that you will receive a prize a (say \$100) if you draw a red ball ('if Red occurs') and a smaller amount b (say, \$0) otherwise ('if not-Red occurs'). One considers the following 4 actions: (I) 'a bet on red', (II) 'a bet on black', (III) 'a bet on red or yellow', (IV) 'a bet on black or yellow', and the pairs of gambles (I, II), (III, IV). Ellsberg found that a very frequent pattern of response is I preferred to II, and IV is preferred to III. Less frequent is: II preferred to I, and III preferred to IV. Both of these violate the STP, which requires the ordering of I to II to be preserved in III and IV (since the two pairs differ only in the pay-off when a yellow ball is drawn, which is constant for each pair). The first pattern, for example, implies that the subject prefers to bet 'on' red rather than 'on' black; and he also prefers to bet 'against' red rather than 'against' black. This contradiction indicates that preferences of subjects are inconsistent with the independence axiom of expected-utility theory.

Approaches have been developed, such as for example the 'info-gap approach', where it is supposed that the considered person implicitly formulates 'info-gap models' for the subjectively uncertain probabilities. The person then tries to satisfy the expected utility and to maximize the robustness against uncertainty in the imprecise probabilities. This robust-satisfying approach can be developed explicitly to show that the choices of decision-makers should display precisely the preference reversal which Ellsberg observed [51].

3   Effects of Quantum Conceptual Thought in Decision Theory

The situation considered in Ellsberg paradox shows a great similarity to situations considered in decision theory with respect to what is called the disjunction effect [52], where also STP is violated, and the conjunction fallacy [16,53]. The disjunction effect occurs when decision makers prefer option x (versus y) when knowing that event A occurs and also when knowing that event A does not occur, but they refuse x (or prefer y) when not knowing whether or not A occurs. A well-known example of such disjunction effect is the so-called Hawaii problem, in which following two situations are considered [52]:

1. Disjunctive Version
Imagine that you have just taken a tough qualifying examination. It is the end of the fall quarter, you feel tired and run-down, and you are not sure that you passed the exam. In case you failed you have to take the exam again in a couple of months—after the Christmas holidays. You now have an opportunity to buy a very attractive 5-day Christmas vacation package to Hawaii at an exceptionally low price. The special offer expires tomorrow, while the exam grade will not be available until the following day. Would you:

x buy the vacation package

y not buy the vacation package

z pay a $5 non-refundable fee in order to retain the rights to buy the vacation package at the same exceptional price the day after tomorrow—after you find out whether or not you passed the exam.

2. Pass/Fail Version

Imagine that you have just taken a tough qualifying examination. It is the end of the fall quarter, you feel tired and run-down, and you find out that you passed the exam (failed the exam. You will have to take it again in a couple of month—after the Christmas holidays). You now have an opportunity to buy a very attractive 5-day Christmas vacation package to Hawaii at an exceptionally low price. The special offer expires tomorrow. Would you:

x buy the vacation package

y not buy the vacation package

z pay a $5 non-refundable fee in order to retain the rights to buy the vacation package at the same exceptional price the day after tomorrow.

In the Hawaii problem, more than half of the subjects who know the outcome of the exam (54% in the passed condition and 57% in the fail condition) choose option x — buy the vacation package, but only 32% do it in the uncertain condition of not knowing the outcome of the exam. This is a crucial demonstration that Tversky and Shafir produced to show the disjunction effect. Here decision makers prefer option x (to buy the vacation package) when they are in a certain condition (passed exam) and they also prefer x when they are not in that condition (failed exam), but they refuse x (or prefer z) when they don't know which condition they are in (they don't know the outcome of the exam).

Next to the disjunction effect in decision theory an effect called the conjunction fallacy was identified. This effect occurs when it is assumed that specific conditions are more probable than a single general one. The most oft-cited example of this fallacy originated with Amos Tversky and Daniel Kahneman [16,53]: *Linda is 31 years old, single, outspoken, and very bright. She majored in philosophy. As a student, she was deeply concerned with issues of discrimination and social justice, and also participated in anti-nuclear demonstrations. Which is more probable? (1) Linda is a bank teller. (2) Linda is a bank teller and is active in the feminist movement.* 85% of those asked chose option (2). However, if classical Kolmogorovian probability theory is applied, the probability of two events occurring together (in 'conjunction') will always be less than or equal to the probability of either one occurring alone.

4   The Guppy Effect in Concept Theories

Situations that can be compared with the foregoing described paradoxes, effects, fallacies, in economics and decision theory have been studied in the field of concepts representation. In [3] the concepts *Pet* and *Fish* and their conjunction *Pet-Fish* are considered, and observed that an item such as *Guppy* turns out to be a very typical example of *Pet-Fish*, while it is neither a very typical example of *Pet* nor of *Fish*. Hence, the typicality of a specific item with respect to the conjunction of concepts can behave in an unexpected way. The problem is often referred to as the 'pet-fish problem' and the effect is usually called the 'guppy effect'. The guppy effect is abundant and appears almost in every situation where concepts are combined [3,4,5,6,7,8,9,10,11,12,13].

The guppy effect was not only identified for the typicality of items with respect to concepts and their conjunctions but also for the membership weights of items with respect to (i) concepts and their conjunction [6], (ii) concepts and their disjunction [7]. For example, the concepts *Home Furnishings* and *Furniture* and their disjunction *Home Furnishings or Furniture* was one of the pair of studied concepts. With respect to this pair, Hampton considered the item *Ashtray*. Subjects estimated the membership weight of *Ashtray* for the concept *Home Furnishings* to be 0.3 and the membership weight of the item *Ashtray* for the concept *Furniture* to be 0.7. However, the membership weight of *Ashtray* with respect to the disjunction *Home Furnishings or Furniture* was estimated only 0.25, less than both weights with respect to both concepts apart. This means that subjects found *Ashtray* to be 'less strongly a member of the disjunction *Home Furnishings or Furniture*' than they found it to be a member of the concept *Home Furnishings* alone or a member of the concept *Furniture* alone. If one thinks intuitively of the 'logical' meaning of a disjunction, this is an unexpected result. Indeed, someone who finds that *Ashtray* belongs to *Home Furnishings*, would be expected to also believe that *Ashtray* belongs to *Home Furnishings or Furniture*. Equally so for someone who finds that *Ashtray* is a piece of *Furniture*.

This deviation from what one would expect of a standard classical interpretation of the disjunction was called 'underextension' in [7]. Although the item *Ashtray* with respect to the disjunction of the concepts *Home Furnishings* and *Furniture* shows underextension, and hence the presence of the effect of underextension, could be interpreted as due to 'ambiguity aversion', also the inverse effect occurs in many occasions. For example with respect to the pair of concepts *Fruits* and *Vegetables* and their disjunction *Fruits or Vegetables*, for the item *Olive* the membership weights with respect to *Fruits*, *Vegetables* and *Fruits or Vegetables* were respectively 0.5, 0.1 and 0.8. This means that *Olive* is estimated by subjects to belong 'more' to *Fruits or Vegetables* than to any one of the concepts apart.

The examples in the different domains of science, economics, decision theory and concept theories, that we gave in the foregoing sections illustrate the presence of the two layers of thought which we called the classical logical layer and the quantum conceptual layer. In the next section we analyze the two layers of thought in detail on a specific example from the domain of concept theory.

5   Classical Logical and Quantum Conceptual Thought

Let us analyze in detail the structure of the two layers of thought on the specific example of the 'disjunction of concepts'. It is indeed the quantum modeling of the disjunction of concepts as elaborated in [30,31,32], and more specifically in [32], section 7, that reveals the presence of the two superposed layers of classical logical thought and of quantum conceptual thought for the situation of the disjunction of two concepts. We will analyze the two layers in detail here now.

Consider the situation of a subject performing one of the experiments of Hampton [6,7], and more concretely the subject is estimating the membership weight of an item X with respect to the disjunction 'A or B' of the two concepts A and B. To make the situation even more concrete, consider as an example the item *Apple* with respect to the pair of concepts *Fruits* and *Vegetables* and their disjunction '*Fruits or Vegetables*'. A subject might go about more or less as follows: 'An apple is certainly a fruit, but it is definitely not a vegetable. But hence it is certainly also a 'fruit or a vegetable', since it is a fruit'. This 'reasoning' fits into a classical Boolean scheme [1], indeed if one proposition is true then also the disjunction of this proposition with another proposition is true. Of course, in general the membership weight which is an average over the yes/no attributions by the individual subjects will not be equal to 1, as it is the case for the items *Apple*, but will be between 0 and 1. Consider for example the item *Pumpkin* with respect to the same pair of concepts. As has been measured in [7], and also to be found in Table 1 of [32], the membership weights of this item with respect to the concepts *Fruits*, *Vegetables* and their disjunction *Fruits or Vegetables* are respectively 0.7, 0.8 and 0.925. We prove in [32], section 2.2, that with these membership weights it is possible for a deterministic logical underlying process to exist, such that these weights are the results of the classical Kolmogorovian chance for a specific subject to choose 'for' or 'against' membership of the item *Pumpkin* with respect to the concepts *Fruits*, *Vegetables* and *Fruits or Vegetables* respectively. This means that for the thought process that enrolls when a subject is deciding 'for' or 'against' membership of *Pumpkin* with respect to the pair of concepts *Fruits*, *Vegetables* and *Fruits or Vegetables* it is possible to find an underlying process that is deterministic and enrolls following classical logic. The manifest process measured in [7] can as a consequence be modeled by means of a classical stochastic process, where the probability, giving rise to the weights, is due to a lack of knowledge of an underlying deterministic classical logic process.

Let us consider now a third item with respect to the same pair of concepts, namely the item *Olive*. In [7] the following weights were measured, 0.5 with respect to *Fruits*, 0.1 with respect to *Vegetables* and 0.8 with respect to *Fruits or Vegetables*. Hence this is a case called overextension in [7]. We prove in [32], section 2.2, that for these weights it is not possible to find a Kolmogorovian representation. This means that these weights cannot be obtained by supposing that subjects reasoned following classical logic and that the weights are the result of a lack of knowledge about the exact outcomes given by each of the individual subjects. Indeed, if 50% of the subjects has classified the item *Olive* belonging to *Fruits*, and 10% has classified it as belonging to *Vegetables* than following a classical reasoning at most 60% of the subjects (corresponding with the set-theoretic union of two mutually distinct sets of subjects) can classify it as belonging to *Fruits or Vegetables*. Hence, this means that these weights arise in a distinct way. Some individual subjects must have chosen *Olive* as a member of *Fruits or Vegetables* and 'not as a member' of *Fruits* and also 'not as a member' of *Vegetables*, otherwise the weights 0.5, 0.1 and 0.8 cannot result. It is here that the second layer of thought, namely quantum conceptual thought, comes into play. Concretely this means that for the item *Olive* the subject does not reason in a logical way, but rather directly wonders whether *Olive* is a member or not a member of *Fruits or Vegetables*. In this quantum conceptual thought process the subject considers *Fruits or Vegetables* as a new concept and not as a classical logical disjunction of the two concepts *Fruits* and *Vegetables* apart. Hence the subject gets influenced in his or her choice in favor or against membership of *Olive* with respect to *Fruits or Vegetables* by the presence of the new concept *Fruits or Vegetables*. This is the reason why we say that within the quantum conceptual thought process it is the emergence of a new concept, i.e. the concept *Fruits or Vegetables* within the landscape of existing concepts, i.e. *Fruits, Vegetables* and *Olive*, that gives rise to the deviation of the membership weight that would be expected following classical logic. And it is the probability to decide for or against membership that is influenced by the presence of this new concept within the landscape of existing concepts. Concretely, in this case, the subject estimates whether *Olive* is characteristic for the new concept *Fruits or Vegetables*, hence whether *Olive* is one of these items where indeed one can doubt whether it is a *Fruit* or a *Vegetable*. And right so, for *Olive* this is typically the case, which is the reason that its weight with respect to *Fruits or Vegetables* is big, namely 0.8, as compared to rather small, 0.5 and 0.1 with respect to the concepts *Fruits* and *Vegetables* apart, and most important 'bigger than the sum of both' (0.8 is strictly bigger than 0.5+0.1), which makes a classical explanation impossible, as we prove in [32]. The proof in [32] that the weights for the item *Olive* with respect to the pair of concepts *Fruits, Vegetables* and *Fruits or Vegetables* cannot be modeled within a Kolmogorovian probability structure, has as a consequence that there cannot exist an underlying deterministic process giving rise to these weights. Hence, it means that the conceptual thought process that takes place when a subject decides 'for' or 'against' membership of the item *Olive* with respect to the concepts *Fruits, Vegetables* and *Fruits or Vegetables* is intrinsically indeterministic. In [32] we also show that the conceptual thought process, hence the weights that it produces with respect to the different conceptual structures, when different possible decisions are considered, can be modeled by means of a quantum mechanical probability structure.

6   Experimental Arguments for Quantum Conceptual Thought

The disjunction effect, one of the effects where following our hypothesis 'quantum conceptual thought is present', can be tested in various experimental settings. We will show in this section that specific situations that have been investigated demonstrate the presence of quantum conceptual thought. More concretely, the experiments that we will describe in this section show that what is crucial to explain the effects measured is; (i) that the whole and overall conceptual landscape is relevant for the situation considered, and; (ii) that the concepts play a role as individual entity, hence different from what this role would be if they were just combinations, i.e. disjunctions or conjunctions, of existing concepts. This is the reason why these experiments show the presence of what we have called 'quantum conceptual thought' in the situations typical of the disjunction effect. This also explains why the often proposed explanation of 'uncertainty aversion', although it can play a role, as long as it fits into the conceptual structure, is not the main cause of the disjunction effect. Let us clarify this concretely by referring to the examples of the foregoing section: *Olive* scores 'higher' than classically expected with respect to *Fruits or Vegetables*, hence contrary to what the disjunction effect would provoke if it were caused by 'uncertainty aversion'. Hence, if one would reflect in terms of 'uncertainty aversion' as explanation of the disjunction effect, it would mean that *Olive* 'likes' uncertainty instead of disliking it. Of course, what is really at work, as we explained in the foregoing section, is that *Olive* is characteristic conceptually for items that are *Fruits or Vegetables*, hence *Olive* indeed 'likes' uncertainty in this sense metaphorically speaking. In the experiments that we consider in the following of this section, we will see proofs of the fact that it is the overall conceptual landscape that is at the origin of the effects, and such that 'concepts influence as individual entities and not as classical logical combinations of other existing concepts' which is exactly what we have called the 'presence of quantum conceptual thought'.

In [21] Maya Bar-Hillel and Efrat Neter explored the possibility of extending the conjunction fallacy [53] to a more general extension fallacy, using natural disjunctive categories and including problems that involve no compound events, which allowed to check whether the fallacy results from incorrect combination rules. Students received brief case descriptions and ordered 7 categories according to the criteria: (a) probability of membership, (b) willingness to bet on membership. Let us give some examples to illustrate the type of test made. A detailed description of a person is given: *Danielle, sensitive and introspective. In high school she wrote poetry secretly. Did her military service as a teacher. Though beautiful, she has little social life, since she prefers to spend her time reading quietly at home rather than partying.* The question is: What does she study? And then the alternatives to choose from are: *Literature*; *Humanities*; *Physics* or *Natural Sciences*. The second person considered is: *Oded: Did his military service as a combat pilot. Was a brilliant high school student, whose teachers predicted for him an academic career. Independent and original, diligent and honest. His hobbies are shortwave radio and Astronomy.* The question is again: What does he study? The alternatives to choose from are again: *Physics*; *Natural Sciences*; *Literature* or *Humanities*.

One of the basic rules of classical probability is violated in all cases tested in [21]. Let us point out more in detail what happens. Consider the *Danielle* case. Following the rules of classical logic, studying *Literature* implies studying *Humanities*. This means that it is 'always more probable that *Danielle* studies *Humanities* than that *Danielle* studies *Literature*'. However, 82% of the subjects indicated *Literature* and not *Humanities* as more probable, and 75% of the subject preferred to bet on *Literature* instead of *Humanities*. This effect is called a 'disjunction fallacy' by the authors. The disjunction fallacy turned out much less pronounced for the less probable choices. Concretely, 45% of the subjects indicated *Physics* and not *Natural Sciences* as more probable for *Danielle*, and 27% of the subjects preferred to bet on *Physics* instead of *Natural Sciences*. A very similar and even more pronounced result for the second example of *Oded*. Subjects indicated *Physics* and not *Natural Sciences* as more probable, and preferred to bet on *Physics* instead of *Natural Sciences*. For the less probable choices again the effect was less. Other but similar situations were tested and all revealed the same disjunction fallacy. The results support the view that the disjunction fallacy in probability judgments is due to representativeness, i.e. the degree of correspondence between an instance and a concept. Concepts that are 'more representative' are preferred, even if such a choice violates the rules of classical probability and logic. More specifically, *Literature* is preferred to *Humanities* for *Danielle* and *Physics* to *Natural Sciences* for *Oded* because as concepts they represent better their respective personalities. On the contrary, *Natural Sciences* is preferred to *Physics* for *Danielle* and *Humanities* is preferred to *Literature* for *Oded* because as concepts both represent less badly their respective personalities. The fact that this inversion of the effect takes place for 'badly representing concept' shows that the effect is not due to some kind of overall preference for basic concepts, which *Literature* and *Physics* are, as compared to superordinate concepts, which *Humanities* and *Natural Sciences* are. Since stimuli used in the experiments could not be judged by combination rules, these results also go against claims that probability fallacies 'stem primarily from the incorrect rules people use to combine probabilities [20]'. We have mentioned earlier 'representativeness' as one of the measurable quantities that reveals the presence of what we have called quantum conceptual thought, and this is what we see at play here. The experiments in [21] show that a concept such as *Humanities*, although originally conceived as the disjunction of *Languages, Literature, History, Philosophy, Religion, Visual and Performing Arts* and *Music*, 'does not behave as the classical logical disjunction of these basic concepts', exactly analogously with the non classical logical behavior we have put forward in detail in the foregoing section of the present article for the disjunction *Fruit or Vegetables* of the two concepts *Fruits* and *Vegetables*. Equally so for the concept *Natural Sciences*, although originally being the disjunction of *Astronomy, Physics, Chemistry, Biology* and *Earth Sciences*, it does not behave as a concept in the way it should when simply being the classical logical disjunction of the different natural sciences. The quantum mechanical approach elaborated in [30,31,32] models this non classical logical behavior.

Another set of experiments related to the disjunction effect was organized by Maria Bagassi and Laura Macchi [22]. Their aim was to show that the disjunction effect does not depend on the presence of uncertainty (pass or fail the exam) but on the introduction into the text-problem of a non-relevant goal. This indicates in a very explicit way that it is the overall conceptual landscape that gives form to the disjunction effect. More specifically Bagassi and Macchi point out that, option z ('pay a \$5 non-refundable fee in order to retain the rights to buy the vacation package at the same exceptional price the day after tomorrow—after you find out whether or not you passed the exam') contains an unnecessary goal, i.e. that one needs to 'pay to know', which is independent of the uncertainty condition. In this sense, their hypothesis is that the choice of option z occurs as a consequence of the construction of the discourse-problem itself ([22], p. 44). Four experiments were performed in which various modifications with respect to option z were considered, ranging from (1) eliminating from the text and option z any connection between the 'knowledge of the outcome' and the 'decision'; (2) eliminating option z, limiting the decision to x ('buy') or y ('not buy'); (3) making option z more attractive; (4) render the procrastination option z more onerous. The experimental results support the view that the disjunction effect does not depend on the uncertainty condition itself, but on the insertion of the misleading goal 'paying to know' in the text-problem. Eliminating it (but maintaining the uncertainty condition) the disjunction effect vanishes (exp. 1 and 2). Also, if choice for z is sensible (exp. 3), most subjects choose it. If option z is onerous (exp. 4), it is substantially ignored. In this sense, option z is not a real alternative to x and y, but becomes an additional premise that conveys information, which changes the decisional context. Hence the crucial factor is the relevance of the discourse-problem of which z is one element, rather than certainty versus uncertainty. These results support the view that the disjunction effect can be realized by applying a *suitable* decisional context rather than an *uncertain* decisional context. If the suitable decision context, or in our terminology 'specific conceptual landscape surrounding the decision situation', is what lies at the origin of the disjunction effect, then this shows that what we have called 'quantum conceptual thought' is taking place during the process that gives rise to the disjunction effect. Following the experimental results of [22] one can argument that 'the specific conceptual landscape surrounding the decision situation' plays a principal role in shaping the disjunction effect.

7   Conclusions

Inspired by a quantum mechanical formalism to model concepts and their disjunctions and conjunctions, we put forward in this paper a specific hypothesis. Namely that two superposed layers exist within human thought:

(i) A layer which we call the 'classical logical layer', and which is given form by an underlying classical deterministic process, giving rise to classical logical thought and its indeterministic manifestation modeled by classical probability theory. We refer to thought in this layer as 'classical logical thought'.

(ii) A layer which we call the 'quantum conceptual layer', and which is given form by the influence and structure of the overall conceptual landscape, where concepts and also all combinations of concepts exercise their influence on an individual basis. This means that combinations of concepts emerge in this layer as new individual concepts and not just logical combinations which is what they are in the classical logical layer. In this quantum conceptual layer the global and holistic effects of the overall conceptual landscape can be experimentally detected by measuring specific quantities that may be different depending on the domain under consideration, and a substantial part of this layer can be modeled by quantum mechanical probabilistic structures. The effects of the presence of conceptual thought can be observed in situations where deviations of logical thought are apparent in a systematic and intersubjective way, i.e. such that the effect can be measured and proven to be not due to chance and be repeated quantitatively. We have considered examples of three specific domains of research where the effects of the presence of conceptual thought and its deviations from logical thought have been noticed and studied, i.e. economics, decision theory, and concept theories. In concept theories quantities such as 'typicality', 'membership', 'similarity' and 'applicability' can be measured and shown to deviate from what their values should be if thought would be classical logical. In decision theory this role is played by quantities such as 'representativeness', 'qualitative likelihood' 'similarity' 'resemblance', and in economics quantities like 'preference', 'utility' and 'presence of ambiguity' put into evidence the presence of quantum conceptual thought by deviating from their classical logical values. We have illustrated in detail the functioning of the two layers of thought on the specific example of the 'disjunction of concepts', and we analyze two experimental investigations on the disjunction effect that put into evidence the presence of quantum conceptual thought with respect to this disjunction effect.

References

1. Boole, G.: An Investigation of the Laws of Thought: On Which Are Founded the Mathematical Theories of Logic and Probabilities. Walton and Maberly, London (1854)
2. Kolmogorov, A.N.: Foundations of the Theory of Probability. Chelsea Publishing Company, New York (1956)
3. Osherson, D.N., Smith, E.E.: On the adequacy of prototype theory as a theory of concepts. Cognition 9, 35–58 (1981)
4. Osherson, D.N., Smith, E.E.: Gradedness and conceptual combination. Cognition 12, 299–318 (1982)
5. Smith, E.E., Osherson, D.N.: Conceptual combination with prototype concepts. Cognitive Science 8, 357–361 (1988)
6. Hampton, J.A.: Overextension of conjunctive concepts: Evidence for a unitary model for concept typicality and class inclusion. Journal of Experimental Psychology: Learning, Memory, and Cognition 14, 12–32 (1988)
7. Hampton, J.A.: Disjunction of natural concepts. Memory & Cognition 16, 579–591 (1988)
8. Storms, G., De Boeck, P., Van Mechelen, I., Geeraerts, D.: Dominance and non-commutativity effects in concept conjunctions: Extensional or intensional basis? Memory & Cognition 21, 752–762 (1993)
9. Rips, L.J.: The current status of research on concept combination. Mind & Language 10, 72–104 (1995)
10. Storms, G., De Boeck, P., Van Mechelen, I., Ruts, W.: The dominance effect in concept conjunctions: Generality and interaction aspects. Journal of Experimental Psychology: Learning, Memory & Cognition 22, 1–15 (1996)

11. Hampton, J.A.: Conceptual combination: Conjunction and negation of natural concepts. Memory & Cognition 25, 888–909 (1997)
12. Storms, G., De Boeck, P., Van Mechelen, I., Ruts, W.: Not guppies, nor goldfish, but tumble dryers, Noriega, Jesse Jackson, panties, car crashes, bird books, and Stevie Wonder. Memory & Cognition 26, 143–145 (1998)
13. Storms, G., de Boeck, P., Hampton, J.A., van Mechelen, I.: Predicting conjunction typicalities by component typicalities. Psychonomic Bulletin and Review 6, 677–684 (1999)
14. Kahneman, D., Tversky, A.: Subjective probability: A judgment of representativeness. Cognitive Psychology 3, 430–454 (1972)
15. Tversky, A.: Features of similarity. Psychological Review 84, 327–352 (1977)
16. Tversky, A., Kahneman, D.: Judgments of and by representativeness. In: Kahneman, D., Slovic, P., Tversky, A. (eds.) Judgment under uncertainty: Heuristics and biases. Cambridge University Press, Cambridge (1982)
17. Morier, D.M., Borgida, E.: The conjunction fallacy: A task specific phenomenon? Personality and Social Psychology Bulletin 10, 243–252 (1984)
18. Wells, G.L.: The conjunction error and the representativeness heuristic. Social Cognition 3, 266–279 (1985)
19. Carlson, B.W., Yates, J.F.: Disjunction errors in qualitative likelihood judgment. Organizational Behavior and Human Decision Processes 44, 368–379 (1989)
20. Gavanski, I., Roskos-Ewoldsen, D.R.: Representativeness and conjoint probability. Journal of Personality and Social Psychology 61, 181–194 (1991)
21. Bar-Hillel, M., Neter, E.: How alike is it versus how likely is it: A disjunction fallacy in probability judgments. Journal of Personality and Social Psychology 65, 1119–1131 (1993)
22. Bagassi, M., Macchi, L.: The 'vanishing' of the disjunction effect by sensible procrastination. Mind & Society 6, 41–52 (2007)
23. Allais, M.: Le comportement de l'homme rationnel devant le risque: critique des postulats et axiomes de l'école Américaine. Econometrica 21, 503–546 (1953)
24. Ellsberg, D.: Risk, Ambiguity, and the Savage Axioms. Quarterly Journal of Economics 75(4), 643–669 (1961)
25. Freud, S.: Die Traumdeutung. Fischer-Taschenbuch, Berlin (1899)
26. James, W.: Some Problems of Philosophy. Harvard University Press, Cambridge (1910)
27. Piaget, J.: Le Langage et la Pensée Chez l'Enfant. Delachaux et Niestlé, Paris (1923)
28. Bruner, J.: Acts of Meaning. Harvard University Press, Cambridge (1990)
29. Gabora, L., Aerts, D.: Evolution of the Integrated Worldview. Journal of Mathematical Psychology (2009)
30. Aerts, D.: Quantum interference and superposition in cognition: A theory for the disjunction of concepts (2007), Archive reference and link:
    http://uk.arxiv.org/abs/0705.0975
31. Aerts, D.: General quantum modeling of combining concepts: A quantum field model in Fock space (2007), Archive reference and link:
    http://uk.arxiv.org/abs/0705.1740
32. Aerts, D.: Quantum structure in cognition. Journal of Mathematical Psychology (2009), Archive reference and link:
    http://uk.arxiv.org/abs/0805.3850
33. Schaden, M.: Quantum finance: A quantum approach to stock price fluctuations. Physica A 316, 511–538 (2002)
34. Baaquie, B.E.: Quantum Finance: Path Integrals and Hamiltonians for Options and Interest Rates. Cambridge University Press, Cambridge (2004)

35. Haven, E.: Pilot-wave theory and financial option pricing. International Journal of Theoretical Physics 44, 1957–1962 (2005)
36. Bagarello, F.: An operational approach to stock markets. Journal of Physics A 39, 6823–6840 (2006)
37. Aerts, D., Aerts, S.: Applications of quantum statistics in psychological studies of decision processes. Foundations of Science 1, 85–97 (1994); Reprinted in: Van Fraassen, B. (ed.): Topics in the Foundation of Statistics, pp. 111–122. Springer, Dordrecht (1994)
38. Gabora, L., Aerts, D.: Contextualizing concepts using a mathematical generalization of the quantum formalism. Journal of Experimental and Theoretical Artificial Intelligence 14, 327–358 (2002)
39. Aerts, D., Gabora, L.: A theory of concepts and their combinations I & II. Kybernetes 34, 167–221 (2005)
40. Busemeyer, J.R., Wang, Z., Townsend, J.T.: Quantum dynamics of human decision making. Journal of Mathematical Psychology 50, 220–241 (2006)
41. Busemeyer, J.R., Matthew, M., Wang, Z.: A quantum information processing theory explanation of disjunction effects. Proceedings of the Cognitive Science Society (2006)
42. Widdows, D.: Orthogonal negation in vector spaces for modelling word-meanings and document retrieval. In: Proceedings of the 41st Annual Meeting of the Association for Computational Linguistics, Sapporo, Japan, pp. 136–143 (2003)
43. Widdows, D., Peters, S.: Word vectors and quantum logic: Experiments with negation and disjunction. In: Mathematics of Language, vol. 8, pp. 141–154. Bloomington, Indiana (2003)
44. Aerts, D., Czachor, M.: Quantum aspects of semantic analysis and symbolic artificial intelligence. Journal of Physics A, Mathematical and Theoretical 37, L123-L132 (2004)
45. Van Rijsbergen, K.: The Geometry of Information Retrieval. Cambridge University Press, Cambridge (2004)
46. Aerts, D., Czachor, M., D'Hooghe, B.: Towards a quantum evolutionary scheme: violating bell's inequalities in language. In: Gontier, N., Van Bendegem, J.P., Aerts, D. (eds.) Evolution- ary Epistemology, Language and Culture - A Non Adaptationist Systems Theoretical Approach. Springer, Dordrecht (2006)
47. Bruza, P.D., Cole, R.J.: Quantum logic of semantic space: An exploratory investigation of context effects in practical reasoning. In: Artemov, S., Barringer, H., d'Avila Garcez, A.S., Lamb, L.C., Woods, J. (eds.) We Will Show Them: Essays in Honour of Dov Gabbay. College Publications (2005)
48. von Neumann, J.: Zur Theorie der Gesellschaftsspiele (On the Theory of Parlor Games). Mathematische Annalen 100, 295–300 (1928)
49. von Neumann, J., Morgenstern, O.: Theory of games and economic behavior. Princeton University Press, Princeton (1944)
50. Savage, L.J.: The Foundations of Statistics. Wiley, New-York (1944)
51. Ben-Haim, Y.: Info-gap decision theory: Decisions under severe uncertainty. Academic Press, London (2006) (section 11.1)
52. Tversky, A., Shafir, E.: The disjunction effect in choice under uncertainty. Psychological Science 3(5), 305–309 (1992)
53. Tversky, A., Kahneman, D.: Extension versus intuitive reasoning: The conjunction fallacy in probability judgment. Psychological Review 90(4), 293–315 (1983)
54. Kahneman, D., Tversky, A.: Prospect theory: An analysis of decision under risk. Econometrica 47, 263–291 (1979)

Quantum and Classical Structures
in Nondeterminstic Computation

Dusko Pavlovic*

Kestrel Institute and Oxford University
dusko@{kestrel.edu,comlab.ox.ac.uk}

Abstract. In categorical quantum mechanics, classical structures characterize the classical interfaces of quantum resources on one hand, while on the other hand giving rise to some quantum phenomena. In the standard Hilbert space model of quantum theories, classical structures over a space correspond to its orthonormal bases. In the present paper, we show that classical structures in the category of relations correspond to direct sums of abelian groups. Although relations are, of course, not an interesting model of quantum computation, this result has some interesting computational interpretations. If relations are viewed as denotations of *nondeterministic* programs, it uncovers a wide variety of non-standard quantum structures in this familiar area of classical computation. Ironically, it also opens up a version of what in philosophy of quantum mechanics would be called an ontic-epistemic gap, as it provides no interface to these nonstandard quantum structures.

1 Introduction

Classical structures came to be a useful algebraic tool for analyzing the conceptual foundations of quantum computation [10,7,12]. They characterize the classical interfaces of quantum resources on one hand, and generate entanglement structures, and other essentially quantum phenomena on the other. In the standard, Hilbert space model of quantum theories, classical structures over a space exactly correspond to its orthonormal bases. In nonstandard models, however, they provide a generic conduit to the classical and the quantum features.

Categorical quantum mechanics, initiated in [2], axiomatizes some basic quantum phenomena in the framework of dagger-compact categories. This remarkably rich yet succinct structure has arisen in part from the experience gathered in semantics of programming languages. The most direct source are probably Abramsky's *interaction categories* [3,25], developed to capture the idea of concurrent programs as *relations extended in time*. As a consequence, categories of relations, in all their various flavors arising from various resources [4,5,13,22,23], provide models of categorical quantum mechanics, albeit degenerate because of the trivial dagger structure. Nevertheless, the notion of a classical structure over

---

* Supported by ONR and EPSRC.

P. Bruza et al. (Eds.): QI 2009, LNAI 5494, pp. 143–157, 2009.
© Springer-Verlag Berlin Heidelberg 2009

relations is well defined. In the present paper, we provide a complete characterization of classical structures over relations.[1]

But what is the relevance and meaning of such a result? Although some relationally based "toy models" of certain quantum phenomena [27] have awaken a lot of interest, a category of relations itself is a rather degenerate model of quantum computation. Its duality and scalar structures in particular seem too simple to accommodate the complex interactions between the quantum and the classical phenomena. — It is therefore only more surprising that, even in this simple framework, classical structures seem to have an interesting story to tell.

Outline of the Paper

In section 2, we summarize the definitions of classical and quantum structures, recall their basic properties, and describe the standard, and some nonstandard examples. In section 3, we describe a rich source of nonstandard examples of classical structures in the category Rel of relations: every Abelian group gives a nonstandard classical structure. In fact, these are exactly the indecomposable classical structures. In section 4, we show that every classical structure in Rel must arise as a direct product of indecomposables. This provides a complete characterisation of classical structures in Rel. In section 5, we summarize the meaning of this characterization, and discuss the questions that it raises.

Notation. To describe relations on finite sets, we often find it convenient to use von Neumann's representation of ordinals, where $0 = \emptyset$ is the empty set, and $n = \{0, 1, \ldots, n-1\}$. Moreover, the pairs $\langle i, j \rangle \in n \times n$ are often abbreviated to $ij \in n \times n$.

2   Algebras for Abstraction and Duality

We begin by introducing classical structures as the classical interface of quantum resources. To justify their algebraic content, we delve into a conceptual reconstruction of their role. A reader only interested in their structure should skip the next subsection.

2.1   Program Abstraction and Quantum Computation

Abstraction is the essence of programming. The first example of program abstraction are probably Gödel's numberings of primitive recursive functions [14]. Gödel's construction demonstrated that recursive programs, specifying entire families of computations (of the values of a function for all its inputs), can be stored as data. Von Neumann later explicated this as the fundamental principle of computer architecture. Kleene, on the other side, refined the idea of program

---

[1] Formally, we work with relations within a given universe of sets. Each of the relational formalisms proposed in the above references will suffice for this.

abstraction into the fundamental lemma of recursion theory: the s-m-n theorem [16]. Church, finally[2] proposed the formal operations of variable abstraction and data application as the driving force of all computation [6]. This proposal became the foundation of functional programming.

But what is variable abstraction? What is a variable? We use them so often that it is sometimes hard to tell. A variable is adjoined to a ring, as an indeterminate element, to generate the ring of polynomials. A programmer denotes by a variable a piece of data that will only be determined when the program is run. The variable captures all the possible values of this piece of data that may arise at run time. The operation of abstraction of a variable binds all of its occurrences (within the declared scope) to the interface where its values will be read, when given. The operation of application substitutes these values for the variable.

A variable is thus a tool for propagating as-yet-unknown data through a program (or through an algebraic structure, etc). The crucial capability of such a tool is that it allows the data to be *copied* wherever they are needed, or *deleted* where they are not needed. While the classical computation, as the above early references show, was built upon this capability as its very foundation — it is a fundamental property of quantum data that they generally cannot be copied or deleted [29,11,20,1].

A classical structure formalizes this distinction: its first feature is a comonoid $X \otimes X \xleftarrow{\Delta} X \xrightarrow{\top} I$, where Δ can be used to copy and $\top$ to delete a piece of data. A datum $I \xrightarrow{\psi} X$ is classical if it can be copied, in the sense that $\Delta \psi = \psi \otimes \psi$, and deleted, in the sense that $\top \psi = \mathrm{id}_I$. This turns out to be exactly what is needed to support variable abstraction in monoidal categories. In the framework of dagger-monoidal categories, the requirement that abstraction preserves the dagger induces the rest of classical structure [24,21].

2.2   Frobenius Algebras

Framework. Let $(\mathcal{C}, \otimes, I)$ be a monoidal category. With no loss of generality, we assume that the tensor is strictly associative and unitary, i.e. that the objects of $\mathcal{C}$ form an ordinary monoid with respect to $\otimes$ and I. Every monoidal category is equivalent to one which is strict in this sense. We call the arrows from I *vectors*, and write $\mathcal{C}(X) = \mathcal{C}(I, X)$.

Definition 2.1. *The structure of a* Frobenius algebra X *in* $\mathcal{C}$ *consists of*

- *an internal monoid* $X \otimes X \xrightarrow{\nabla} X \xleftarrow{\perp} I$, *and*
- *an internal comonoid* $X \otimes X \xleftarrow{\Delta} X \xrightarrow{\top} I$,

---

[2] Although Church's paper appeared three years earlier than Kleene's, Church's proposal is the final step in the conceptual development of function abstraction as the foundation of computation.

such that the following diagram commutes

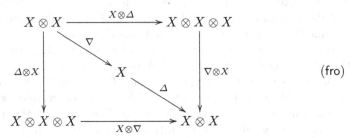

(fro)

A *Frobenius algebra* $(X, \nabla, \Delta, \bot, \top)$ is special *if its monoid and comonoid structures are normalized, in the sense that the diagram*

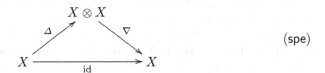

(spe)

also commutes.

Proposition 2.2. *In every monoidal category, being a special Frobenius algebra is a property of the monoid* $(X, \nabla, \bot)$. *More precisely, if both* $(X, \nabla, \Delta_1, \bot, \top_1)$ *and* $(X, \nabla, \Delta_2, \bot, \top_2)$ *are special Frobenius algebras, then* $\Delta_1 = \Delta_2$ *and* $\top_1 = \top_2$.

Dually, being a special Frobenius algebra can be viewed as a property of the comonoid $(X, \Delta, \top)$.

The monoid part $(X, \nabla, \bot)$ *of a special Frobenius algebra is abelian if and only if the corresponding comonoid part* $(X, \Delta, \top)$ *is.*

Much of the power of the Frobenius algebra structure arises from the way in which it gives rise to dualities.

Duality. A *duality* in a monoidal category $\mathcal{C}$ consists of two objects and two arrows, written $(\eta, \varepsilon) : X \dashv X^*$, where

- the copairing $I \xrightarrow{\eta} X^* \otimes X$ and
- the pairing $X \otimes X^* \xrightarrow{\varepsilon} I$

are required to satisfy the equations

$$(\varepsilon \otimes X)(X \otimes \eta) = X \qquad (X^* \otimes \varepsilon)(\eta \otimes X^*) = X^*$$

If every object $X \in \mathcal{C}$ has a chosen dual X^*, then the duality can be extended to a functor $(-)^* : \mathcal{C}^{op} \longrightarrow \mathcal{C}$, which maps $A \xrightarrow{f} B$ to

$$f^* : B^* \xrightarrow{\eta B^*} A^* A B^* \xrightarrow{A f B^*} A^* B B^* \xrightarrow{A^* \varepsilon} A^*$$

Frobenius algebras and dualities. Every Frobenius algebra X induces

- a pairing $\varepsilon : X \otimes X \xrightarrow{\nabla} X \xrightarrow{\top} I$, and
- a copairing $\eta : I \xrightarrow{\perp} X \xrightarrow{\Delta} X \otimes X$,

which together make X into a self-dual object, with $X^* = X$. Categorically, this means that X is adjoint to itself at the same time on the left and on the right, if the monoidal category $\mathcal{C}$ is viewed as a bicategory with a single 0-cell. E.g., if this 0-cell is a category $\mathbb{D}$ and $\mathcal{C} = \mathbb{D}^{\mathbb{D}}$ is the category of endofunctors $F, G \ldots : \mathbb{D} \longrightarrow \mathbb{D}$, with the natural transformations as the arrows between them, and with the composition playing the role of the tensor $F \otimes G = F \circ G$, then

- the monoid $F \otimes F \xrightarrow{\nabla} F \xleftarrow{\perp} Id$ makes the functor F into a monad,
- the comonoid $F \otimes F \xleftarrow{\Delta} F \xrightarrow{\top} Id$ makes it into a comonad, and
- condition (fro) makes the pairing $\varepsilon : FF \longrightarrow Id$ and copairing $\eta : Id \longrightarrow F$ into an adjunction $F \dashv F$.

This functorial setting was first described by Lawvere [18], who characterized it by requiring the commutativity of the following diagrams of natural transformations

$$
\begin{array}{ccc}
FFF \xleftarrow{F\Delta} FF \xrightarrow{\Delta F} FFF \\
{\scriptstyle \nabla F}\downarrow \qquad \downarrow{\scriptstyle \nabla} \qquad \downarrow{\scriptstyle F\nabla} \\
FF \xrightarrow[\top F]{} F \xleftarrow[F\top]{} FF
\end{array}
\qquad
\begin{array}{ccc}
FF \xleftarrow{\perp F} F \xrightarrow{F\perp} FF \\
{\scriptstyle \Delta F}\downarrow \qquad \downarrow{\scriptstyle \Delta} \qquad \downarrow{\scriptstyle F\Delta} \\
FFF \xrightarrow[F\nabla]{} FF \xleftarrow[\nabla F]{} FFF
\end{array}
$$

and attached the name of Frobenius to such structures. The equivalent but simpler condition (fro) first appeared in Carboni and Walters' work [5], characterizing relations as a cartesian bicategory. The geometric meaning of (fro) in the category of cobordisms brought the same condition to prominence in the categorical version of Topological Quantum Field Theory [17]. Finally, its role in supporting a generic form of abstraction, on which the interface between the classical and the quantum computation turns out to be based [21], made it into an important piece in categorical Quantum Mechanics [2,10].

2.3 Classical and Quantum Structures

Framework. Categorical quantum mechanics actually requires a slight extension of monoidal categories: besides the monoidal structure, the category $\mathcal{C}$ should come equipped with a functor $(-)^{\ddagger} : \mathcal{C}^{op} \longrightarrow \mathcal{C}$, which is identity on the objects, and involutive on the arrows, i.e. satisfying $f^{\ddagger\ddagger} = f$. The arrows u such that $u^{\ddagger}u = \mathrm{id}$ and $uu^{\ddagger} = \mathrm{id}$ are called unitary. All monoidal coherences in a dagger monoidal category are required to be unitary. In the strict case, this boils down to the requirement that the symmetries are unitary, since the other coherences are already identities.

The abstract structure of a *symmetric dagger-monoidal* category $(\mathcal{C}, \otimes, I, \ddagger)$ turns out to support the main constructions of quantum mechanics, normally presented using Hilbert spaces [2,26,10].

Definition 2.3. *A special Frobenius algebra* $(X, \triangledown, \vartriangle, \bot, \top)$ *in a symmetric dagger-monoidal category* $\mathcal{C}$ *is called a* classical structure *if its monoid and its comonoid parts are*

- *adjoint, i.e.* $\triangledown = \vartriangle^{\ddagger}$ *and* $\bot = \top^{\ddagger}$
- *abelian, i.e.* $\triangledown \circ (a \otimes b) = \triangledown \circ (b \otimes a)$.

Interpretation. In categorical quantum mechanics, classical structures can be used to distinguish the classical resources from the quantum resources. On one hand, each classical structure extracts the *classical elements*. On the other hand, it supports the entanglement phenomena, implemented through *quantum structures*. We now recall these concepts from [10].

Definition 2.4. *A quantum structure in a dagger-monoidal category is a pair* (X, η), *such that* $\eta : I \longrightarrow X \otimes X$ *and* $\eta^{\ddagger} : X \otimes X \longrightarrow I$ *make* X *self-dual, i.e.* $(\eta, \eta^{\ddagger}) : X \dashv X$.

Proposition 2.5. *Every classical structure induces a quantum structure, with the pairing* $\varepsilon : X \otimes X \xrightarrow{\triangledown} X \xrightarrow{\top} I$, *and the copairing* $\eta : I \xrightarrow{\bot} X \xrightarrow{\vartriangle} X \otimes X$.

Several classical structures may induce the same quantum structure. Some quantum structures do not arise from a classical structure.

Definition 2.1. Classical elements[3] *for a classical structure* X *in* $\mathcal{C}$ *are the arrows* $\varphi \in \mathcal{C}(X)$ *such that* $\vartriangle \varphi = \varphi \otimes \varphi$ *and* $\top \varphi = \mathrm{id}_I$.

Classical elements are thus just those vectors that can be copied and deleted. On the other hand, the entanglement capability of quantum structures is obtained by applying the copying facility of a classical structure to non-classical elements, such as the monoid unit of the classical structure itself.

2.4 Examples

The trivial example of a classical structure, present in every monoidal category, is the tensor unit I: the canonical isomorphisms $I \otimes I \cong I$ make it into a special Frobenius algebra. In the categories $(\mathsf{FVec}, \otimes, \mathsf{K})$ of finitely dimensional vector spaces and $(\mathsf{FHilb}, \otimes, \mathsf{K})$ over any field K, a choice of base $|0\rangle, |1\rangle, \ldots |n\rangle \in X$ makes each space X into a classical structure, defined by the linear operators

$$\vartriangle |i\rangle = |ii\rangle \qquad\qquad\qquad \top |i\rangle = 1$$

---

[3] In the Hopf algebra theory, the elements that satisfy the same conditions are called *set-like*.

In the monoidal category $(\mathsf{Rel}, \times, 1)$, every object comes with a similar classical structure

$$\Delta(i) = \{ii\} \qquad\qquad \top(i) = 0$$

where ij abbreviates the pair $\langle i, j \rangle \in X \times X$, and 0 is the unique element of 1. Both of these families of classical structures are induced by the cartesian structure of the category FSet of finite sets, canonically embedded in FVec and FRel. We call them *standard* classical structures. They are characterized and analyzed in detail in [5]. In [8], it has been shown that all classical structures in FHilb are standard. Very recently [9], though, Bill Edwards and Bob Coecke noticed a nonstandard classical structure $2 \times 2 \xleftarrow{\oslash} 2 \xrightarrow{\curlyvee} 1$ over the set $2 = \{0, 1\}$, defined as follows

$$\oslash(0) = \{00, 11\} \qquad\qquad \curlyvee(0) = \{0\}$$
$$\oslash(1) = \{01, 10\} \qquad\qquad \curlyvee(1) = 0$$

Note that both the standard classical structure $2 \times 2 \xleftarrow{\Delta} 2 \xrightarrow{\top} 1$ and the nonstandard classical structure $2 \times 2 \xleftarrow{\oslash} 2 \xrightarrow{\curlyvee} 1$ induce the same quantum structure $1 \xrightarrow{\eta} 2 \times 2$, relating 0 with 00 and 11. But this turns out to be an exception. E.g., a little trial and error leads to the following nonstandard classical structure $3 \times 3 \xleftarrow{\oslash} 3 \xrightarrow{\curlyvee} 1$, where $3 = \{0, 1, 2\}$

$$\oslash(0) = \{00, 12, 21\} \qquad\qquad \curlyvee(0) = \{0\}$$
$$\oslash(1) = \{22, 01, 10\} \qquad\qquad \curlyvee(1) = 0$$
$$\oslash(2) = \{11, 02, 20\} \qquad\qquad \curlyvee(2) = 0$$

The induced quantum structure $1 \xrightarrow{\eta} 3 \times 3$ now relates 0 with $00, 12$ and 21, whereas the standard one relates 0 with 00, 11 and 22. Soon we shall see how this comes about.

2.5   Representations of Classical Structures

By definition 2.1, classical structures are given as *internal* algebras. They are thus defined in any dagger monoidal category $\mathcal{C}$. However, some parts of the analysis of classical structures is simper with a more concrete representation.

According to proposition 2.2, a classical structure $(X, \nabla, \Delta, \bot, \top)$ is completely determined by the monoid part $(X, \nabla, \bot)$. This internal monoid can be represented as a monoid of endomorphisms on X in $\mathcal{C}$, as follows: proceeding as follows:

- first *externalize* the internal monoid $(X, \nabla, \bot)$ in $\mathcal{C}$ as an ordinary monoid of vectors $(\mathcal{C}(X), \cdot, \bot)$, by setting

$$\varphi \cdot \psi = \nabla \circ (\varphi \otimes \psi)$$

— then represent every vector $\varphi \in \mathcal{C}(X)$ as an action $\Upsilon\varphi$ over X, by

$$\Upsilon : \mathcal{C}(X) \longrightarrow \mathcal{C}(X,X)$$
$$\left(I \xrightarrow{\varphi} X\right) \longmapsto \left(X \xrightarrow{\varphi \otimes X} X \otimes X \xrightarrow{\nabla} X\right)$$

This second step can be viewed either as a generalization of Cayley's group representation to monoids, or as a special case of Yoneda's embedding of categories.

Proposition 2.6. *The monoid* $(\mathcal{C}(X), \cdot, \bot)$ *is isomorphic with the submonoid of* $(\mathcal{C}(X,X), \circ, \mathrm{id})$ *which consists of the endomorphisms* $f : X \to X$ *such that* $f \circ (a \cdot x) = (f \circ a) \cdot x$ *holds for all* $a, x \in \mathcal{C}(X)$.

This allows representing any monoid as a monoid of endomorphisms. But those monoids that come from a classical structure carry more. As observed in [8], and further explored in [28], the externalisation of every Frobenius algebra, and hence every classical structure, is also a $\star$-algebra. The categorical presentations [15,28] of the antilinear operation $\star$ involve the formal duals, as spelled out in sec. 2.2.

Definition 2.7. *An internal $\star$-monoid in a monoidal category* $\mathcal{C}$ *is a structure* $(X, X^*, \nabla, \bot, \star)$ *where*

- *$(X, \nabla, \bot)$ is internal monoid*
- *X^* is a dual of X, and*
- *$\star : X \cong X^*$ is an isomorphism (always unitary).*

We write $\varphi^* = \star \circ \varphi \in \mathcal{C}(X^*)$ *for* $\varphi \in \mathcal{C}(X)$.
A $\star$-monoid homomorphism $f : X \longrightarrow Y$ *is a monoid homomorphism which moreover preserves the $\star$, in the sense that*

$$\begin{array}{ccc} X^* & \xrightarrow{f_*} & Y^* \\ \star \downarrow & & \downarrow \star \\ X & \xrightarrow{f} & Y \end{array}$$

commutes.

Proposition 2.8. *The monoid* $(\mathcal{C}(X), \cdot, \bot)$ *induced by a classical structure* $(X, \nabla, \bot)$ *comes with an involution*

$$(-)^\star : \mathcal{C}(X) \longrightarrow \mathcal{C}(X)$$
$$\left(I \xrightarrow{\varphi} X\right) \longmapsto \left(I \xrightarrow{\eta} X \otimes X \xrightarrow{\varphi^\ddagger \otimes X} X\right)$$

This involution is preserved by the representation $\Upsilon : \mathcal{C}(X) \longrightarrow \mathcal{C}(X,X)$, *in the sense that* $\Upsilon(\varphi^\star) = (\Upsilon\varphi)^\ddagger$ *holds.*

3   Simple Classical Structures in Rel

In the rest of the paper, we explore and characterize classical structure in a category Rel of sets and relations. Any of its formalizations (some mentioned in the Introduction) will do. Computationally, relations are usually viewed as denotations of nondeterministic programs: a binary relation $R : A \longrightarrow B$ is the input/output relation of a program, which may output b when given an input a whenever aRb holds [19].

3.1   Meaning of (spe) in Rel

On the other hand, the isometry condition (spe) here means that the relation $\triangledown : X \times X \longrightarrow X$ is single-valued and surjective on X, i.e.

$$\forall x, y, u, v \in X.\ x, y \in \triangledown(u, v) \implies x = y$$
$$\forall x \in X \exists uv \in X.\ x = \triangledown(y, z)$$

Equivalently, (spe) means that $\Delta = \triangledown^{op} : X \longrightarrow X \times X$ injects X into parts of $X \times X$ and is total on X.

$$\forall x, y \in X.\ \Delta(x) \cap \Delta(y) \neq 0 \implies x = y$$
$$\forall x \in X.\ \Delta(x) \neq 0$$

3.2   Meaning of (fro) in Rel

In the monoidal category (Rel, $\times$, 1), the monoid action $\triangledown : X \times X \longrightarrow X$ is a relation, which assigns to every pair $x, y \in X$ a set $\triangledown(x, y) \subseteq X$. The Frobenius condition (fro) becomes

$$\{\langle x, y \rangle \mid \triangledown(i, j) \cap \triangledown(x, y) \neq 0\} = \{\langle x, \triangledown(y', j) \rangle \mid i \in \triangledown(x, y')\}$$
$$= \{\langle \triangledown(i, x'), y \rangle \mid j \in \triangledown(x', y)\}$$

This must be satisfied for all $i, j \in X$.

3.3   Meaning of (fro) $\wedge$ (spe) in Rel

Notation. Since $\triangledown(u, v)$, according to (spe), has at most one element, $\triangledown$ is a partial operation. It is thus convenient to write it in the infix form whenever it is defined, i.e. $u\triangledown v = \triangledown(u, v) \neq 0$.

The condition $\triangledown(i, j) \cap \triangledown(x, y) \neq 0$ now becomes $i\triangledown j = x\triangledown y$ and $i \in x\triangledown y$ means that $i = x\triangledown y$. The Frobenius condition thus boils down to

$$\{\langle x, y \rangle \mid i\triangledown j = x\triangledown y\} = \{\langle x, y'\triangledown j \rangle \mid i = x\triangledown y'\} \tag{1}$$
$$= \{\langle i\triangledown x', y \rangle \mid j = x'\triangledown y\}$$

This characterisation provides a rich source of classical structures.

Proposition 3.1. *Every abelian group $(X, \triangledown, \bot)$ in* Set *induces a classical structure in* Rel.

Proof. If $(X, \triangledown, \bot)$ is a group, then (1) is satisfied by $x' = j\triangledown y^{-1}$ and $y' = x^{-1}\triangledown i$. $\qquad\square$

Remark. The nonstandard classical structures described in section 2.4 are easily seen to arise from the groups $\mathbb{Z}_2$ and $\mathbb{Z}_3$.

3.4 Simplicity

In general, an object in a category is said to be *simple* if it has no nontrivial subobjects. All classical objects derived from abelian groups, along the lines of proposition 3.1, turn out to be simple.

Proposition 3.2. *A comonoid monomorphism between two classical structures, induced in* Rel *by abelian groups, must be either the empty relation, or a group isomorphism. Such classical structures are thus simple.*

Towards an easy proof, note that a relation $R : A \longrightarrow B$ is a monomorphism in Rel if and only if $R^{op} : B \longrightarrow A$ is a partial surjection.

Corollary 3.1. *Classical structures induced in* Rel *by nontrivial[4] abelian groups do not have any classical elements.*

4   Classification of Classical Structures in Rel

4.1   ⋆-algebras in Rel

In section 2.5, we saw that every classical structure induces a ⋆-algebra. In Rel, this restricts them to a very small family. The decomposition of Frobenius algebras into simple subalgebras follows.

Proposition 4.1. *The representation* $\Upsilon : X \longrightarrow$ Rel(X, X) *maps the elements of any classical structure X in* Rel *to partial bijections.*

Proof. We saw in section 3 that classical structures in Rel are partial monoids, in the sense that $x\triangledown y$ has at most one element. This means that for every $y \in X$ $\Upsilon y : X \longrightarrow X$ is a partial map.

Since $\Upsilon : X \longrightarrow$ Rel(X, X) is a ⋆-representation, $\Upsilon(y^{\ddagger}) = (\Upsilon y)^{op}$ is also a partial map. But a relation $R \in$ Rel(X, X) such that both R and R^{op} are partial maps

$$xRy \wedge xRy' \Longrightarrow y = y'$$
$$xRy \wedge x'Ry \Longrightarrow x = x'$$

must be a partial bijection. In words, for every x there is at most one y such that xRy; and for every y there is at most one x such that xRy.   □

---

[4] A group is said to be trivial if it has a single element.

4.2   The Main Results

Proposition 4.2. *Every special Frobenius algebra in* Rel *is a biproduct of special Frobenius algebras where the unit is a singleton.*

Terminology. The biproduct of sets A and B in Rel is simply their disjoint union $A+B$. This means that it is at the same time their product, and their coproduct.

Proof of proposition 4.2. Let the unit $\perp \in \mathsf{Rel}(X)$ of the special Frobenius algebra X be $\perp = \{\phi_j\}_{j\in J}$. We claim that there is a partition

$$X = \bigcup_{j\in J} X_j$$

such that

$$X_k \cap X_\ell \neq 0 \Longrightarrow k = \ell$$

and such that for every $j \in J$ the partial bijection $\Upsilon\phi_j : X \longrightarrow X$ is just the identity on X_j.

To prove this, note that

- $\Upsilon\perp = \mathrm{id}_X$,
- $\Upsilon\phi_j \subseteq \Upsilon\perp$
- if $a \in X_i \cap X_j$, then $a\Upsilon\phi_j = a = a\Upsilon\phi_k \Longrightarrow j = k$, because $a\Upsilon : X \longrightarrow X$ must also be a partial bijection, as demonstrated in proposition 4.1.

Thus the domains of $\Upsilon\phi_j$ must cover X, and they must be disjoint.

Now we claim that $(X_j, \triangledown_j, \perp_j)$ is a submonoid of $(X, \triangledown, \perp)$. This means that for every $x \in X_j$, the partial bijection $\Upsilon x : X \longrightarrow X$ restricts to a bijection $\Upsilon_j x : X_j \longrightarrow X_j$.

Suppose that for $x, y \in X_j$ happens that $x\triangledown y \in X_k$. That would mean that $y\triangledown\phi_k$ must be defined, because $x\triangledown y = (x\triangledown y)\triangledown\phi_k = x\triangledown(y\triangledown\phi_k)$. But then $y = y\triangledown\phi_k \in X_k$, and we get $y \in X_j \cap X_k$. We have seen above that this implies $j = k$.   □

Proposition 4.3. *Suppose that* $(X, \triangle, \triangledown, \top, \perp)$ *is a classical structure in* Rel, *such that the unit* $\perp : 1 \longrightarrow X$ *is a function, i.e. a single element of* X. *Then the monoid* $(X, \triangledown, \perp)$ *must be an abelian group in* Set.

Proof. We first show that the monoid part every classical structure X in Rel must admit the inverses, as soon as it satisfies the assumptions of the proposition. More precisely, the claim is that condition (1) from section 3.3, together with the assumption that the unit is a singleton $\perp \in X$, implies that for every $k \in X$ there is $k^{-1} \in X$ such that $k\triangledown k^{-1} = k^{-1}\triangledown k = \perp$.

First consider condition (1) for $i = k$ and $j = \perp$. For the pair $\langle x, y \rangle = \langle \perp, k \rangle$, the second equation gives $x' \in X$ such that $\perp = x'\triangledown k$. Dually, (1) also holds for $i = \perp$ and $j = k$. For the pair $\langle x, y \rangle = \langle k, \perp \rangle$, the first equation gives $y' \in X$

such that $\bot = k\triangledown y'$. Since $x' = x'\triangledown\bot = x'\triangledown(k\triangledown y') = (x'\triangledown k)\triangledown y' = \bot\triangledown y' = y'$, we can set $k^{-1} = x' = y'$.

To see that $(X, \triangledown, \bot)$ is a group, it remains to be shown that the operation $\triangledown$ is total, i.e. that $k\triangledown\ell$ is defined for all $k, \ell \in X$. To see that this is the case, note that in each of the equations $\ell = (k^{-1}\triangledown k)\triangledown\ell = k^{-1}\triangledown(k\triangledown\ell)$, the left-hand side is defined if and only if the right-hand side is defined. Hence $k\triangledown\ell$ must be defined.

Since the monoid operation $\triangledown : X \times X \longrightarrow X$ is total, and every element $k \in X$ has an inverse k^{-1}, we conclude that $(X, \triangledown, \bot)$ is indeed a group.   □

Given an arbitrary classical structure $(X, \vartriangle, \triangledown, \top, \bot)$ in Rel, we can now first apply proposition 4.2 to decompose it as a biproduct

$$X = \sum_{j \in J} X_j$$

of classical structures $(X_j, \vartriangle_j, \triangledown_j, \top_j, \bot_j)$ where each $\bot_j$ is a singleton. By proposition 4.3, each of these classical structures is a group. Hence the final result:

Theorem 4.4. *Every special Frobenius algebra in* Rel *is a biproduct (disjoint union) of groups. Every classical structure in* Rel *is a biproduct of abelian groups.*

Using this result, we can now effectively enumerate all classical structures in Rel with a given number of elements.

4.3   Examples of Classical Structures

Any classical structure over a set of n elements can thus be constructed by choosing

- a partition $n = \sum_j n_j$, where $j \geq 1$,
- an abelian group X_j of order n_j for each n_j.

For $n = 2$, there are just two partitions: $n = 1 + 1$ and $n = 2$. Since there is just one group with a single element, and just one group with 2 elements, these two partitions each determine a unique classical structure. They were described in section 2.4.

For $n = 3$, besides $n = 1 + 1 + 1$ and $n = 3$, we can also write $n = 1 + 2$. The first two partitions give the classical structures described in section 2.4. The nonstandard one comes from $\mathbb{Z}_3$. The third classical structure is the disjoint union $\mathbb{Z}_1 + \mathbb{Z}_2$.

For $n = 4$ there are five partitions. It is easy to see the pattern: e.g., $n = 2 + 2$ induces the classical structure $\mathbb{Z}_2 + \mathbb{Z}_2$, whereas $n = 1 + 3$ induces $\mathbb{Z}_1 + \mathbb{Z}_3$. Since there are two groups with 4 elements, $\mathbb{Z}_4$ and the Kleinian group $D_4 = \mathbb{Z}_2 \times \mathbb{Z}_2$, the trivial partition $n = 4$ induces two different classical structures. They both have the same classical element, consisting of all of $n = 4$; but they induce different quantum structures, entangling each element with its group inverse. Since each element of D_4 is its own inverse, its quantum structure coincides with the one induced by the standard classical structure $\mathbb{Z}_1 + \mathbb{Z}_1 + \mathbb{Z}_1 + \mathbb{Z}_1$. In any case, there are exactly 6 different classical structures with 4 elements.

For $n = 5$, there are 7 different partitions, and they induce 8 different classical structures. E.g., the partition $n = 2 + 3$ corresponds to the classical structure $\mathbb{Z}_2 + \mathbb{Z}_3$, with the classical elements $\{0, 1\}$ and $\{2, 3, 4\}$. The quantum structure is $\eta = \{00, 11, 22, 34, 43\}$.

For $n = 6$, there are 11 partitions. There are 2 groups with 6 elements, but only the cyclic one is abelian. . .

In all cases, the classical elements of a classical structure are just the underlying sets of its constituent groups. They do not depend on the actual structure of the groups. This structure is, however, reflected in the induced quantum structure, which entangles each element with its group inverse.

5   Conclusions and Future Work

We classified classical structures in Rel, and found that many are nonstandard. They also induce many nonstandard quantum structures $I \xrightarrow{\eta} X \times X$ in Rel. If X is a group, then η entangles each $a \in X$ with its inverse a^{-1}. Moreover, each of the nonstandard classical structures induces a nonstandard abstraction operator κx, binding the variable x in the polynomial relations $\varphi(x) \in \mathsf{Rel}[x]$. For monoidal categories in general, such operations and their meaning were analyzed in [24]. In Rel in particular, the situation seems rather curious. While the nonstandard classical structures support specifying relational polynomials, as nondeterministic programs with nonstandard variables — corollary 3.1 says that there are few classical elements to be substituted for these variables. The distinctions of the elements belonging to the same group within a nonstandard classical structure turn out to be classically indistinguishable. However, this indistinguishability, observed through a different classical structure, can be used as a computational resource. Indeed, switching between the different classical structures in order to use this resource is the essence of some of the most important quantum algorithms. The interesting structural repercussions of this method within the relational view of nondeterministic computation need to be further explored in future work.

References

1. Abramsky, S.: No-cloning in categorical quantum mechanics. In: Gay, S., Mackie, I. (eds.) Semantical Techniques in Quantum Computation, 32 p. Cambridge University Press, Cambridge (2008) (to appear)
2. Abramsky, S., Coecke, B.: A categorical semantics of quantum protocols. In: Proceedings of the 19th Annual IEEE Symposium on Logic in Computer Science: LICS 2004, pp. 415–425. IEEE Computer Society, Los Alamitos (2004), arXiv:quant-ph/0402130
3. Abramsky, S., Gay, S., Nagarajan, R.: Interaction categories and the foundations of typed concurrent programming. In: Broy, M. (ed.) Proceedings of the 1994 Marktoberdorf Summer Sxhool on Deductive Program Design, pp. 35–113. Springer, Heidelberg (1996)

4. Barr, M., Grillet, P., van Osdol, D. (eds.): Exact Categories and Categories of Sheaves. Lecture Notes in Mathematics, vol. 236. Springer, Heidelberg (1971)
5. Carboni, A., Walters, R.F.C.: Cartesian bicategories, I. J. of Pure and Applied Algebra 49, 11–32 (1987)
6. Church, A.: A formulation of the simple theory of types. The Journal of Symbolic Logic 5(2), 56–68 (1940)
7. Coecke, B., Duncan, R.: Interacting quantum observables. In: Aceto, L., Damgård, I., Goldberg, L.A., Halldórsson, M.M., Ingólfsdóttir, A., Walukiewicz, I. (eds.) ICALP 2008, Part II. LNCS, vol. 5126, pp. 298–310. Springer, Heidelberg (2008)
8. Coecke, B., Pavlovic, D., Vicary, J.: A new description of orthogonal bases. Math. Structures in Comp. Sci., 13 (2008) (to appear), arXiv:0810.0812
9. Coecke, B., Edwards, W.: Toy quantum categories. In: Coecke, B., Panangaden, P. (eds.) Proceedings of the 2008 QPL-DCM Workshop, pp. 25–35. Springer, Heidelberg (2008), arXiv:0808.1037
10. Coecke, B., Pavlovic, D.: Quantum measurements without sums. In: Chen, G., Kauffman, L., Lamonaco, S. (eds.) Mathematics of Quantum Computing and Technology. Taylor and Francis, Abington (2007), arxiv:quant-ph/0608035
11. Dieks, D.: Communication by EPR devices. Physics Letters A 92(6), 271–272 (1982)
12. Coecke, B., Paquette, É.O., Pavlovic, D.: Classical and quantum structuralism. In: Gay, S., Mackie, I. (eds.) Semantical Methods in Quantum Computation, 42 p. Cambridge University Press, Cambridge (2008) (to appear)
13. Freyd, P.J., Scedrov, A.: Categories, Allegories. North Holland Publishing Company, Amsterdam (1991)
14. Gödel, K.: Über formal unentscheidbare Sätze der Principia Mathematica und verwandter Systeme. I. Monatshefte fr Mathematik und Physik 38, 173–198 (1931)
15. Joyal, A., Street, R.: An introduction to Tannaka duality and quantum groups. In: Carboni, A., Pedicchio, M.C., Rosolini, G. (eds.) Category Theory Proceedings, Como 1990. LNM, vol. 1488, pp. 411–492. Springer, Berlin (1991)
16. Kleene, S.C.: Recursive predicates and quantifiers. Transactions of the American Mathematical Society 53(1), 41–73 (1943)
17. Kock, J.: Frobenius Algebras and 2D Topological Quantum Field Theories. London Mathematical Society Student Texts, vol. 59. Cambridge University Press, Cambridge (2004)
18. William Lawvere, F.: Ordinal sums and equational doctrines. In: Seminar on Triples, Categories and Categorical Homology Theory. Lecture Notes in Mathematics, vol. 80, pp. 141–155. Springer, Heidelberg (1969)
19. Moggi, E.: Notions of computation and monads. Inf. Comput. 93(1), 55–92 (1991)
20. Pati, A.K., Braunstein, S.L.: Impossibility of deleting an unknown quantum state. Nature 404, 164–165 (2000)
21. Pavlovic, D.: Geometry of abstraction in quantum computation. In: TANCL 2007 (2007) (manuscript)
22. Pavlovic, D.: Maps I: relative to a factorisation system. J. Pure Appl. Algebra 99, 9–34 (1995)
23. Pavlovic, D.: Maps II: Chasing diagrams in categorical proof theory. J. of the IGPL 4(2), 1–36 (1996)
24. Pavlovic, D.: Categorical logic of names and abstraction in action calculus. Math. Structures in Comp. Sci. 7, 619–637 (1997)

25. Pavlovic, D., Abramsky, S.: Specifying interaction categories. In: Moggi, E., Rosolini, G. (eds.) CTCS 1997. LNCS, vol. 1290, pp. 147–158. Springer, Heidelberg (1997)
26. Selinger, P.: Dagger compact closed categories and completely positive maps. Electron. Notes Theor. Comput. Sci. 170, 139–163 (2007)
27. Spekkens, R.W.: Evidence for the epistemic view of quantum states: A toy theory. Phys. Rev. A 75, 30 p. (2007)
28. Vicary, J.: Categorical formulation of quantum algebras, 37 p. (2008)
29. Wootters, W.K., Zurek, W.H.: A single quantum cannot be cloned. Nature 299, 802–803 (1982)

A Symbolic Classical Computer Language for Simulation of Quantum Algorithms

Peter Nyman

International Center for Mathematical Modeling
in Physics, Engineering and Cognitive science
MSI, Växjö University, S-35195, Sweden

Abstract. Quantum computing is an extremely promising research combining theoretical and experimental quan tum physics, mathematics, quantum information theory and computer science. Classical simulation of quantum computations will cover part of the gap between the theoretical mathematical formulation of quantum mechanics and the realization of quantum computers. One of the most important problems in "quantum computer science" is the development of new symbolic languages for quantum computing and the adaptation of existing symbolic languages for classical computing to quantum algorithms. The present paper is devoted to the adaptation of the Mathematica symbolic language to known quantum algorithms and corresponding simulation on the classical computer. Concretely we shall represent in the Mathematica symbolic language Simon's algorithm, the Deutsch-Josza algorithm, Grover's algorithm, Shor's algorithm and quantum error-correcting codes. We shall see that the same framework can be used for all these algorithms. This framework will contain the characteristic property of the symbolic language representation of quantum computing and it will be a straightforward matter to include this framework in future algorithms.

Keywords: Deutsch-Josza algorithm, Grover's algorithm, Quantum computing, Quantum error-correcting, Shor's algorithm, Simon's algorithm, Simulation of quantum algorithms.

1   Introduction

In [1] Richard Feynman pointed out that it is totally impossible to efficiently simulate quantum mechanics on classical computers.[1] Consider a system consisting of N quantum particles. According to quantum formalism it is described by the tensor product $H_N = H \otimes H \otimes \ldots \otimes H$ of the N copies of the state space H for a single particle. It is evident that the dimension of H_N grows exponentially with N. Feynman's observation was the first step toward the creation of quantum computers. Nowadays the main aim of the quantum computing is not the simulation of quantum mechanical structures, but the execution of quantum algorithms for solving NP problems in polynomial time.

---

[1] Simulation is efficient if the execute time is in polynomial time.

P. Bruza et al. (Eds.): QI 2009, LNAI 5494, pp. 158–173, 2009.

A number of quantum algorithms have been developed since David Deutsch presented the first algorithm [2]. This algorithm determines whether a Boolean function $f : \{0,1\} \rightarrow \{0,1\}$ is balanced or constant[2]. The Deutsch-Jozsa algorithm [3] is the generalization of Deutsch's algorithm to a Boolean function $f : \{0,1\}^n \rightarrow \{0,1\}$. The well-known classical algorithm should query the function more than $2^{n-1} + 1$ times, while the quantum algorithm needs only one interaction with the oracle which implements the function.

Simon's algorithm [4,5] is similar to the Deutsch-Jozsa algorithm, but, instead of determining whether the Boolean function is balanced or constant, it is designed to find the period of a Boolean function $f : \{0,1\}^n \rightarrow \{0,1\}^n$. A comparison with the well-known classical algorithm shows that the application of Simon's algorithm [4,5] (of course, on a quantum computer) would imply exponential speedup.

These first quantum algorithms have no direct practical applications. However, their creation played an important role in quantum computing. It became evident that quantum computers might increase the speed of calculations tremendously.

Two algorithms with a greater potential for direct implementation to practical application are Grover's search algorithm [3] and Shor's factorization algorithm [6]. Grover's algorithm is a quantum search algorithm which searches through an unsorted list with square roots less queries than the most effective classical algorithm. Shor's algorithm factorizes integers exponentially faster than any known classical algorithm. It has obvious applications to cryptography.

At the first stage of quantum computing research expectations of quick progress dominated in the quantum community. However, it seems that such high expectations were not totally justified. Numerous foundational and technological problems such as the decoherence of quantum bits and the instability of quantum structures with an already sufficiently small number of registers induced doubts about the quick elaboration of really working quantum computers. Although it could not be denied that great progress had been made in quantum technologies, it is clear that there is still a huge gap between the creation of toy quantum computers with 10-15 quantum qubits and satisfying e.g. the technical conditions of the project announced a few years ago in the Canadian [7]: 100 quantum qubits. It is also evident that difficulties increase nonlinearly with an increase in the number of registers.

Therefore the simulation of quantum computations on classical computers became an important part of the quantum computing. Of course, one could not expect that quantum algorithms would help to solve NP problems for polynomial time on classical computers. This is not at all the aim of classical simulation, however. The classical simulation of quantum computations will cover part of the gap between the theoretical mathematical formulation of quantum mechanics and the realization of quantum computers. One of the most important problems in "quantum computer science" is the development of new symbolic languages

---

[2] A Boolean function $f : \{0,1\}^n \rightarrow \{0,1\}$ is constant if $f(x) = 1$ or $f(x) = 0$ for all inputs x and balanced if $f(x) = 1$ for half the inputs.

for quantum computing and the adaptation of existing symbolic languages to quantum algorithms.

The present paper is devoted to the adaptation of the Mathematica symbolic language to known quantum algorithms and a corresponding simulation on a classical computer. Concretely we shall represent in the Mathematica symbolic language Simon's algorithm, Deutsch-Josza's algorithm, Grover's algorithm, Shor's algorithm and quantum error-correcting codes.

We shall see that the same framework can be used for all these algorithms. This framework will express the characteristic property of the symbolic language representation of quantum computing. It will be a straightforward matter to include this framework in future algorithms.

There are great numbers of results in simulations of quantum computing using wide numbers of classical programming languages to implement quantum algorithms. These quantum programming languages will together cover a large field of simulations. Generally, all quantum programming languages may be separated into three group's imperative, functional and other quantum programming language.

The survey of quantum programming languages by Sofge [8] will give deeper analysis of the quantum programming languages. One of the first [9] imperative languages introduced Knill 1996 the QRAM model [10]. Two years later present Ömer an imperative language the quantum computing language QCL and develop it further [11,12,13]. Bettelli's, Calarco's and Serafini's quantum computing language [14] is also an imperative language. Both this quantum computing language written in C++ may be downloading and test in classical computers. Juliá-Díaz,Burdis and Tabakin developed imperative languages QDENSITY [15] using Mathematica which shares some common points with the approach presented in this article. But still is QDENSITY different from this approach in many ways and one of the essential different is that this language operators not are represented as matrices.

The first functional quantum programming languages springs out from the Lambda-Calculus and the QRAM model [16]. Moreover present Selinger [17] also a programming language together with Valiron with is based on Valiron's Master's thesis [18]. Altenkirch, and Grattage introduced linear logic functional quantum programming language QML with contraction [19,20]. They are several papers using Haskell to investigate quantum language [21,22,23,24,25].

2   Symbolic Language for Simulation

In this paper it will be demonstrated that the Mathematica symbolic language can be used for all known quantum algorithms. This is a consequence of the realization of all basis laws of quantum mechanics with the help of Mathematica: the superposition of quantum states, the representation of the state of a composite system in the tensor product of Hilbert spaces describing the states of the components of the system, Schödinger's unitary evolution and the measurement process based on von Neumann's projection postulate.

Some blocks of the program code can be shared in all algorithms. The simulation procedure will adhere to the following pattern:

(a) definition (the framework),
(b) the flow through the quantum circuit,
(c) the measurement of a specially chosen observable on the output state.

Thus we start with the formalization in the Mathematica symbolic language of all basic elements of the mathematical formalism of quantum mechanics which are used in quantum computing, e.g., qubits. Then we represent the quantum circuits specific to the algorithm under investigation. At the end of the programming of the algorithm we need the symbolic language representation of the measurement.

Coming back to (a) we point out that in the symbolic language the representation corresponding to Dirac's formalism with bra and ket vectors should be constructed [26]. After this, superposition and tensor product, etc. are expressed in symbolic Mathematica notations. Then, by moving to (b) we express quantum gates.

To be more precise, an arbitrary one-qubit quantum state

$$\phi = c_1 \left|0\right\rangle + c_2 \left|1\right\rangle$$

will be implemented as

$$c_1 \, e[0] + c_2 \, e[1],$$

where $e[j], j = 0, 1$, are symbolic Mathematica representations of the basis vectors and $c_j, j = 0, 1$ are complex numbers. An arbitrary one-qubit quantum operator is symbolized as

$$U := \{ \, e[0] \rightarrow c_3 \, e[0] + c_4 \, e[1], \, e[1] \rightarrow c_5 \, e[0] + c_6 \, e[1] \}.$$

Thus:

$$U \, | \, (c_1 \, e[0] + c_2 \, e[1]) = (c_1 c_3 + c_2 c_5) \, e[0] + (c_1 c_4 + c_2 c_6) \, e[1]$$

3   The Simulation Framework

Let us introduce the part of the program that will be the framework for the simulation of quantum algorithms and error correction. Several well-known quantum algorithms have been implemented in this framework by the author. We will point out that there is a symbolic similarity between our framework and the mathematical foundation of quantum computing. For this reason we will represent the code by a simple modification of Dirac's notation. A quantum state in n dimensions can be represented by a linear combination of n numbers of basis vectors as $\{ \, e[0], e[1], \ldots, e[n] \} = \{ \, e[0]^{\otimes n}, e[0]^{\otimes n} \otimes e[1], \ldots, e[1]^{\otimes n} \}$. In the two-dimensional case a quantum state $|\phi\rangle$ is represented as a superposition of two basis vectors, say $|0\rangle$ and $|1\rangle$, known as computational basis (computational basis, see [27,28]). In this basis a quantum state $|\phi\rangle$ is represented as

$$|\phi\rangle = \alpha|0\rangle + \beta|1\rangle, \tag{1}$$

where α and β are complex numbers such as $|\alpha|^2 + |\beta|^2 = 1$. We will introduce some new symbols for the states of the computational basis as follows:

e[0] = |0⟩ and e[1] = |1⟩. This is the foundation for the structure of the program code. For more than one qubit we will use the computational basis states e[$x_1, \ldots, x_n$] = |$x_1 \ldots x_n$⟩, where $x_j \in \{0, 1\}$ or the more compact notation e[y] = |y⟩, where $y = x_n 2^0 + \cdots + x_1 2^{n-1}$. We will write the state ϕ as e[ϕ] = αe[0] + βe[1], in analogy to (1). The operator A acts on the state ϕ and is usually written as $A|\phi⟩$ in the quantum mechanical literature. To match these symbols we will use the computational symbols A|e[ϕ] for this operation. One might regard |$x_1 \ldots$|$y_1 \ldots y_m$⟩$\ldots x_n$⟩ as |$x_1 \ldots y_1 \ldots y_m \ldots x_n$⟩ in order to simplify the program code. This will be a computational problem, since Mathematica will distinguish between e[$x_1, \ldots$, e[$y_1, \ldots, y_m$], $\ldots, x_n$] and e[$x_1, \ldots, y_1, \ldots, y_m, \ldots, x_n$]. The computer must regard these expressions as equal even if the notations are not identical with each other. As an example the expression e[0, e[1], 1] must be equal to e[0, 1, 1] in the code. We can bring in the command e[0, e[1], 1] := e[0, 1, 1] or the more general e[a_-, e[b_-], c_-] := e[a, b, c] to solve this problem. Moreover, the program code must be able to handle the linearity of the tensor product. Let e[.] be vectors and α a complex number. We define the tensor product as

$$\alpha(\text{e}[v] \otimes \text{e}[w]) = (\alpha\text{e}[v]) \otimes \text{e}[w] = \text{e}[v] \otimes (\alpha\text{e}[w]) \tag{2}$$

$$(\text{e}[v_1] + \text{e}[v_2]) \otimes \text{e}[w] = \text{e}[v_1] \otimes \text{e}[w] + \text{e}[v_2] \otimes \text{e}[w] \tag{3}$$

$$\text{e}[v] \otimes (\text{e}[w_1] + \text{e}[w_2]) = \text{e}[v] \otimes \text{e}[w_1] + \text{e}[v] \otimes \text{e}[w_2]. \tag{4}$$

Two short commands in the program code will implement this definition of the tensor product. The command

```
e[a___,α_.e[x__],b___]:=αe[a,x,b]
```

will transform e[a]⊗αe[x]⊗e[c] into αe[$a \otimes x \otimes b$] = αe[a, x, b]. This command is the computational dual to the tensor expression in Dirac's notation |a⟩⊗α|x⟩⊗|b⟩ = α|$a\,x\,b$⟩. The other command

```
e[a___,ξ_.(α_.e[x__]+β_.e[y__]),b___]:=
      ξαe[a,x,b]+ξβe[a,y,b]
```

will transform e[a] ⊗ ξ(α e[x] + β e[y]) ⊗ e[b] to $\xi\alpha$e[a, x, b] + $\xi\beta$e[a, y, b]. Let U be an arbitrary unitary one-qubit quantum gate. Then U will transform a one-qubit state e[ϕ], which is represented in the computational basis states as e[ϕ] = a e[0]+b e[1], into the state U|e[ϕ] → $a(c_1$ e[0]+c_2 e[1])+$b(c_3$ e[0]+c_4 e[1]), where a, b, c_i are complex numbers. We add the *Mathematica* gate U to the program code as follows: U|e[0] → c_1 e[0] + c_2 e[1] and U|e[1] → c_3 e[0] + c_4 e[1]. For example, the Hadamard gate H will be added in Mathematica as the command H:={ e[0] → $1/\sqrt{2}$(e[0] + e[1]), e[1] → $1/\sqrt{2}$(e[0] − e[1])}. We will define a one-qubit gate O_i as an operator which acts on the qubit in position i and leaves the other qubits unchanged. The program code must be able to operate with a gate on an arbitrary qubit. Consequently, we will define an operator O_i in the Mathematica code. Defined the operator O_i as $O_i = I^{\otimes i-1} \otimes U \otimes I^{\otimes n-i}$, which acts on n-qubits where I is the one-qubit unit operator and U is an arbitrary one-qubit operator. Then operator O_i is a function of O_i|e[v] → e[ψ]. Similarly, we will define $O_{i,j}$ as an operator which operates as the two-qubit operator on the qubits in positions i, j and leaves the other qubits unchanged. Now we have the tools to build the quantum circuit for quantum algorithms and error correction. The definitions for quantum algorithms using one or two qubits will be as in following listing 1.1 where we also define Hadamard gate

Listing 1.1. Definition of register and quantum gates in Mathematica

```
1  e[a___,α_.e[x__],b___]:=αe[a,x,b]
2  e[a___,ξ_.(α_.e[x__]+β_.e[y__]),b___]:=
3     ξαe[a,x,b]+ξβe[a,y,b]
4  O_i_|v_:=Chop[Expand[v/.
5     (e[x__]:→ReplacePart[e[x],e[{x}[[i]]]/.O,i])]]
6  O_i_,j_|v_:=Chop[Expand[v/.O[i,j]]]
7  H:={e[0]:→ 1/√2(e[0]+e[1]),e[1]:→ 1/√2(e[0]-e[1])}
```

The definitions in listing 1.1 will be used in all simulations in this article and we will assume this part executed and omit it from now on. Before introducing the simulations for the quantum algorithms, let us consider one simple example of quantum circuit represented in the symbolic language. Defined the Pauli-X gate in mathematica as

Listing 1.2. Pauli-X gate

```
1  X:={e[0]:→e[1],e[1]:→e[0]}
```

In general, one qubit state is represented in the symbolic language as

$$e[\psi] = \alpha\, e[0] + \beta\, e[1],$$

where α and β are complex numbers in the way that the sum of squares of their absolute values is equal to one. Here $e[0]$ and $e[1]$ are representations of Dirac's notations in Mathematica. Generalization to the case of multi-qubits is evident. Let us regard the definitions part as executed as in listings 1.1 and 1.2, then apply following quantum gates to a state $e[00]$.

Consider the Pauli operator $X = |1\rangle\langle 0| + |0\rangle\langle 1|$ and the Hadamard operator $H = 1/\sqrt{2}(|0\rangle\langle 0| + |1\rangle\langle 0| + |0\rangle\langle 1| - |1\rangle\langle 1|)$ and the quantum circuit

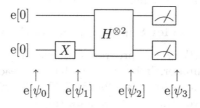

Fig. 1. A quantum circuit example

Let the initial state be e[00]. We apply X_2, H_1 and H_2 to the first and second qubit. Thus the new state will be

$$H_1|H_2|X_2| \, e[00] = 1/2 \, (\, e[00] - e[01] + e[10] - e[11])$$

where we leaving out the parentheses.

This is straightforward in simulation in Mathematica;

$$In[7] := H_1| \, (H_2| \, (X_2| \, e[0,0]))$$

The result is given by

$$Out[7] := \frac{1}{2}e[0,0] - \frac{1}{2}e[0,1] + \frac{1}{2}e[1,0] - \frac{1}{2}e[1,1]$$

where In[7] and Out[7] denote input and output number 7.

4   Simulation of Shor's Algorithm

Shor's algorithm contains a part of classical algorithm and the quantum Fourier transform (QFT). Let us focus on the quantum part. The QFT is a quantum analogous to the classical discrete Fourier transform. QFT contains the following quantum gates: Hadamard operator H, Controlled Not operator and Rotation operator R. The implementation of QFT will follow from the QFT Circuit:

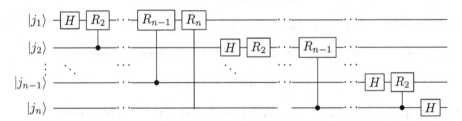

Fig. 2. The QFT Circuit

where we omit the Controlled Not gates which swap the order of the qubits at the end of the circuit. This circuit is directly implemented by the code

$$H_1|R_{1,2}| \cdots |R_{1,n-1}|R_{1,n}|H_2|R_{2,3}| \cdots |R_{2,n-1}| \cdots |H_{n-1}|R_{n-1,n-2}|H_n| \, e[\psi]$$

in this simulation language (leaving out the parentheses).

4.1   Mathematica Code for Simulation of Shor's Algorithm

Let us introduce the *Mathematica* program code which implements Shor's algorithm in a classical computer. It is assumed that the reader already has some familiarity with Shor's algorithm (see[29,28,27,30]) We will follow the Mathematica program code evolutions and compare this with Shor's algorithm. The program will try to find two factors to N, where N is an odd prime factorization and has at least two different prime factors.

Listing 1.3. Shor's Algorithm in Mathematica

```
1  N=3*5;q=⌈Log[2,N^2]⌉;
2  Do[x=Random[Integer,{2,N-1}];
3  If[GCD[x,N]==1,SecondStep;QFT;OutPrint,
4   Print["Chosen x=",x,"a multiplier of",GCD[x,N]],"."];
5    ];,{⌈(160Log[Log[N]])/9⌉}]
```

The algorithm will choose $q= \lceil Log_2(N^2) \rceil$ so that the algorithm will find a factor with large probability, i.e. if it selects q to satisfy $N^2 \leq 2^q < 2N^2$, the two factors will then be found with a probability of at least $\frac{9}{160 \log \log N}$. The program will choose a random integer $x \in \{2, 3, \cdots, N-1\}$.

Listing 1.4. Rotation and Swap gate in Mathematica

```
1  R_d:={e[1,1]:→ e^{\frac{\pi i}{2^d}}e[1,1]}
2  Swap:={e[i_, j_]:→e[j, i]}
3  SecondStep:=(Secstep[q_,x_,N_]:=Expand[\frac{1}{\sqrt{2^q}} \sum_{2^q-1}^{c=0} e[
4   Sequence@@IntegerDigits[c,2,q],Sequence@@
5   IntegerDigits[Mod[x^c,N],2,q]]];u=Secstep[q_,x_,N_])
```

To compute QFT the algorithm requires three gates, Hadamard (**H**), Rotation (**R**) and Swap (**Swap**), where the Hadamard gate already is defined in listing 1.1. **Secstep** calculates $x^c \pmod{N}$ in the second register, where q is the number of qubits.

Listing 1.5. Quantum Fourier transform

```
1  QFT:=(For[i=1,i≤q,i++,u=H_2|u;For[j=i+1,j≤q,j++,
2   u=(R_{(j-i)})_{i,j}|u]];
3  For[i=1,i≤IntegerPart[q/2],i++,u=Swap_{i,q+1-i}|u];
4  OutQFT=(u//.a_.e[y___]+b_.e[y___]:→
5   Together[(a+b)]e[y]);
6  Probability=List@@OutQFT/.α_.e[y___]→
7   {Abs[α]$^2$.,e[y]};
8  Probability=Probability/.{a_,e[y__]}→a;
9   b={Probability[[1]]};
10  For[i=2,i@\leq$@Length[Probability],i++,
11   b=AppendTo[b,b[[i-1]]+Probability[[i]]]];
12  r=Random[];
13  For[i=1,i≤Length[b],i++,
14   If[r≤b[[i]],MeasureQFTStep=i;Break[]]];
15  p=(List@@OutQFT/._.e[x__]:→
16  FromDigits[{Sequence@@Take[{x},q]},2]))
```

The **QFT** will act on the state u by means of the three gates in the following order: $H_q(R_{q,q-1}H_{q-1}(\cdots(R_{q,2}\cdots R_{3,2}H_2(R_{q,1}\cdots R_{2,1}H_1\,e[\mathbf{u}]))))$. The third line in the program code will swap the qubits. All terms with identical computational basis states will be collected in the command **OutQFT**. **Probability** is a list of the probabilities used to measure one of the terms in the register. One of the terms will be randomly chosen taking into consideration of probability to measure the state. The position of the chosen term will be saved in **Measure-QFTStep**. Finally, the list **p** of decimal numbers is derived from the binary list **OutQFT**.

```
1  OutPrint:=(CFD:=Denominator[
2    Convergents[p[[MeasureQFTStep]]/2^q]];
3  Do[{If[Mod[x^CFD[[i]],N]==1&&EvenQ[CFD[[i]]]&&
4    Mod[x^(CFD[[i]]/2),N]≠N-1,
5  Print["Factors a1=",GCD[N,x^CFD[[i]]/2 +1],"
6    and","a2=",GCD[N,x^CFD[[i]]/2 +1],
7    " have been found."];]},{i,1,Length[CFD]}];)
```

The randomly chosen value in the register is in p[[**MeasureQFTStep**]]. In **CFD** the program saves the denominator of convergents p[[**MeasureQFTStep**]]$/2^q$. From this we can select all even denominators, where $\mathbf{x^{CFD}} \equiv 1 \pmod N$ and $\mathbf{x^{\frac{CFD}{2}}} \neq N-1 \pmod N$. If any of the denominators satisfies these three conditions, it will give us two factors.

The entire simulation is present in [31].

5   Simulation of Deutsch-Jozsa Algorithm

The commission for Deutsch-Jozsa's algorithm is easy: we need to decide whether a Boolean function is balanced or constant. Our aim is to make a simulation of this quantum algorithm in the symbolic language. The definition section contains one gate and a so-called quantum oracle. A quantum oracle is an operator in a black box defined as:

$$Uf|x\rangle^{\otimes n}|y\rangle = |x\rangle^{\otimes n}|y \oplus f(x)\rangle$$

Let us give a simplified picture of this task. Bob selects a balanced or constant Boolean function and Alice's task is to determine function property. Alice uses a quantum computer to implement the algorithm by means of the quantum circuit:

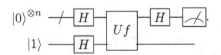

Fig. 3. Deutsch-Jozsa algorithm circuit

Alice prepares the initial state $e[0, 0, 0, \ldots, 1]$ and the quantum gate according to the scheme in the circuit

$$H_{n-1}|\ldots(H_2|(H_1|(U_f|(H_n|\ldots(H_2|(H_1|\,e[0,0,0,\ldots,1])))))).$$

Finally, the measurement is performed. She should apply the oracle once, instead of using the classical algorithm and query the function at least $2^{n-1} + 1$ times. The results of this simulation is presented in [32].

5.1   The Deutsch-Jozsa Algorithm in Mathematica

The Deutsch-Jozsa algorithm is implemented as follow. We will begin to define characteristic properties of quantum computers in the program code.

Listing 1.6. Definition of register and quantum gates in mathematica

```
1  Uf:={e[i___,j_]↦e[i,Mod[j+f[i],2]]}
2  Enlarge[ψ_,i_]:=ψ/.ψ->(ψ/.e[x_]↦e[
3    Sequence@@Table[0,{i-1}],x])
```

The Deutsch- Jozsa algorithm only need to apply the two operators H and Uf. The last line will enlarge a state of the one qubit $e[1]$ to the n-qubit state $e[0, 0, 0, \ldots, 1]$. The operators, linearity, superposition and tensor products are defined in listing 1.1. The implementation of the algorithm will follow from the listing 1.7.

Listing 1.7. Deutsch-Jozsa algorithm in mathematica

```
1  q = 8; φ = Enlarge[e[1],q]; Do[φ =
2  (H_i | φ), {i,q}]; φ = Uf_{q-1,q} | φ; φ = φ /. {f[1,
3  x__]↦0,f[0, x__]↦1}; Do[φ =
4  (H_i | φ), {i,q - 1}]; φ
```

There are certain advantages to compare this listing 1.7 with the circuit in figure 3. By using the command **Enlarge** on state $e[1]$, defined in line 2, will the register will be prepared in the q-qubit state $e[0, 0, 0, \ldots, 1]$. At line 3-4 in listing 1.7 the Hadamard operator and the unitary function are applied on all the qubits in the register. In line 5 we will be able to choose the function property. In this example we have chosen a balanced function. The algorithm will output the result when the Hadamard operator have been applied on the $q - 1$ first qubits. The output will contain zeros in $q - 1$ first qubits if and only if f is constant, otherwise it is balanced.

6   Simulation of Simon's Algorithm

The structure of this algorithm is similar to that of Deutsch-Jozsa's algorithm. It also contains an oracle. The essential difference is the function given by the oracle. Bob selects a periodic function and Alice's task is to find the period. Alice prepares the state e[0, 0, ...], applies the quantum gates and the oracle and finally, Alice performs the measurement. She continues by restarting the algorithm until she obtains a sufficient number of values to solve a linear equation giving the answer.

6.1 Simon's Algorithm in Mathematica

The Simon's algorithm is implemented as in Mathematica as following. The First part of the program code is a framework for simulation of quantum algorithms in Mathematica and this part define characteristic properties of quantum computers.

Listing 1.8. Definition of quantum gates in mathematica

```
1  Uf:={e[i___,j_]:→e[e[x][[1;;i]],
2    e[Sequence@@Mod[List@@e[x][[i+1;;j]]
3      +f[Sequence@@e[x][[1;;i]]],2]]]}
4  Enlarge[ψ_,i_]:=ψ/.ψ->(ψ/.e[x_]:→e[
5    Sequence@@Table[0,{i-1}],x])
```

For Simon's algorithm apply the two operators H and Uf. The last line will enlarge a state of the one qubit e[0] to the n-qubit state $e[0]^{\otimes n}$. The implementation of the Simon's algorithm will follow from the listing 1.9.

Listing 1.9. Simon's algorithm in mathematica

```
1       q=5;Measure={};
2  While[Length[Union[Measure]]≤q,
3    φ=Enlarge[e[0],q];
4    Do[φ=(H_i|φ),{i,q}];
5    a={1,0,1,1,0};
6    f[0,x__]:=Mod[{0,x},2];
7    f[1,x__]:=Mod[{1,x}+a,2];
8    φ=Uf_{q,2q}|φ;
9    Do[φ=(H_i|φ),{i,q}];
10   Probability=
11   List@@Expand[φ/.α_.e[y___]->Abs[α]^2];
12     Table[Probability[[i+1]]=
13     Plus@@Take[Probability,{i,i+1}],
14     {i,1,Length[φ]-1};r=Random[];
15       AppendTo[Measure,Take[φ[[1+LengthWhile[
16       Probability,#<r&],2]],{1,q}]];]]
17  Union[Measure]
18
19  Out[11]={e[0,0,0,0,1],e[0,0,1,1,0],e[0,1,0,0,1],
20          e[0,1,1,1,0],e[1,0,1,0,0],e[1,0,1,0,1]}
```

Numbers of qubits in the two registers will be selected and an empty list of measured states creates in the first line. The algorithm iterates with use of a **while** loop until it has measured q-number linearly independent states. The command **Enlarge** applied on state e[0], defined in line 3, will prepare the two register in q-qubit states e[0]$^{\otimes q}$. In line 4 the Hadamard operator are applied to the first register. Take the unknown period a to be $\{1, 0, 1, 1, 0\}$ in this example. In this specific example will the function be defined as in line 6 and 7, here must mention that it is conceivable too explicit defined every function value. After that the code will apply Uf the both registers where the first register is the control qubits and the second is targets qubits. Then apply the Hadamard gate to the first register before measurement. Moreover will a measurement of the first register means that one state of all the q-qubit states in superposition will be randomly chosen where the measurement probability is equal to square of the absolute value of the phase. Consequently, must the simulation of a measurement depend on the phase. As a first step to make a measurement will line 10 to 14 create a list called **Propability**, with contains the probability to measure the states. The algorithm chose a randomly $r \in [0, 1]$. This random r decides which of the element (state) that will be measured in the list **Propability**. The end of the **while** loop in line 15-16 will measure an element and add it to the list **Measure**. Finally line 17 will output all measured elements in the list **Measure**. It remains to solve the linear equation to find a, but since it will be done in a classical computer will we leave this besides. We can easy verify that $y \cdot a = 0$ for all measured states in the list of output(line 18-19).

7   Simulation of Grover's Algorithm in Mathematica

Grover's algorithm is a search algorithm for unsorted lists. This quantum algorithm needs a $O(\sqrt{N})$ query as an alternative to $O(N)$ for the classical algorithm, where N is the number of elements in the list. Moreover, this algorithm uses an oracle and, in fact, repeats the query of the oracle (Grover iteration) until the probability to obtain the searched element approaches max.

We prepare the initial state e[0, 0, 0, ..., 1], and then apply the Hadamard gate to all qubits. After this, we repeat the Grover iteration and perform the measurement. This flow for Grover's algorithm is given by the following circuit:

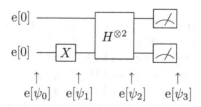

Fig. 4. A quantum circuit example

7.1   Grover's Search Algorithm in the Simulation Framework

The Grover's Search Algorithm algorithm is implemented as in Mathematica as following. The first part (see the four first code lines in listing 1.1) of the program code is a framework for simulation of quantum algorithms in Mathematica and this part define characteristic properties of quantum computers. The follow code lines in the listing 1.10 defined the operators X, U_f and U_0. Notice that it is straight forward to introduce new gates in this program from the mathematical representation of gates.

Listing 1.10. Definition of quantum gates in mathematica

```
1  InitialState[q_]:=e[Sequence@@Table[0,{q}]]
2  X:={e[0]:→e[1],e[1]:→e[0]}
3  Uf[i_,j_]:={e[x__]:→e[e[x][[1;;i]],
4    e[Sequence@@Mod[List@@e[x][[i+1;;j]]
5      +f[Sequence@@e[x][[1;;i]]],2]]]}
6  U0[i_,j_]:={e[x__]:→e[e[x][[1;;i]],
7    e[Sequence@@Mod[List@@e[x][[i+1;;j]]
8      +g[Sequence@@e[x][[1;;i]]],2]]]}
```

Let us choose an example to describe the part of the program which represents the quantum circuit for Grover algorithm. Assume that we will search for x_0 in a list of 120 element where x_0 have four solutions $19, 29, 39, 79$ (i.e. $M = 4$). The register needs to contain $q + 1$ numbers of qubits where $q = \log_2(N) = 7$ and $N = 128 \geq 120$. In this special case will numbers of iterations be

$$k = \text{round}\left(\frac{\arccos(\sqrt{4/128})}{2\arcsin(\sqrt{4(128-4)/128})}\right) = 4.$$

In next part the register will be prepared in the initial state $e[0]^{\otimes q+1}$, then algorithm will follow the circuit in figure 4 and apply the X gate and the other gates. The last part of the program will simulate a measurement of the register where probability to measure a basis state depends on its corresponding phase.

Listing 1.11. A search in a list of 120 elements

```
1  Clear[f,g]
2  q=Length[f[Sequence@@IntegerDigits[120,2]]]
3  f[Sequence@@IntegerDigits[79,2,q]]:=1;
4  f[Sequence@@IntegerDigits[39,2,q]]:=1;
5  f[Sequence@@IntegerDigits[19,2,q]]:=1;
6  f[Sequence@@IntegerDigits[29,2,q]]:=1;f[x__]:=0;
7  g[Sequence@@IntegerDigits[0,2,q]]:=1;g[x__]:=0;
8  M=4;N=2^q;
9  K[M_,N_]:=
```

```
10    Round[ArcCos[Sqrt[M/N]]/(ArcSin[2Sqrt[M(N-M)]/N])];
11    Φ=InitialState[q+1];
12    Φ=(H_{q+1}|Φ);
13    Do[Φ=(H_i|Φ),{i,q+1}];
14    Do[Φ=(Uf_{q,q+1}|Φ);
15    Do[Φ=(H_i|Φ),{i,q}];
16    Φ=(UO_{q,q+1}|Φ);
17    Do[Φ=-(H_i|Φ);,{i,q}],
18     {j,K[M,N]}];
19    Φ=(H_q+1|Φ);
20    Probability=List@@Expand[Φ]/.α_.
21     e[y___]->Abs[α]^2;
22    Table[Probability[[i+1]]=Plus@@Take[Probability,
23     {i,i+1}],{i,1,Length[Φ]-1}];
24    r=Random[];
25    Take[Φ[[1+LengthWhile[Probability,#<r&],-1]]]
```

8   Simulation of Quantum Error Correcting Code

Error correction will be necessary in quantum computing and quantum information. We have also implement Shor code as an example of error corrections code,
this simulation language will be found in [33]. The Shor code is a development
of the classical error correcting code known as majority voting. There are some
great differences between quantum and classical error correcting. Measurements
destroy the quantum states and another problem in quantum computing is continuous errors. Moreover, it is impossible to clone an arbitrary quantum state. In
classical computing errors indicate that bits have flipped, but continuous errors
in quantum computing indicate that states phases flips or qubits flips or some
combination of this errors. Shor code will overcome these problems in quantum
computing.

Acknowledgments. I am grateful to my supervisor Professor Andrei Khrennikov for many discussions on the formulations of quantum mechanics and for
introducing me into this field of research.

I am very thankful to Guillaume Adenier for all the discussions we have had
about the subject of quantum mechanics.

References

1. Feynman, R.P.: Simulating Physics with Computers. International Journal of Theoretical Physics 21, 467 (1982)
2. Deutsch, D.: Quantum theory, the Church-Turing principle and the universal quantum computer. Proceedings of the Royal Society of London. Series A, Mathematical
   and Physical Sciences (1934-1990) 400(1818), 97–117 (1985)
3. Deutsch, D., Jozsa, R.: Rapid solution of problems by quantum computation. Proc.
   Roy Soc. Lond A 439, 553–558 (1992)

4. Simon, D.R.: On the power of quantum computation. In: Proceedings of the 35th Annual Symposium on Foundations of Computer Science, Los Alamitos, CA, pp. 116–123. Institute of Electrical and Electronic Engineers Computer Society Press (1994)
5. Simon, D.R.: On the power of quantum computation. SIAM J. Comput. 26(5), 1474–1483 (1997)
6. Shor, P.W.: Polynomial-time algorithms for prime factorization and discrete logarithms on a quantum computer. SIAM Journal on Computing 26(5), 1484–1509 (1997)
7. http://www.dwavesys.com
8. Sofge, D.A.: A survey of quantum programming languages: History, methods, and tools. In: Second International Conference on Quantum, Nano and Micro Technologies, February 2008, pp. 66–71 (2008)
9. Gay, S.J.: Quantum programming languages: survey and bibliography. Mathematical Structures in Computer Science 16(04), 581–600 (2006)
10. Knill, E.: Conventions for quantum pseudocode. Technical report, Technical Report LAUR-96-2724, Los Alamos National Laboratory (1996)
11. Ömer, B.: A procedural formalism for quantum computing. Master's thesis, Department of Theoretical Physics (1998)
12. Ömer, B.: Structured Quantum Programming. Ph.D thesis, Technical University of Vienna (2003)
13. Ömer, B.: Classical Concepts in Quantum Programming. International Journal of Theoretical Physics 44(7), 943–955 (2005)
14. Bettelli, S., Calarco, T., Serafini, L.: Toward an architecture for quantum programming. The European physical journal. D, Atomic, molecular and optical physics 25(2), 181–200 (2003)
15. Juliá-Díaz, B., Burdis, J.M., Tabakin, F.: QDENSITY-A Mathematica Quantum Computer simulation. Computer Physics Communications 174(11), 914–934 (2006)
16. Selinger, P.: Towards a quantum programming language. Mathematical Structures in Computer Science 14(04), 527–586 (2004)
17. Selinger, P., Valiron, B.: A lambda calculus for quantum computation with classical control. Mathematical Structures in Computer Science 16(03), 527–552 (2006)
18. Valiron, B.: A Functional Programming Language for Quantum Computation with Classical Control. Master's thesis, University of Ottawa (2004)
19. Altenkirch, T., Grattage, J.: A functional quantum programming language. In: Proc. 20th Annual IEEE Symposium on Logic in Computer Science LICS 2005, pp. 249–258 (2005)
20. Grattage, J.: QML: A functional quantum programming language. Ph.D. thesis, The University of Nottingham (2006)
21. Shin-Cheng, M., Bird, R.: Functional quantum programming. In: Proceedings of the 2nd Asian Workshop on Programming Languages and Systems (2001)
22. Sabry, A.: Modeling quantum computing in Haskell. In: Proceedings of the 2003 ACM SIGPLAN workshop on Haskell, pp. 39–49. ACM, New York (2003)
23. Vizzotto, J.K., da Rocha Costa, A.C.: Concurrent quantum programming in Haskell. In: VII Congresso Brasileiro de Redes Neurais, Sessão de Computação Quântica (2005)
24. Vizzotto, J., Altenkirch, T., Sabry, A.: Structuring quantum effects: superoperators as arrows. Mathematical Structures in Computer Science 16(03), 453–468 (2006)
25. Grattage, J.: An overview of QML with a concrete implementation in Haskell (2008)

26. Dirac, P.A.M.: The Principles of Quantum Mechanics. Clarendon Press, Oxford (1995)
27. Hirvensalo, M.: Quantum Computing, 1st edn. Springer Series on Natural Computing. Springer, Heidelberg (2001)
28. Nielsen, M.A., Chuang, I.L.: Quantum Computation and Quantum Information. Cambridge University Press, Cambridge (2000)
29. Chen, G., Staff, C.G., Kauffman, L.H., Lomonaco, S.J.: Mathematics of Quantum Computation and Quantum Technology. Chapman & Hall/CRC, Boca Raton (2007)
30. Kaye, P., Laflamme, R., Mosca, M.: An Introduction to Quantum Computing. Oxford University Press, Oxford (2007)
31. Nyman, P.: Simulation of Quantum Algorithms on a Symbolic Computer. In: AIP Conference Proceedings. AIP, vol. 889, p. 383 (2007)
32. Nyman, P.: Simulation of deutsch-jozsa algorithm in mathematica. In: AIP, vol. 962, p. 312 (2007)
33. Nyman, P.: Simulation of Quantum Error Correcting Code (2008), Arxiv preprint arXiv:0809.3306

Quantum Theory, the Chinese Room Argument and the Symbol Grounding Problem

Ravi V. Gomatam

Bhaktivedanta Institute
2334 Stuart Street, Berkeley, CA 94705
rgomatam@bvinst.edu
http://www.bvinst.edu

Abstract. I offer an alternative to Searle's original Chinese Room argument which I call the Sanskrit Room argument (SRA). SRA distinguishes between syntactic TOKEN and semantic SYMBOL manipulations and shows that *both* are involved in human language understanding. Within classical mechanics, which gives an adequate scientific account of TOKEN manipulation, a symbol remains a subjective construct. I describe how an objective, quantitative theory of semantic symbols could be developed by applying the Schrodinger equation directly to macroscopic objects independent of Born's rule and hence independent of current statistical quantum mechanics. Such a macroscopic quantum mechanics opens the possibility for developing a new theory of computing wherein the Universal Turing Machine (UTM) performs semantic symbol manipulation and models *macroscopic* quantum computing.

Keywords: Artificial Intelligence, Chinese Room Argument, Symbol Grounding Problem, Sanskrit Room Argument, Quantum theory, Universal Turing Machine, Topology, Exotic manifolds.

1 Introduction

This paper seeks to relate quantum mechanics to artificial intelligence, human language, and cognition. It is a position paper, reporting some relevant details of a long-term research program currently in progress.

In his justly famous Chinese Room Argument (CRA), Searle argued that "we could not give the capacity to understand Chinese or English [that humans evidently have] to a machine where the operation of the machine is solely defined as the instantiation of a computer program." ([1], p. 422) Searle avoided the need to positively define 'human understanding' by relying on an operator (such as himself) who does not understand Chinese in any sense. Skipping the details of this well-known paper, Searle's main conclusions were as follows:

1. At present, digital computers only perform symbol manipulation.
2. The CRA shows that symbol manipulation alone will never lead the operator to understand Chinese; thus an equivalent digital computer cannot be said to understand Chinese.

P. Bruza et al. (Eds.): QI 2009, LNAI 5494, pp. 174–183, 2009.

3. Hence, the thesis of Strong AI is false. An appropriately programmed digital computer may be able to simulate human thinking and intelligence, but it cannot be said to literally possess cognitive states.

Searle identifies the Chinese language letters only in terms of their shape, and refers to them as symbols. But the term "symbol" implies something more. Henceforth I shall distinguish between syntactic TOKENS (letters qua shapes) and semantic SYMBOLS (letters qua components of a language, generating meaning).[1]

Under this terminology, Searle's argument is that computers, by virtue of performing *token* manipulation alone, cannot demonstrate understanding of Chinese.

I wish to show that the CRA does not preclude in any way a more general form of the strong-AI thesis, namely that present digital computers can be said to possess some portion of the abilities constituting human 'understanding'. To this end, without much ado, I offer an alternative to the CRA, which I shall call the "Sanskrit Room argument".

2   The Sanskrit Room Argument (SRA)

The Sanskrit Room is similar to the Chinese Room in all respects but two. The language spoken is Sanskrit[2], and the person inside (say, myself) understands Sanskrit. There are now two possibilities:

Scenario 1: I ignore my ability to understand the questions, merely decomposing the queries into a sequence of shapes. I then string together replies from baskets containing a good supply of these same shapes, following a hypothetical program. As with the CRA, let us say that I pass the Turing test in this mode without difficulty.

Scenario 2: I discard the program, and use my Sanskrit skills to construct an appropriate response from the store of *language symbols*. In this case I should also pass the Turing test.

To an outsider, my behavior is the same in both cases. However, I hold it as self-evident that I am doing something fundamentally different in each scenario. In the first, I am mimicking Searle's operator in the Chinese room (who doesn't know the language) and am doing solely token manipulation. In the second, I *understand* the input and answer on my own (without reference to the program).

---

[1] TOKENS are also referred to in the literature as physical symbols.

[2] Although for the purposes of this paper I could simply replace Searle by a native Chinese speaker, I shift to Sanskrit for a reason. It is a well-discussed topic in the field of natural language processing (NLP) that Panini's grammar can generate semantically valid sentences in Sanskrit to an extent that has so far been impossible for other natural languages [2]. I expect the idea that the *individual letters* in Sanskrit carry semantic content, in a manner possibly unique to that language, to become relevant to the issue at hand in future publications.

I am therefore manipulating semantic symbols. But I am *also* doing token manipulation at the level of physically stringing together letters. Thus, unlike the CRA, the SRA gives some insight into the nature of human language processing. It involves both token and symbol manipulation at the level of the subject.

I now turn to a key question: Are there any differences between token and symbol manipulation at the objective level?

3   TOKENS and Classical Physics

I shall assume metaphysical realism to start with, and refer to the external identity of objects I manipulate as the Kantian things-in-themselves (or THINGS). From this viewpoint, the terms token and symbol *both* refer to our subjective conceptions of the THING. In standard philosophical parlance, tokens and symbols are phenomenal objects given in our experience.

We objectify phenomenal tokens as *res extensa* in everyday thinking and in physics. That is, we endow them with an absolute location, mass and derived extensional attributes such as velocity, acceleration, etc. We believe that there *is* an object out there with features that correspond in every detail to our conception of the phenomenal token. We might call this belief the classical version of direct realism. Classical direct realism is referred to as naive realism in everyday thinking, and as correspondence realism in physics.[3]

Let us refer to the phenomenal token objectified in classical realism as a TOKEN. It is a physical object defined entirely in terms of its primary physical properties. On this basis, we could say that I am manipulating objective TOKENS in Scenario-1. Classical physics suffices to explain how I can recognize and manipulate objects qua TOKENS. The operation of the digital computers is presently understood in similar classical terms, rendering the computer a token manipulating system.

From this perspective, any symbol manipulation (and thus 'understanding') attributed to the computer (or, to the human operator in Scenario-1 of SRA) becomes *our interpretation* of operations that objectively must be considered solely syntactic. However, this is more a limitation of our classical point of view than an intrinsic limitation of the computer (or the human operator). We must find a way to overcome this limitation, since classical physics itself has been superseded by quantum theory. This paper will propose that present-day computers can be understood as objectively performing, not just TOKEN manipulation

---

[3] It is now recognized that a logical defense of scientific realism is highly problematic, even with regard to pre-quantum theories. (See reference [[3]] for an easy introduction to the topic.) The underdetermination of theory by data and pessimistic meta-induction are but two reasons. Nevertheless, physicists maintain a realist view of their theories based on pragmatic success. Likewise, it has been impossible to logically justify naive version of direct realism in everyday thinking too. Yet this viewpoint is embraced by all of us uncritically, again on pragmatic grounds. Physicists too rely on naive realism, at the level of reporting their laboratory observations using ordinary language.

but *also* SYMBOL manipulation, provided we apply the Schrodinger equation to macroscopic objects in a new manner.[4]

To develop this proposal further, we need to move toward a direct realist view of phenomenal symbols in everyday thinking and physics, one which is complementary to our direct realist view of the phenomenal tokens. That is to say, just as classical physics and its associated ordinary language-based thinking link our sense experiences to objective TOKENS, phenomenal symbols can be given objective counterparts in a new physics and its associated range of everyday thinking. I call these objective counterparts SYMBOLS. I will now argue that the quantum formalism, if properly applied to the macroscopic world, can serve as the basis for a scientific theory of SYMBOLS in physics.

4   SYMBOLS and Macroscopic Quantum Mechanics

Quantum theory arose out of the failure of classical physics to give an account of certain phenomena (blackbody radiation, the photoelectric effect, and atomic spectra). In statistical quantum mechanics, atomic objects acquire context sensitive properties reminiscent of symbols. This feature suggests that quantum mechanics, if it can be directly applied to the macroscopic world, could provide a basis for physically treating phenomenal symbols on a non-classical footing, i.e. as physical objects other than TOKENS.

Indeed, Schrodinger's equation *in principle* applies to the world of tables and chairs and measuring devices just as much as it applies to the world of atomic particles. The modern practice of statistical quantum mechanics (SQM), however, presupposes the classicality of the everyday world in general, and our measurement devices in particular. Although the quantum mechanical wave function, a superposed state in the most general case, evolves as per the Schrodinger equation, Born's rule is required to link the same function statistically to the classical observed states of the measuring devices. Trying to extend SQM to obtain a quantum description of the macro world is not just inconsistent but impossible; this is one way of understanding the famous cat paradox of Schrodinger [5].

Quantum formalism does provide a basis for a physics of macroscopic objects; it is just that SQM itself is not up to this task. Any successful quantum theory of the macroscopic world would have to be *logically independent* of SQM. Let me refer to this putative theory as macroscopic quantum mechanics (MQM). I shall now argue that MQM can be our physics of macroscopic objects qua semantic SYMBOLS.

4.1   Against 'Position'

The central idea behind my vision of MQM is the following. In quantum mechanics, we know that the classical idea of a determinate trajectory is lost. It seems

---

[4] While it is true that present SQM has nothing to offer to neuroscience [see [4]],the move being outlined here would be of significance to neuroscience, and for our scientific understanding of symbol manipulation at the human level.

inadvisable and contradictory to think of a quantum object as existing at successive classical locations in actual observations, while denying it a trajectory. We would have a better chance of understanding the quantum nature of objects if we could completely forego interpreting the x-observable in terms of the classical kinematical notion of position. Ideally, such an approach would recover SQM as a limiting case.

Why do we infer from a localized click experience that we observe an electron at a definite position? There are *two* reasons. One is that that we think of the electron as causing the click. The other reason is even more fundamental; we think that our localized observation experience corresponds to a real detector which has a localized position in the world and which clicked prior to our experience. This latter notion is a consequence of our interpreting the observation experience in terms of physical TOKENS in the real world. A main thrust of my research is that such an interpretation is *not necessary*; we could also conceive the real detector as a physical SYMBOL.

It should be emphasized that renouncing the classical kinematical idea of position to interpret the x-observable in a theory about SYMBOLS would require denying an even more visceral assumption: that our localized observation experiences must be interpreted as occurring at absolute locations in the physical world.

The reason is, as von Neumann pointed out, our direct sense experiences allow us to make statements only of the type, 'an observer has made a certain (subjective) observation'; never any like 'a physical quantity has a certain value' [6]. Thus, there are generally three stages in the full development of a physical theory.[5]

1. Design and conduct an experiment, then report the observations qua our sense experiences using ordinary language.
2. Develop a theory along with its mathematical formalism that can verifiably predict these sense experiences.
3. Describe the physical content of the formalism, if possible, using ordinary language words (and associated space-time pictures).

Ordinary language (OL) is invoked by physicists in stages 1 and 3. If the use of OL creates serious problems in visualizing the physical content of a theory (stage-3), as is the case with quantum mechanics, one can choose to discuss the physical content of the theory in entirely abstract, mathematical terms. This is arguably the preferred mode of thinking amongst quantum physicists [7].

We thus have a choice to invoke OL or not in stage-3. However, the use of OL in stage 1 is mandatory and unavoidable.

Thus, the renunciation of position at the formal level within quantum theory would require macroscopic objects to be identified, repeatedly observed and reported using OL in a manner that does not involve their absolute locations in stage-1. This is not a whimsical suggestion. Consider the case of a spin-$\frac{1}{2}$ particle being made to interact with a macroscopic apparatus in such a way that a meter

---

[5] The first two stages can be interchanged.

points to either $+1$ or -1 according to the particle's observable eigenstate $|\uparrow\rangle$ or $|\downarrow\rangle$. If SQM is extended to describe the joint electron + meter system, we obtain the following quantum state, $|J\rangle$, prior to observation:

$$|J\rangle = \frac{1}{\sqrt{2}}\left[|\uparrow\rangle|+\rangle + |\downarrow\rangle|-\rangle\right]$$

Even according to the standard textbook interpretation, SQM formally rules out the possibility that the macroscopic pointer has a determinate location prior to and independent of an observation carried out on the joint state $|J\rangle$.

4.2   SYMBOL as a Relational Object

Clearly, to recover macroscopic realism, we need a different conception of the macroscopic world which does not involve absolute location. Ideally, this conception should be complementary to the token-based view which has provided SQM with much pragmatic success. To this end, I have argued elsewhere [5] that we already maintain two mutually exclusive conceptions of macroscopic entities in OL and everyday thinking: generic TOKENS and specific KINDS. The haecceity of a TOKEN is operationally defined in terms of an absolute location at any given time, which any other object can occupy at a subsequent instant in time. Thus TOKEN is a universal class to which all THINGS belong.

STONE, PAPERWEIGHT, DOORSTOP and WEAPON, on the other hand, are examples of KINDS that correspond to the same token. Clearly, any token can be classified under many different kinds; but no one KIND can apply to all TOKENS.

What is the difference between a TOKEN and a SYMBOL at the physical level? Where a TOKEN ultimately derives all its properties from its constituent matter, a KIND does not. A piece of wood and a piece of stone can both be paperweights. There is at least one well-known KIND that we can immediately connect to quantum theory: the 'blackbody'. A blackbody is of course a classical mechanical object (a TOKEN), but it also has a distinctive observable property (its emission spectrum) that is independent of its constituent classical matter. Quantum mechanics, the theory which accounts for this emission spectrum, therefore ought to be taken as already treating the blackbody as a SYMBOL. However, SQM effectively models the blackbody as a TOKEN because all the states of measuring devices that interact with the blackbody are still interpreted classically (i.e., involving absolute locations). The theory of MQM aims to *reverse* this course by re-interpreting our sensory observations in relation to the state of macroscopic devices qua SYMBOLS.

What would such a theory look like? A TOKEN is defined primarily in terms of the absolute position it occupies. A KIND's identity, on the other hand, can only be defined in terms of external referents. Thus, an object qua KIND consists essentially of relational properties. A KIND then is a relational object.

Although KINDS are generally identified by their relation to a conscious subject, this need not be always the case; a SYMBOL is a particular KIND defined essentially by virtue of its properties in relation to another object. In other words, a SYMBOL is defined in terms of a physical relation.[6] For this reason, I believe that symbols can be naturalized and dealt with scientifically. I envision MQM as the theory of a particular set of KINDS called SYMBOLS, which are to be defined in terms of objective relational properties that describe their kinematics as well as their dynamic interactions with other SYMBOLS.

5 Digital Computers as Semantic SYMBOL Manipulating Systems

I have already noted that digital computers are seen as performing classical computation. There is a fast developing alternative field known as quantum computation, which is based on SQM. However, within its framework, a quantum computer would have to be built from scratch using qubits. The design and production of such a computer is not yet commercially feasible. There are extreme practical difficulties in producing qubits that can be isolated from the environment at room temperature.

Could the existing digital computer be re-interpreted as functioning according to quantum mechanical principles *without* delving into SQM? An affirmative response would clearly be of immense value, both theoretically and practically. One might think this task impossible, since digital computers were designed and developed without considering the quantum aspects of electronic chips. It would seem more reasonable to develop quantum computers from scratch by building qubits based on quantum physical principles, as is being currently attempted.

However, I believe it is equally reasonable to try to give current digital computers two complementary (i.e. mutually exclusive) treatments: one based on the classical theory of computing, and another based on MQM. An example of complementary classical and quantum treatments given to the same physical system can be found in optical polarizers. Consider a 45-degree polarized light incident on an HV polarizer. Classically, the polarizer is understood as splitting the incident light into two equal parts that pass through both channels simultaneously. After quantum theory was developed, the same polarizers received a second and more fundamental treatment in terms of individual photons. Individual photons are said to pass randomly through one or the other channel. These two ways of speaking about the functioning of the polarizer are complementary, or mutually exclusive. Similarly, if we are trying to understand token manipulation, the classical theory of computing is sufficient. To understand how a digital computer can perform symbol manipulation, we shall have to invoke a quantum theory of computing which I expect will be based on MQM.

---

[6] In this regard, it is useful to note that Bohr differed from the Copenhagenists by holding that the wave function describes the quantum system and the macroscopic measuring device simultaneously, as a *joint system*. Feyerabend and Jammer have pointed out that this idea implies that quantum mechanical observables correspond to *irreducible relations*, not primary properties that acquire definite values just at the point of measurement [8].

What would a theory of computing based on MQM entail? The basic operations of an Universal Turing Machine (UTM) involve motion: a tape sliding in either direction, a print head being raised or lowered onto the tape. Currently we treat these basic operations in terms of classical mechanics. Our notion that the UTM is carrying out computation is simply *our interpretation* of purely mechanical or token-manipulating operations (as Searle and other opponents of strong-AI rightly argue). This argument retains its force at the level of digital computers, where the mechanical motions of the UTM are replaced by mechanical motions of electrons.

Thus, to defend the strong-AI thesis, we need to go past the classical mechanical view of the basic operations of the UTM.

Now, the Schrodinger equation is a non-classical equation of motion. Yet when this equation is applied to the phenomenon of electron motion via SQM, we continue to invoke the classical kinematical notions of position and momentum. This has led to deep, abiding conceptual problems in reconciling quantum dynamics at the formal level with the very word "motion" as we understand it in everyday thinking. We therefore *cannot* expect SQM to provide a non-classical theory of motion at the macroscopic level.

I have suggested that a macroscopic quantum theory (MQM) can be developed which would renounce the idea of absolute position (and hence classical motion) starting from the level of observation itself. If successful, such an MQM would form a suitable basis for interpreting the basic operations of the UTM in terms of a quantum theory of macroscopic objects qua symbols.

This is the potential connection I see between MQM and a quantum theory of computation. There are two complementary ways of thinking about macroscopic motion: one (token manipulation) clearly related to classical physics and the other (symbol manipulation) related to MQM.

6   Macroscopic Quantum Mechanics - The Road Ahead

We are now in a position to outline in broad strokes the stages by which I envision such an MQM could be developed more fully.

1. Develop a range of ordinary language (OL)-based thinking that can serve as the foundation for objective, physical notions of macroscopic objects as SYMBOLS. To this end I have elsewhere developed a detailed set of arguments distinguishing between the P-mode of OL (statements that describe our sense experiences) and R-mode of OL (statements that describe a real-world situation underlying our sense experiences) [9]. Most pertinently, given the same P-mode statement, there will be one R-mode statement relating to the usual TOKEN view of the real macroscopic world and a separate R-mode statement relating to the SYMBOL view. In related work I have introduced the idea of "relational properties", which occupy a middle ground between primary and secondary properties [10].

2. Use these notions to develop an alternative range of kinematical properties corresponding to the quantum observables. I expect this step to involve the notion of objective semantic information, which I have yet to publish.

3. Develop a mathematical framework for MQM based on the aforementioned quantum kinematical concepts. Here, I am considering a line of attack involving "exotic manifolds" from 4-dimensional topology [11]. Exotic manifolds feature alternate smoothness, and thus contain room for developing an alternative calculus. This could permit the Schrodinger equation to provide a quantum mechanical description of the observed (i.e., the phenomenal) motion of macroscopic objects qua SYMBOLS that does not reduce to that of classical physics. Instead of interpreting our observation experiences as objective physical events in the standard 4-dimensional space-time, the idea is to try and interpret the same observation experiences as objective physical events in an alternative exotic 4-dimensional space-time.

Clearly, the development of MQM is a long-term program. Further details of this ongoing research must await future publications.

7   Conclusion

The human ability to 'understand' has many facets: self-awareness, specific mental capacities, intentionality, physical organs to sense the world and manipulate objects, etc. Among these myriad aspects, this paper has argued that our ability to work with the objects of the world consists of at least two complementary physical processes: syntactic token manipulation and semantic symbol manipulation.

The CRA is a negative argument. It demonstrates that token manipulation alone cannot lead to human 'understanding', but does not provide any insight into what such understanding actually involves. Thus, we cannot check to see whether digital computers possess this capability. The SRA, on the other hand, is a positive argument. It shows that at a minimum, human understanding requires symbol manipulation in addition to token manipulation. This argument shifts the focus of the strong-AI project to finding a way to naturalize symbols.

In this regard, I have tried to offer a well-posed practical motivation for developing a macroscopic version of quantum mechanics. Such a theory would physically characterize macroscopic objects in terms of their relational properties, as SYMBOLS with semantic content. MQM would open up the possibility for an alternative quantum theory of the UTM allowing even present computers to be seen as performing symbol manipulation and thus demonstrating human-like understanding to a limited extent. This is different from the current approach to developing quantum computers from scratch, based on SQM.

References

1. Searle, J.R.: Minds, Brains, and Programs. Behavorial and Brain Sciences 3, 417–424 (1980)
2. Kiparsky, P.: On the architecture of Panini grammar. Lectures delivered at the Hyderabad Conference on the Architecture of Grammar (2002),
   http://www.stanford.edu/~kiparsky/Papers/hyderabad.pdf
3. Ladyman, J.: Understanding Philosophy of Science. Routledge, London (2002)
4. Koch, C., Hepp, K.: Quantum Mechanics in the Brain. Nature 440, 611–612 (2006)
5. Gomatam, R.: Quantum Realism and Haecceity. In: Ghose, P. (ed.) Materialism and Immaterialism in India and the West: Varying Vistas. CSC, New Delhi (2008),
   http://bvinst.edu/gomatam/pub-2008.pdf
6. von Neumann, J.: Mathematical Foundations of Quantum Mechanics. Princeton University Press, Princeton (1955)
7. Fuchs, C., Peres, A.: Quantum Theory Needs No Interpretation. Physics Today, 70–71 (March 2000)
8. Gomatam, R.: Niels Bohr's Interpretation and the Copenhagen Interpretation – Are the Two Incompatible? Philosophy of Science 74(5), 736–748 (2007),
   http://www.bvinst.edu/gomatam/pub-2007-01.pdf
9. Gomatam, R.: Physics and Commonsense - Relearning the connections in the light of quantum theory. In: Chattopadhyaya, D.P., Sengupta, A.K. (eds.) Philosophical Consciousness and Scientific Knowedge. CSC, Indian Council of Philosophical Research, New Delhi (2004), http://www.bvinst.edu/gomatam/pub-2004-01.pdf
10. Gomatam, R.: Quantum Theory and the Observation Problem. Journal of Consciousness Studies 6(11-12), 173–190 (1999),
    http://www.bvinst.edu/gomatam/pub-1999-01.pdf
11. Asselmeyer-Maluga, T., Brans, C.H.: Exotic Smoothness and Physics: Differential Topology and Spacetime Models. World Scientific, Singapore (2007)

Conservation of Information: A New Approach to Organizing Human-Machine-Robotic Agents under Uncertainty

William F. Lawless[1], Donald A. Sofge[2], and H.T. Goranson[3]

[1] Paine College, 1235 15th Street, Augusta, Georgia, USA
lawlessw@mail.paine.edu
[2] Naval Research Laboratory, 4555 Overlook Avenue SW, Washington, DC, USA
donald.sofge@nrl.navy.mil
3 Earl Research, 500 Crawford St., Suite 402, Portsmouth, VA 23704, USA
tedg@earlresearch.com

Abstract. For many years, social scientists have struggled to make sense of the shift between individual and group perception, the difference between observation and action, and the meaning of interdependence. Interactions between these factors produce stable worldviews that contain more illusory than actual connections to reality. We attribute these struggles to scientists embedded in the social fabric, the lack of a measurement theory, and the difficulty of testing new theory with human subjects, groups and organizations. Our work includes field research with observations of citizen organizations advising the Department of Energy (DOE) on its environmental cleanup; laboratory simulations of DOE field results; stock market data; and computational modeling (coupled differential equations, control theory, AI, Gaussian distributions, uncertainty models, Fourier transform pairs, continuous and discrete wavelets). Results from laboratory experiments and stock markets agree with our theory, but many questions remain, forming a high-risk research plan. Our objective is to incorporate interdependent uncertainty into computational intelligence to better instantiate autonomy or decentralized control for mixed human-machine systems.

Keywords: interdependence, uncertainty, observation-action conjugation, quantum model, autonomous systems, social interaction.

1 Introduction

Two obstacles confront researchers modeling the dynamics of behavior in homogenous or heterogeneous (mixed) groups and organizations composed of humans, machines or robots: human agent observations, known as self-reports, and interdependent uncertainties in social interaction. We have known for some time that observing human workers affects their performance — the Hawthorne effect (Roethlisberger & Dickson, 1939). And we know that collecting data directly from humans changes their responses (Carley, 2002). To overcome these obstacles to model mixed systems of human and virtual agents or robots requires a theory encompassing these effects under uncertainty, which we consider high-risk research.

P. Bruza et al. (Eds.): QI 2009, LNAI 5494, pp. 184–199, 2009.
© Springer-Verlag Berlin Heidelberg 2009

1.1 Measurement

Individuals. The primary obstacle to a successful computational model and application of social interdependence for human and agent-based models (ABMs) has been the reliance by researchers on agent self-reports, a problem that persists in the new field of applying quantum models to human decision-making. Busemeyer and his team (2008) illustrate it with elegantly crafted mathematics that nonetheless rely on traditional self-reports, violating the central tenet of the quantum model: measurement disturbs what is measured. Evidence supporting theories of human self-reports and dynamic behavior is virtually non-existent. This evidence is lacking for self-reported preferences and actual choices in game theory (Kelley, 1992), in the very first (but not yet efficaciously (Sanfey, 2007)) mathematical model of interdependence (Von Neumann & Morgenstern, 1953), and for decision-making by groups or organizations (Levine & Moreland, 1998) (this work has yet to be established theoretically (Pfeffer & Fong, 2005)). And for self-reported self-esteem with actual academic or work performance, one of the most studied of psychological phenomena, Baumeister and his colleagues (2005) in a meta-analysis found only weak correlations, leaving Baumeister to conjecture that there were no other means to collect this data. In business management, Bloom and colleagues (2007) found only a weak correlation between self-assessments by managers and the performance of their firms. The likely cause in all of these cases is that self-reports are assumed to measure phenomena independent of human observers; however, a computer's ability to capture the full knowledge of the social objects interacting in its virtual computational space, known as the "God's eye" view displayed on monitors, recreates this obstacle for ABMs and robots (moving computational spaces) (Lawless et al., 2000). The assumption that humans or ABMs can be exactly located on a psychological measure or other conceptual grid, even when the reliability of the measure is well-established, ignores the evidence that cognitive dissonance orders experience and illusions into stable worldviews which interact with the illusions and experiences of others to form bi-stable (conjugate) interdependent effects (Lawless et al., 2007).

Groups and Organizations. The lack of a theory of measurement at the individual level is amplified at the group (Levine & Moreland, 1998), organization (Pfeffer & Fong, 2005), and virtual organization levels (Lawless et al., 2008d). As Kohli and Hoadley (2006) conclude, this absence leads to ineffective metrics to improve organizational performance and effectiveness or business redesign; i.e., currently, Business Process Restructuring (BPR) does not guarantee success (Sommer, 2004). Smith and Tushman (2005) propose the need to maintain a state of tension between optimizing performance (mission) and adapting to change (vision). We have been applying their idea with a web-based eIRB (Institutional Review Board) designed to better train military physicians in research methods (Wood et al., 2009).

1.2 Interdependence

According to Kenny and colleagues (1998), the general strategy of working with interdependence in social science creates confounds. The general goal for the analyses of social data becomes removal or control of the effects of interdependence in data.

Dawes and his colleagues (1989) found similarly that, compared to the estimates by experts, actuarial estimates of human behavior were less immune to the effects of interdependence; as a corrective, Dawes found that actuarial data were more reliable and valid than clinical or forensic (expert) diagnoses of human pathology. Nonetheless, our goal is the opposite: instead of controlling for observer dependencies, we plan to remove observer independencies to better establish the science of interdependence in organization, system and social processes. For example, measuring deindividuation among the citizens of Burma imposed by the reigning military dictatorship may indicate over time and geospatially the variability in the amounts of power expended to oppress Burmese citizens. By a similar method, tradeoffs in the amount of energy expended and duration to coordinate a mixed system of human-robot-machine agents could indicate computationally the effectiveness of an organization's plan. And the interdependence in emergency responses, as with the fire that gutted the upper house of Egypt's Parliament (Slackman, 2008), should indicate for censorship, even when mathematically justified (May, 1973), the cost of dictatorship.

1.3 Perturbations

Financial interdependence may indicate the spread of market volatility. Based on Fama's (1964) random walk model, modern finance assumes that randomization among independent pricing events increases the value of distributing risk in diversified portfolios (Markowitz, 1952). Long-Term Capital Management's rational options pricing model, built by the firm's two Nobel Laureates Merton and Scholes, the cornerstone of modern finance, estimated random variations in the valuations of options that could be exploited. However, these randomizations break down then as in 2008 when economic stresses generate long-range interdependences characterized as correlations (waves of panic generating waves of deleveraging) across a stock market (see news analysis by Lowenstein, 2008).

Historically, during periods of normal corporate and financial activities, the spread between BAA and AAA bond interest rates has ranged randomly between 0.85 and 0.95 interest points (Ranson, 2007). Spreads over 1.15 points forecast perturbations across a market of impending GNP creating financial instability that threatens stock values (Figure 1; the collapse and merger of Countrywide Financial Corporation by Bank of America on July 1, 2008. Bear Sterns was taken over by J.P. Morgan Chase on 3/17/08 [#110 on x-axis]; IndyMac collapsed on July 17, 2008 [#198]; Fannie and Freddie went into receivership on September 5, 2008 [#234], producing a brief respite; and on September 15, 2008, Lehman Brothers filed for bankruptcy, Merrill Lynch was taken over by Bank of America, and American International Group struggled to recapitalize. The spread was 1.61 at end of the chart; on January 5, 2009, the spread was 3.35 points). On October 23, 2008, the spread at 2.92 indicated significant new incoming waves of threats to corporations, motivating mergers to reduce the threats, thereby creating a measurement problem. We believe that modeling perturbations is essential to computing organizational dynamics in the stock market and, by extension, military organization dynamics or the dynamics of military effects on a target country (e.g., actions by the U.S. government on a target country to produce desired effects; in Lawless, 2008). For example, we have written that mergers

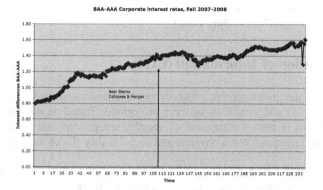

Fig. 1. Daily BAA-AAA bond interest rate spreads from 10/16/07 to 9/11/08, http://research.stlouisfed.org/fred2/categories/119 (retrieved 9/15/08)

between business organizations, even terrorist organizations, can be better understood mathematically by considering uncertainty, producing the measurement problem (Lawless et al., 2007; see the computational example with Equation 2).

From a self-report perspective of the measurement problem, in April 2008 Lehman Brothers' Chief Financial Officer E. Callan stated in an interview with CNBC that rumors by short-sellers had hurt the firm: "Unfortunately, we're in a market where perception trumps reality" (news by Motley Fool, 2008). But after Lehman was successful when its public offer of preferred convertible stock was oversubscribed, raising $4 billion in capital on April 1st, Callan reversed himself to conclude that Lehman had learned its lesson about the 1998 credit crunch (news by Bloomberg, 2008). However, illustrating the power of an illusion, since April the collapse in the value of Lehman's stock held just by its employees alone had fallen by over $10 billion, signaling that management had lost control: "Everyone knows that [Lehman] has been on the ropes since May" (interview of T. Truman (2008), former Fed official). Lehman filed for bankruptcy on September 15, 2008 (end of Figure 2).

Our approach to the study of interdependence in measurement is not to ignore or replace self-reports, but to constrain them. Our overall goal is to predict computationally the outcomes of an interaction, including for ABMs, which have been ruled unlikely (Bankes, 2002; Conzelman et al., 2004). Instead of asking organizational responders — human and artificial — to report on their effectiveness, we want more objective measures combined with predictions of what will be self-reported to constrain the interpretations of agent reports (Lawless et al., 2007). Our view has important ramifications for consensual constructs of situation awareness in military combat; e.g., the divergent views held in 2006 by General G. Casey on the "surge" in Iraq compared with those represented by W.J. Luti, reflected in a report by the American Enterprise Institute written by Gen. J. Keane and F.W. Kagan, and supported by Generals R. Odierno and D. Petraeus (http://www.washingtonpost.com/wp-dyn/content/article/2006/12/26/AR2006122600773.html). Despite extraordinary negative political fallout, the Keane and Kagan report led the White House to commit to a surge of American troops by concluding that "our center of gravity — public support — was in jeopardy because of doubts that our Iraq efforts are on a trajectory leading to success ..." (see news accounts by Gordon, 2008).

2   Conservation of Information (COI)

At the heart of the measurement problem exists interdependent uncertainties over space (geospatially) and time that conserve information based on the existence of bistable perceptions in reality composed of variable exogenous and endogenous mixtures of facts and illusions about physical and social reality (Lawless et al., 2005).

Bistability. Bistability is best explained with an example of a simple illusion (see Figure 2). It occurs when data produces mutually exclusive interpretations. When a entire data set is processed cognitively by an individual, two interpretations of a bistable illusion cannot be held in awareness simultaneously (Cacioppo et al., 1996). According to Bohr (1955), multiple interpretations support the existence of different cultures. Further, given the importance of feedback to social dynamics (Lawless et al., 2007), the possibility of shifts ($j\omega$, where j is the imaginary number $\sqrt{-1}$ and ω the discrete frequency in radians per second) between bistable interpretations increases uncertainty in the excluded interpretation which not only creates interpretation barriers between social groups (e.g., courtrooms, politics), but supports the existence of COI, characterized as tradeoffs between incommensurable views (e.g., between action and observation). We propose that measurements of bistable phenomena collapse interdependence to produce the classical information found in self-reports, decreasing uncertainty in the observed aspect of a bistable phenomenon while increasing uncertainty in its non-observed aspect (the insider-outsider minimal inter-group effect; Tajfel, 1970). Figure 2 could reflect the electorate in the U.S. composed of liberals, conservatives and undecided neutrals, with the "faces" *arbitrarily* representing the incommensurable beliefs of liberals (or others, such as Shiite) and the "vase" <u>*arbitrarily*</u> representing those incommensurable beliefs of conservatives (or others, such as Sunni). A winning plan occurs when neutrals are persuaded to endorse a belief, decreasing conflict (Kirk, 2003) by promoting compromise, but making it unlikely that a story can capture the original state of interdependence (Lawless et al., 2008a).

Fig. 2. An example of bistability. In viewing the simple two-faces vase illusion, an observer is incapable of holding both interpretations in awareness simultaneously.

2.1 Illusions

The brain has two independent systems for action and observation (Rees et al., 1997). From Bohr (1955), observation and action form interdependent uncertainty tradeoffs between images constructed by the mind. With his gray square checker-shadow illusion, Adelson (2000) concluded that if the eye system worked only as a signal detector, it could not distinguish a white surface in dim light from a dark surface in bright light, which humans accomplish easily ("lightness constancy"). Three brain processes combine and constrain signals to form a cognitive image: lower level image processing (signals perceived), mid-level processing (Gestalt groupings), and high-level processing (experiences and worldview constructs). Adelson illustrated that illusions are not based on random processes but constructed by combining signal perceptions, groupings, and experiences with the properties of objects to form a visual context. Statistical processes on signals from configurations map light reflectance to illuminate surfaces as an individual estimates the inverse transfer function imperfectly but adaptively with a subsystem that operates like a filter (convolution). Low-pass filters produce long-term signal averages in images, while high-pass filters add details (Lawless & Sofge, 2008c).

Generalizing to social situations led us to predict and find with Citizen Advisory Boards advising the Department of Energy (DOE) on the cleanup of sites across its complex (Lawless et al., 2008a) that risk perceptions (illusions) abound unchallenged under cooperative decision strategies (consensus rules) compared to competitive strategies (majority rule). In stark contrast to scientific peer review, we concluded that DOE managers acted as command decision-makers to force a consensus that created the widespread contamination of its sites by picking winners and losers in funding research programs. Consensus also reduces the likelihood of political compromises that accelerate action; e.g., DOE Hanford and its consensus Citizens Advisory Board in Washington State are "gridlocked" (similarly with global warming negotiations, see Oppenheimer et al., 2007). Moreover, under majority rule with significant numbers of neutrals among the deciders, conflict and ill-will are moderated to promote compromises and what we have named, counterintuitively, an "action consensus" (Lawless et al., 2008a).

Experimental Evidence. We have collected preliminary data from the laboratory with college students making recommendations to improve college experiences. The results appear to have partially replicated the DOE CAB study (with triads, decision duration, but not decision quality; the latter improves as predicted with larger groups).

2.2 The Mathematics of COI

We propose the following mathematical relationships (Lawless & Sofge, 2008c).

Geospatial Interdependence. Multitasking degrades performance at the individual level (Wickens, 1992). In contrast, the function of a group or organization is to multitask (Ambrose, 2001), underscoring the existence of interdependence in business models (Jervis, 1997), chains of command, and organizational centers of gravity (Arnold, 2006). Multitasking coordinates one activity with several others, an interdependence say between localizing an event, Δx_{COG}, or the spatial frequencies of a chain of events, Δk, where $k = 1/\lambda$ and λ is the distance between spatially controlled events, giving

$$\Delta x_{COG} \Delta k \geq c \qquad (1)$$

In support of Equation (1), the defensive pattern against suicide bombings in Jerusalem "is always a tradeoff between model accuracy and area reduction that we can describe for any given model" (Willis, 2007, p. 2). Social interdependencies form geospatial wave patterns to transmit information, like traffic congestion occurring at a critical density when traffic is sufficiently interdependent to transmit waves (Helbing, 2001). Other examples are Wal-Mart's distribution system during the holidays; air-traffic patterns at major airports during busy cycles; and the launch of military operations against Baghdad identified by General Tommy Franks as the center of gravity during Operation Iraqi Freedom (Arnold, 2006, p. 13). Table 1 illustrates geospatial reasoning by thieves in burglary frequencies collected over time across three contiguous counties in Georgia and South Carolina. No information exists in the uniformly high correlations over time (columns). Geospatially (rows), the correlation between Richmond and Columbia, GA is negligible, and weak between Columbia and Aiken, SC. The surprise found in the strong correlation between Richmond, GA, and Aiken, SC, is removed by interpreting that police do not chase thieves across state lines. These counties are across state lines separated by the Savannah River but connected by five bridges. It indicates that thieves are swapping territories and information to steal in different police jurisdictions and unload ill-gotten gains in their own states.

Table 1. Burglary data from the Augusta Chronicle, 10/09/2008

Burglaries by County in Georgia (GA) and South Carolina (SC), USA			
	2008	2007	2006
Richmond, GA	340	222	244
Columbia, GA	33	24	45
Aiken, SC	116	79	74
Correlating columns 1:2	0.99996783		
Correlating columns 1:3	0.991677525		
Correlating columns 2:3	0.990612918		
Correlating rows R:C	0.093784077		
Correlating rows R:A	0.95954857		
Correlating rows C:A	-0.190311882		

Our research indicates that risk perceptions (illusions), as opposed to risk determinations (science), interfere with achieving practical organizational decisions (Lawless et al., 2008a). We have found that illusions increase in number and virulence under enforced cooperative decision-making, but are better managed under a competition for the best among a series of ideas (Holmes, 1919). But Adelson suggests the existence of a social cognitive "screen" where entangled constructive and destructive interferences from competing ideas play out under consensus or competitive decisions. In the future, this new model may allow us to calculate the entropy shifts between mutually exclusive (classical) or entangled stories (illusions) or decisions and their execution (Lawless et al., 2007), where cognitive uncertainty in organizational stories or business models (ΔBM_{COG}) and the conjugate uncertainty in executing these models, Δv (see Lawless et al., 2007; Goranson & Cardier, 2007) gives:

$$\Delta BM_{COG}\, \Delta v \geq c \tag{2}$$

Based on COI, Equation (1) can be reconfigured for energy, ΔE, time, Δt:

$$\Delta E\, \Delta t \geq c \tag{3}$$

Equation (3) agrees with findings for the brain: greater expenditures of energy (power) in the brain are associated with higher cognitive functions, leading to an increase in the ability to resolve mental maps of reality (Hagoort et al., 2004). Words spoken in anger expend about twice the energy of regular voices (Lawless, 2002). When performing a complex military exercise, the brains of novices, compared to experts, light up like a Christmas tree, an increase in the energy wasted by novices (Milton et al., 2007). The goal of the brain becomes to minimize wasted energy to maximize free energy, ΔA (Friston & Stephan, 2007), at a given temperature, T, and entropy S, and where

$$-\Delta A = \Delta E - T\Delta S \tag{4}$$

Based on Equations (3) and (4), organizations attempt to increase free energy by reducing waste and temperature (panic, excessive debate, emotional displays). Increasing organizational size with mergers reduces the perturbations experienced (Andrade et al., 2001). From stock market data, volatility (beta) decreases with increased organizational size (see Table 2 in Lawless et al., 2008b). The interpretation is two-fold: first, as mentioned before, organizations attempt to grow, organically or via mergers, to increase organizational stability; second, COI exists as a tradeoff between firm size relative to a market and volatility. Moreover, innovation, while costly, increases free energy by reducing wasted efforts to better compete (Sood & Telles, 2008), including acquiring a smaller, more innovative firm that has developed break-though technology (e.g., Oracle-BEA merger). Mergers also reduce the energy expended from competing in a market.

Equation (2) exposes a hidden cognitive reason for the pursuit of mergers and strategic alliances, contradicting what is popularly considered to be a poor business decision. Assuming the equality aspect of Equation (2) pertains gives

$$\Delta_1 BM_{COG}\, \Delta_1 v = \Delta_2 BM_{COG}\, \Delta_2 v$$

Next, assume that organization #1 has ½ the uncertainty in its business model as organization #2, resulting in $\Delta_1 BM_{COG} = \frac{1}{2} \Delta_2 BM_{COG}$. Merging gives $\frac{1}{2} \Delta_2 BM_{COG}\, \Delta_1 v = \Delta_2 BM_{COG}\, \Delta_2 v$, then $\frac{1}{2}\, \Delta_1 v = \Delta_2 v$, and finally, $\Delta_1 v = 2\, \Delta_2 v$. Thus, a reason to merge based on mathematics is being able to execute twice as fast, if and only if the new entity retains the same level of uncertainty in its business model (ΔBM). For example, the 1989 merger between US Air and Piedmont degraded ΔBM for the merged firm (US Airways).

A different approach arises from assuming that information signals control behavior at the individual, group or organizational levels, revising Equation (3):

$$\sigma_f \sigma_t \geq \tfrac{1}{2}, \tag{5}$$

where $\sigma_f \sigma_t$ represents a Fourier transform pair, with σ_f as the standard deviation for frequency and σ_t for time. See Cohen (1995) for the derivation of Equation (5).

Applications of Equation (5). Among others, Cohen (1995) recognized tradeoffs in signal detection. Assuming a normalized signal with zero mean for duration, t, and bandwidth, f, Fourier transforms produce variances of σ_t^2 and σ_f^2 for duration and frequency (where $\omega = 2\pi f$), respectively, resulting in a Fourier pair of $\sigma_t \sigma_f \geq \tfrac{1}{2}$ (Table 2). COI means that a signal of "narrow waveform yields a wide spectrum and a wide waveform yields a narrow spectrum and that both the time waveform and frequency spectrum cannot be made arbitrarily small simultaneously." It means that a limited duration signal produces a spectrum with indistinct, poorly-resolved frequencies, a signal characterized by a spike in time yields a broad spectrum, and a signal which "rings" for a long duration (e.g., sine wave) yields a spectrum with only one distinct frequency.

Table 2. An illustration of arbitrarily selected Gaussian distributions that form "Fourier pairs". Notice that as σ_f increases, σ_F decreases; note also that $\sigma_f \sigma_F$ is greater than 1/2 in all cases (from Lawless et al., 2008b). This represents COI.

Function	σ_f	Fourier Transform	σ_F	$\sigma_f \sigma_F$
$f(t) = e^{-2t^2}$	0.558	$F(s) = \dfrac{\sqrt{2\pi}}{2} e^{-\frac{1}{8}s^2}$	5.013	2.806
$f(t) = e^{-\frac{1}{2}t^2}$	1.583	$F(s) = \sqrt{2\pi} e^{-\frac{1}{2}s^2}$	2.507	3.969
$f(t) = e^{-\frac{t^2}{8}}$	4.478	$F(s) = 2\sqrt{2\pi} e^{-2s^2}$	1.253	5.612

As an exercise with a Fourier transform and tradeoffs between Fourier pairs, assume that a system, with pe`riod τ and amplitude $A = 1$, experiences a pulse P described by the following forcing function $f(t)$:

$$f(t) = \begin{cases} A, & -\tau/2 \leq t \leq \tau/2 \\ 0, & |\tau| > \tau/2 \end{cases}$$

The Fourier transform becomes:

$$F(f(t)) = \int_{-\infty}^{\infty} f(t)\, e^{-i\omega t}\, dt = -A\, e^{-i\omega t}/\omega t\, |_{-\tau/2}^{\tau/2} = 2A \sin(\omega\, \tau/2)/\omega = A\, \tau\, sinc\,(\omega\, \tau/2)$$

For each time pulse, letting $\tau = 8$ and 2 for $A = 1$ produces the graphs in Figure 3.

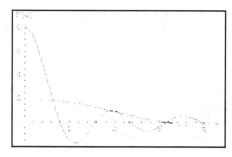

Fig. 3. Sketches of $F(f(t))$ for $\tau = 8$ s and 2 s with $A = 1$. Find $\omega = 0$ Hz on the ω axis with l'Hôpital's rule: $lim_{\omega \to 0}$ $\tau \sin \theta/\theta = \tau \cos \theta/1$, $\tau = 8$. The $\tau = 8$ curve goes to zero on the $F(\omega)$ axis when $4\omega = n\pi$ to give $\omega \approx 0.393 n$, where $n=1,2,...$

Table 3. Standard deviations from the exercise above for $f(t)$. Notice that the single duration signal pulses of widths 2 and 8 seconds do not conserve information. But by estimating a Gaussian distribution approximated by σ_f, COI is recovered.

Function	σ_f	FT	σ_F	$\sigma_f \sigma_F$	Est. GD σ_f	Est. FT σ_f	Rev. $\sigma_f \sigma_F$
$f(t)$	2.31	$F(\omega)$	2.10	4.86	2.0	2.2	4.4
$f(t)$	18.48	$F(\omega)$	8.99	166.23	17.69	.45	9
				$\neq$ COI			$\approx$ COI

From this exercise, as noted by Riefel (2007), despite the lack of a Gaussian distribution, we conclude that as the time pulse shortens, the bandwidth broadens, but also the frequency amplitude decreases. In turn, as the period of the time-domain pulse broadens, the bandwidth narrows and frequency amplitude increases. But by estimating Gaussian distributions, we arrive at Fourier pairs as predicted by COI.

Our research with multiple regressions has established that the value of mergers is that an increase in size decreases volatility from market threats. While we believe that a mathematical model of mergers holds promise even among terrorist organizations (Lawless et al., 2007), we have found support only with multiple regressions (Lawless et al., 2008b), but not with Fourier transform pairs, possibly because we were using discrete rather than continuous transforms (Hubbard, 1998), and because we did not account for market volatility (Frieden, 2004). Going forward, we plan Gaussian fits to distributions, along with semblance wavelet techniques to compare multiple time-series data sources (Cooper & Cowan, 2008).

As a tentative test, we considered the merger of Washington Mutual (WAMU) on 10/26/2008 by J.P. Morgan Chase (JPMC). We did not expect to find COI in the distributions because they were not Gaussian and had trends in them. But by using volatility (beta) along with estimated Gaussian distributions, tentatively and encouragingly, we have confirmed the existence of COI in the tradeoffs between the two.

2.3 Control

To control an organization requires a reference baseline or goal. Based on the control theory of Csete and Doyle (2002), we predict that cooperation will decrease variance with the worldview baseline based on an organization's effort to enforce a consensus worldview among its members (Lawless et al., 2008a), but making it less able to adapt compared to a baseline hammered out under competition. In the future, we plan to convert these metrics into four interrelated standard deviations (Wood et al., 2009). For example, duration data (t) is transformed into frequencies (ω) for which we can calculate energy (resources) and the standard deviations for both, where $\sigma_t \sigma_\omega \geq \frac{1}{2}$. As we reported above, initial results are supportive: organizations make decisions regarding mergers and acquisitions to gain stability. We plan to work with coupled Lotka-Volta equations where control of the contrasting consensus-competitive social images on a group's or organization's community "screen" is lost or gained (May, 1973).

Technology's Effect on Free Energy. Organizational size limits are raised by technology (Mattick & Gagen, 2005); i.e., as size increases, competitiveness decreases, but that effect is countered by new technology (e.g., "back office" integration software). Technology helps to manage tradeoffs in performance, but they still remain (Csete & Doyle, 2002). However, the value on the return on investment (ROI) for technology is subjective (Rouse & Boff, 2003). But mergers often occur to acquire the technology of a firm owning an R&D breakthrough to maintain market leadership or to survive (Bell South's inability to upgrade its technology led to its takeover by AT&T). Technology limits a business model (Carley, 2002). But technology integration can change a business model, which it did for J.P. Morgan Chase, to better manage costs and productivity and to better compete, as in our field research with training physicians in research methods for the Army. Technology can produce its own shocks (Andrade et al., 2001) if new companies unexpectedly take the leadership to control a market (e.g., Google's leadership threatened Yahoo and Microsoft).

2.4 Interdependent Information in State Merging

To test computational models of interdependent uncertainty on alliances among organizations with mixed systems of human-machine-robot agents, we first review the correlation between subsystems in an interdependent state. Merging organizations (interdependent states) gains knowledge. Consider a merged system whose state space can be represented by the tensor product, $I = I_A \otimes I_B$, with ρ_{AB} as the density matrix acting on I, giving Von Neumann entropy $I(\rho_{AB}) = -Tr\, \rho_{AB}\, log\, \rho_{AB}$. If $|\psi\rangle$ is in I_A and $|\phi\rangle$ in I_B, the tensor product $|\psi\phi\rangle$ is the complex scalar $\psi(x) = \langle x|\psi\rangle = ce^{-ikx}$. This function allows us to cross between irreversible and cyclic Lotka-Volterra equations to the entropy of Gaussian distributions and the mathematics of interdependence borrowed from joint quantum entropy (e.g., Horodecki et al., 2005).

Lewin (1951) first proposed that the whole is greater than the sum of its parts; i.e., decohered subsystems have more entropy than the whole. To us this means that having more information makes state mergers cheaper. In the case where a target, Organization B, has prior knowledge of Organization A, the acquiring organization, the partial information needed from A based on prior information by B is the conditional entropy. In a classical two-player game or information network, the conditional entropy occurs with correlations between B with prior information Y merging with acquirer A holding the desired missing information X. In classical situations, this relatively insensitive Shannon partial information ranges from zero when all is known to maximum 1 when nothing is known, but is never negative.

From Horodecki and his colleagues, we propose for interdependent situations that the missing information is Von Neumann entropy ($-Tr\ \rho\ log\ \rho$) having states with density operators ρ_A for the acquirer and ρ_B for the target with sensitive joint entropy ρ_{AB}. Following the classical model, with $I(A|B)$ as the interdependent conditional entropy and $I(AB)$ the joint interdependent entropy:

$$I(A|B) = I(AB) - I(B)$$

where $I(B)$ is the state of ρ_B and $I(A)$ is that of ρ_A. $I(A|B)$ is not an entropy in the traditional sense; it is the meaning gained from preserving interdependence (e.g., "culture"). Interdependent entropy is positive, zero or negative. If positive, the sender must transmit the missing bits; if negative, receiver and sender gain full knowledge from classical knowledge alone at no extra cost by sharing pure maximally interdependent states. Conditional negative entropy reflects excess knowledge held. With a maximally interdependent state like a Bell state, the total system is in a pure state of entropy 0 even when each subsystem is maximally mixed with entropy 1, giving ($-I(A|B)$) EPR pairs (where each reduces entropy by 1 unit) to transmit interdependence using only classical information:

$$\tfrac{1}{2}\,(\,|\,00 > + |\,11 >\,)$$

A maximally mixed state has maximum Von Neumann entropy; the pure state (a rank one projection) has zero Von Neumann entropy. We plan the following test: prior information makes state merging cheaper (similar cultures make merging easier as when HP and Compaq merged, versus Bank of America and Merrill Lynch). We plan to contrast measures of the classical information transferred with that actually transferred between poorly organized subsystems coordinating with each other (classical limits) and well-organized subsystems (interdependent limits).

3   Summary

For future research into COI, we plan to incorporate computational interdependent uncertainty into intelligent and mixed human-machine systems to more robustly instantiate autonomy or to better decentralize mixed systems. Our plan to achieve this goal is to develop and test new theory, to produce computational models, and to search for evidence in the field and laboratory to accept or reject and then revise these models. We have proposed that current social science models relied on by social scientists are flawed. They often do not incorporate uncertainty, nor do they take a fundamental approach to the effects of interdependent (conjugate) uncertainties on the basic collection of data reported by agents involved in interactions. Our theory and research is a small step towards overcoming these obstacles.

For COI we are considering Fourier models, wavelet models, and Gabor or short-time windowed Fourier models, along with Gaussian fitting techniques to work with discrete data. In continuing Lewin's (1951) research, we plan to model the propagation of elastic social wave fields with coupled partial differential equations to account for the virtual displacement of beliefs and their behavior velocity or social embeddedness and its geospatial wave numbers across two or three dimensions with computational intelligence based on evolutionary algorithms (Lawless & Sofge, 2008c).

Acknowledgment

Donald Sofge's contribution to this effort was supported by the Naval Research Laboratory under JONs 55-6207-A8 and 55-6329-09.

References

Adelson, E.H.: Lightness perceptions and lightness illusions. In: Gazzaniga, M. (ed.) The new cognitive sciences, 2nd edn. MIT Press, Cambridge (2000)

Ambrose, S.H.: Paleolithic technology and human evolution. Science 291, 1748–1753 (2001)

Andrade, G.M.-M., Mitchell, M.L., Stafford, E.: New Evidence and Perspectives on Mergers. Harv. Bus. Sch. Working Paper No. 01-070 (2001),
http://ssrn.com/abstract=269313

Arnold, K.A.: PMESII and the non-state actor. Questioning the relevance. US Army Monograph, Fort Leavenworth, KA. School of Advanced Military Studies (2006)

Bankes, S.C.: Perspective. Agent-based modeling. Proceedings National Academy of Sciences 99(3), 7199–7200 (2002)

Baumeister, R.F., Campbell, J.D., Krueger, J.I., Vohs, K.D.: Exploding the self-esteem myth. Scientific American (January 2005)

Bloom, N., Dorgan, S., Dowdy, J., Van Reenen, J.: Management practice and productivity. Quarterly Journal of Economics 122(4), 1351–1408 (2007)

Bloomberg (2008) (retrieved September 14, 2008), http://www.bloomberg.com

Bohr, N.: Science and the unity of knowledge. In: The unity of knowledge, pp. 44–62. L. Leary, New York (1955)

Busemeyer, J.R., Santuy, E., Mogiliansky, A.L.: Distinguishing quantum and Markov models of human decision making. In: Bruza, P., Lawless, W.F., van Rijsbergen, K., Sofge, D., Coecke, B., Clark, S. (eds.) Quantum Interaction, Proceedings of the Second Quantum Interaction Symposium (QI 2008), King's College, London (2008)

Cacioppo, J.T., Berntson, G.G., Crites Jr., S.L. (eds.): Social neuroscience: Principles, psychophysiology, arousal and response. In: Social Psychology Handbook of Basic Principles. Guilford, New York (1996)

Carley, K.M.: Simulating society: The tension between transparency and veridicality. In: Social Agents: ecology, exchange, and evolution, University of Chicago, Argonne National Laboratory (2002)

Cohen, L.: Time-frequency analysis: theory and applications. Prentice Hall Signal Processing Series (1995)

Conzelmann, G., Boyd, Cirillo, R., Koritarov, V., Macal, C.M., North, M.J., Thimmapuram, P.R., Veselka, T.: Analyzing the Potential for Market Power Using ABM Approach. In: Conference on Computing, Communication and Control Technologies, Austin, TX (2004)

Cooper, G.R.J., Cowan, D.R.: Comparing time series using wavelet-based sem-blance analysis. Computers and Geosciences 34, 95–102 (2008)

Csete, M.E., Doyle, J.C.: Science 295, 1664–1669 (2002)

Dawes, R.M., Faust, D., Meehl, P.E.: Clinical versus actuarial judgement. Science 243(4899), 1668–1674 (1989)

Fama, E.F.: Random walks in stock market prices. Financial Analysts Journal 51(1), 75–80 (1965) (reprinted, January-February 1995)

Friston, K., Stephan, K.E.: Free energy and the brain. Synthese 159, 417–458 (2007)

Frieden, B.R.: Science from Fisher information. Cambridge U. Press, New York (2004)

Goranson, H.T., Cardier, B.: Scheherazade's Will: Quantum Narrative Agency. In: Proceedings, AAAI Spring Series Workshop on Quantum Interaction, Stanford (2007)

Gordon, M.R.: Troop 'Surge'. New York Times, p. A1,10 (August 31, 2008)

Hagoort, P., Hald, L., Bastiaansen, M., Petersson, K.M.: Integration of word meaning and world knowledge in language comprehension. Science 304, 438–441 (2004)

Helbing, D.: Traffic. Reviews of Modern Physics 73, 1067–1141 (2001)

Holmes, O.W.: U.S. Supreme Court, Abrams et al. v. United States, No. 316. The dissent was written by Justice Holmes and concurred by Justice Brandeis (1919)

Horodecki, M., Oppenheim, J., Winter, A.: Quantum information can be negative (May 9, 2005) (Retrieved on January 10, 2008), arXiv:quant-ph/0505062v1

Hubbard, B.B.: The world according to wavelets, 2nd edn. A. K. Peters, Natick (1998)

Jervis, R.: Systems effects: Complexity in political and social life. PUP, Princeton (1997)

Kelley, H.H.: Lewin, situations, and interdependence. J. Social Issues 47, 211–233 (1992)

Kenny, D.A., Kashy, D.A., Bolger, N.: Data analysis in social psychology. In: Gilbert, D.T., Fiske, S.T., Lindzey, G. (eds.) The handbook of Social Psychology, pp. I: 233–268. McGraw-Hill, Boston (1998)

Kirk, R.: More terrible than death. Massacres, drugs, and America's war in Columbia. Public Affairs (2003)

Kohli, R., Hoadley, E.: Towards developing a framework for measuring organizational impact of IT–tneabled BPR: case studies of theree firms. ACM SIGMIS Database 37(1), 40–58 (2006)

Lawless, W.F., Castelao, T., Ballas, J.A.: Virtual knowledge: Bistable reality and solution of ill-defined problems. IEEE Systems, Man, & Cybernetics 30(1), 119–124 (2000)

Lawless, W.F.: Adversarial cooperative collaboration: An overview of social quantum logic. In: Proceedings Collaborative Learning Agents. AAAI 2002 Spring Symposium, Stanford University, pp. 122–123 (2002),
http://www.aaai.org/Symposia/symposia.html

Lawless, W.F.: Draft Final Report: The requirements for social and cultural metrics for DIME actions on PMESII effects. A report prepared for the Naval Research Laboratory, funded by a Summer Fellowship award from ASEE, Washington, DC (2008)

Lawless, W.F., Bergman, M., Feltovich, N.: Consensus-seeking versus truth-seeking. ASCE Practice Periodical of Hazardous, Toxic, and Radioactive Waste Management 9(1), 59–70 (2005)

Lawless, W.F., Bergman, M., Louçã, J., Kriegel, N.N., Feltovich, N.: A quantum metric of organizational performance: Terrorism and counterterrorism. Computational & Mathematical Organizational Theory 13, 241–281 (2007)

Lawless, W.F., Whitton, J., Poppeliers, C.: Case studies from the UK and US of stakeholder decision-making on radioactive waste management. ASCE Practice Periodical of Hazardous, Toxic, and Radioactive Waste Management 12(2), 70–78 (2008a)

Lawless, W.F., Poppeliers, C., Grayson, J., Feltovich, N.: Toward a classical (quantum) uncertainty principle of organizations. In: Bruza, P., Lawless, W.F., von Rijsbergen, K., Sofge, D., Coecke, B., Clark, S. (eds.) Quantum Interaction, Proceedings of the Second Quantum Interaction Symposium (QI 2008), King's College, Strand (2008b)

Lawless, W.F., Sofge, D.: Conservation of information (COI). A concept paper on virtual organizations and communities. In: NSF Computational Workshop: Building CIML Virtual Organizations, Fairfax, VA, October 24 (2008c)

Lawless, W.F., Howard, C.R., Kriegel, N.N.: A quantum real-time metric for NVO's. In: Putnik, G.D., Cunha, M.M. (eds.) Encyclopedia of Networked and Virtual Organizations. Information Science Reference, IGI Global, Hershey (2008d)

Levine, J.M., Moreland, R.L.: Small groups. In: Gilbert, D.T., Fiske, S.T., Lindzey, G. (eds.) Handbook of Social Psychology, pp. II: 415-469. McGraw-Hill, Boston (1998)

Lewin, K.: Field theory in social science. Harper (1951)

Lowenstein, R.: Essay: Long-Term Capital, p. BU1, 9. New York Times (September 7, 2008)

Markowitz, H.M.: Portfolio Selection. Journal of Finance 7(1), 77–91 (1952)

Mattick, J.S., Gagen, M.J.: Accelerating networks. Science 307, 856–858 (2005)

May, R.M.: Stability and complexity in model ecosystems. Princeton University Press, Princeton (1973/2001)

Milton, J., Solodkin, A., Hluštík, P., Small, S.L.: The mind of expert motor performance is cool and focused. J. Neuroimage (2007) (retrieved August 25, 2008),
http://www.sciencedirect.com

Motley Fool (2008) (Retrieved September 14, 2008),
http://www.fool.com/investing/dividends-income/
2008/04/04/take-that-lehman-short-sellers

Oppenheimer, M., O'Neill, B.C., Webster, M., Agrawala, S.: Climate change: The limits of consensus. Science 317, 1505–1506 (2007)

Pfeifer, R., Lungarella, M., Lida, F.: Review: Self-organization, embodiment, and biologically inspired robots. Science 318, 1088–1093 (2007)

Ranson, D.: The Fed's Predicament. Op-Ed, Wall Street J, p. A-11 (December 24, 2007)

Rees, G., Frackowiak, R., Frith, C.: Two modulatory effects of attention that mediate object categorization in human cortex. Science 275, 835–838 (1997)

Rieffel, E.G.: Certainty and uncertainty in quantum information processing. In: Quantum Interaction: AAAI Spring Symposium, Stanford U. AAAI Press, Menlo Park (2007)

Roethlisberger, F.J., Dickson, W.J.: Mgt. and the worker. HUP, Cambridge (1939)

Rouse, W.B., Boff, K.R.: Cost-benefit analysis for human systems integration. In: Handbook of Human Systems Integration, ch. 17, pp. 631–657. Wiley, Chichester (2003)

Sanfey, A.G.: Social decision-making: Insights from game theory and neuroscience. Science 318, 598–602 (2007)

Slackman, M.: The New York Times, p. 10 (September 7, 2008)

Smith, W.K., Tushman, M.L.: Organizational Science 16(5), 522–536 (2005)

Sommer, R.A.: Architecting cross-functional business processes: new views on traditional business process reengineering. Int'l. J. Mgt. Enterprise Dev. 1(4), 345–358 (2004)

Sood, A., Tellis, G.J.: Do innovations really pay off? (forthcoming, 2008); Forthcoming in Marketing Science (Retrieved September 15, 2008), SSRN:
http://ssrn.com/abstract=1121005

Tajfel, H.: Experiments in intergroup discrimination. Sci. Am. 223(2), 96–102 (1970)

Trueman, T.: Former Fed Off. Lehman Bro., The Wall Street Journal, p. A1,14 (September 12, 2008)

Von Neumann, J., Morgenstern, O.: Theory of games and economic behavior. Princeton University Press, Princeton (1953)

Wickens, C.D.: Engineering psychology and human performance, 2nd edn. Merrill Publishing, Colombus (1992)

Willis, R.P.: The counterinsurgency pattern assessment (CIPA) Program 2007. NRL (2007)

Wood, J., Tung, H.-L., Marshall-Bradley, T., Sofge, D.A., Grayson, J., Lawless, W.F.: Applying an Org. Uncertainty Principle: Semantic Web-Based Metrics. In: Manuela Cunha, M., Oliveira, E., Tavares, A., Ferreira, L. (eds.) Handbook of Research on Soc. Dimensions of Semantic Techn. and Web Services. IGI, Hershey (forthcoming, 2009)

On Voting Process and Quantum Mechanics

François Dubois

Conservatoire National des Arts et Métiers, Department of Mathematics, Paris,
France, and Association Française de Science des Systèmes
duboisf@cnam.fr
11 January 2009

Abstract. In this communication, we propose a tentative to set the
fundamental problem of measuring process done by a large structure on
a microscopic one. We consider the example of voting when an entire
society tries to measure globally opinions of all social actors in order to
elect a delegate. We present a quantum model to interpret an operational
voting system and propose an quantum approach for grading step of
Range Voting, developed by M. Balinski and R. Laraki in 2007.

[Quantum Interaction 2009, Saarbruecken, 25-27 March 2009].

Keywords: Fractaquantum hypothesis, Range Voting, Information Retrieval, Gleason theorem.

1 Measure Process between Different Scales

Matter is constituted by discrete quanta and this fact was empirically put in evidence by E. Rutherford in the beginning of 20th century. Microscopic quanta as
classical atoms or photons are not directly perceptible by our senses, as pointed
out by M. Mugur-Schächter [MMS08]. In consequence, any possible knowledge
for a human observer of a microscopic quantum is founded on experimental
protocols. The mathematical framework constructed during the 20th century describes unitary "free evolution" through the Schrödinger equation and "reduction
of the wave packet" associated to measure process through a projection operator
in Hilbert space. We refer the reader *e.g.* to the book of C. Cohen-Tannoudji
et al [CDL77]. The philosophical consequences of this new vision of Nature are
still under construction; in some sense, an *a priori* or an external description
of Nature is not possible at quantum scale. We refer to B. D'Espagnat [DE02]
and M. Bitbol [Bi96]. Independently of the development of this renewed physics,
the importance of scale invariance have been recognized by various authors as
B. Mandelbrot [Ma82] and L. Nottale [No98]. The word "fractal" is devoted to
figures and properties that are self-similar whatever the refering scale.

We have suggested in 2002 the fractaquantum hypothesis [Du02], founded
on two remarks: Nature develops a scale invariance and quantum mechanics
is completely relevant for small scales. In order to express this hypothesis, we
have introduced (see *e.g.* [Du05, Du08a]) the notion of "atom", in fact very
similar to the way of vision of Democrite and the ancient Greek philosophers

P. Bruza et al. (Eds.): QI 2009, LNAI 5494, pp. 200–210, 2009.

(see *e.g.* J. Salem [Sa97]). To fix the ideas, an "atom" can be a classical atom, or its nucleus, or a molecule, or a micro-organism like a cell, or an entire macro-organism as a human being or till an entire society! If we divide an "atom" into two parts, its qualitative properties change strongly at least in one of these parts. With this framework, elementary components are supposed to exist in Nature at different scales. A classical atom is a "micro state" relative to a Human observer. In this particular case, a ℓittle "atom" ℓ is a classical atom and a Big "atom" B is a human observer. More generally, two "atoms" ℓ and B have different scales when "atom" ℓ is not directly perceptible to "atom" B. In other words, a direct interaction between B and ℓ can not be controlled by B himself. In this case, the direct interaction between little "atom" ℓ and big "atom" B can be neglected as a first order approximation.

In this contribution, we suggest to revisit this classical quantum formalism when little and big "atoms" are nonclassical ones. In fact, this research program is tremendous! For similar programs, we refer *e.g.* to the works of G. Vitiello [Vi01], P. Bruza *et al* [BKNE08], A. Khrennikov and E. Haven [KH07], P. La Mura *et al* [LMS07]. The phenomenology of possible measurement interactions should be reconstructed. What is a big "atom" B that can measure some quantities on little "atom" ℓ? Does the classical framework of quantum mechanics operates without any modification? Of course all these questions motivate our communication. Due to the lack of knowledge of what can be a measure done by "atoms" at mesoscopic or microscopic scales, we restrict ourselves in this contribution to measures done by human society considered as a whole on individual human beings.

We consider here a particular example of the measurement process associated with voting. In this case, "atom" ℓ is a social actor and "atom" B is the entire society. We first introduce the scientific problem of voting process and in the following section, we present a preliminary quantum model for voting. In the two following sections we describe with the help of fractaquantum hypothesis the range voting procedure ("vote par valeurs") developed independently by M. Balinski and R. Laraki [BL07a] at Ecole Polytechnique (Paris) and by W.D. Smith [Sm07, RS07] at the "Center of Range Voting" (Stony Brook, New York).

2   On the Voting Process

We consider a macroscopic "atom" B composed by an entire social structure. For example, B is a state like France to fix the ideas. The social actors of society B are the little "atoms" ℓ in our model. We write here

$$\ell \in B \tag{1}$$

even if the expression (1) does not take precisely into account the detailed structure of society B. The numbers of such indistinguable individuals are quite important (10^6 to 10^9 typically). The democratic life in society B suppose that social responsabilities are taken by elected representants of social corpus. Thus a voting process has the objective to determine one particular social actor among all for accepting social responsabilities. This kind of position is supposed to be attractive and a set Γ of candidates γ among the entire set of "atoms" ℓ is supposed to be given in our framework.

The problem is to determine a single "elected" candidate γ_1 among the family Γ thanks to the synthesis of all opinions of different electors ℓ. The social objective of society B is the determination of one candidate among others through a social process managed by the entire society, modelized here as a macro "atom" B. This problem is highly ill posed and we refer to the pioneering works of J.C. de Borda [1781] and N. de Condorcet [1785] followed more recently by the theorem of non existence of a social welfare function satisfying reasonable hypotheses, proved by K. Arrow [Ar51]. We describe this result in the following of this section.

With K. Arrow, we suppose that each elector ℓ determines some ordering denoted by $\succ_{\sigma_\ell}$ (or simply by σ_ℓ) among the candidates $\gamma \in \Gamma$:

$$\gamma_{\sigma_l(1)} \succ_{\sigma_\ell} \gamma_{\sigma_l(2)} \succ_{\sigma_\ell} \cdots \gamma_{\sigma_l(i)} \succ_{\sigma_\ell} \gamma_{\sigma_l(i+1)} \cdots \succ_{\sigma_\ell} \gamma_{\sigma_l(K)}, \quad \ell \in B.$$

We consider now the set σ of **all** orderings σ_l for all the electors ℓ

$$\sigma = \{\sigma_\ell, \ \sigma_\ell \text{ ordering of candidates } \Gamma, \ \ell \in B\}.$$

A so-called social welfare function f determines a particular social ordering $\sigma^* = f(\sigma)$ as a global synthesis of all orderings σ_ℓ in order to construct a commun and socially coherent position. Some democratic properties are *a priori* required for this function f:

(i) *Unanimity*
If everybody thinks that candidate γ is better than γ' the social choice must satisfy this property:

$$\text{If } (\forall \ell \in B, \ \gamma \succ_{\sigma_\ell} \gamma') \quad \text{for some } \gamma, \gamma' \in \Gamma, \quad \text{then } (\gamma \succ_{\sigma^*} \gamma'). \qquad (2)$$

(ii) *Independance of irrelevant alternatives*
Consider two orderings σ and τ grading in a similar way the two candidates γ and γ':

$$((\gamma \succ_{\sigma_\ell} \gamma') \text{ and } (\gamma \succ_{\tau_\ell} \gamma')) \text{ or } ((\gamma \prec_{\sigma_\ell} \gamma') \text{ and } (\gamma \prec_{\tau_\ell} \gamma')), \quad \forall \ell \in B. \qquad (3)$$

Then the social orderings $\sigma^* = f(\sigma)$ and $\tau^* = f(\sigma)$ must satisfy the corresponding property:

$$\begin{cases} \gamma \succ_{\sigma^*} \gamma \quad \text{when } ((\gamma \succ_{\sigma_\ell} \gamma') \text{ and } (\gamma \succ_{\tau_\ell} \gamma')) \\ \qquad \text{or} \\ \gamma \prec_{\sigma^*} \gamma \quad \text{when } ((\gamma \prec_{\sigma_\ell} \gamma') \text{ and } (\gamma \prec_{\tau_\ell} \gamma')). \end{cases} \qquad (4)$$

The social welfare function depends only on the relative ranking and not on the intermediate candidates.

Then the Arrow impossibility theorem (proven elegantly by J. Geanakoplos in [Ge01]) implies that under conditions (2) of unanimity and (3)-(4) of independance of irrelevant alternatives, the social welfare function is simply a constant:

(iii)   **Dictatorship**

$$\exists d \in \Gamma, \quad f(\{\sigma_\ell, \ell \in B\}) \equiv \sigma_d \tag{5}$$

and the result is a dictature! In other terms, it is impossible to construct a social welfare function that has the two first properties of unanimity and independance of irrelevant alternatives and the non-dictatorship property, obtained by negation of (5).

3   A Preliminary Quantum Model for Voting

We describe in this Section a quantum model presented in [Du08b]. We restrict here to the so-called "first tour" process as implemented in a lot of situations. In this process, each elector ℓ has to transmit the name of at most **one** candidate γ. Then an ordered list of candidates is obtained by counting the number of expressed votes for each candidate. Introduce the space H_Γ of candidates generated formally by the finite family Γ of all candidates:

$$H_\Gamma = \bigoplus_{\gamma \in \Gamma} \mathbb{C}\,|\gamma> \tag{6}$$

where $\mathbb{C}$ denotes the field of complex numbers. This decomposition (6) is supposed to be orthogonal:

$$<\gamma|\gamma'> = \begin{cases} 0 \text{ if } \gamma \neq \gamma' \\ 1 \text{ if } \gamma = \gamma', \end{cases} \quad \gamma, \gamma' \in \Gamma.$$

The "wave function" associated with an elector ℓ is represented by a state denoted by $|\ell>$ in this space H_Γ:

$$|\ell> = \sum_{\gamma \in \Gamma} |\gamma><\ell|\gamma> . \tag{7}$$

The scalar product $<\ell|\gamma>$ in relation (7) is the component of elector ℓ relative to each candidate γ. This number represents the political sympathy of elector ℓ relative to the candidate γ. We suppose here that the norm $\|\ell\|$ of state $|\ell>$ *id est*

$$\|\ell\| \equiv \sqrt{\sum_{\gamma \in \Gamma} |<\ell|\gamma>|^2}$$

is **inferior or equal** to unity. We follow the Born rule and suggest that the probability for elector ℓ to give its vote to candidate γ is equal to $|<\ell|\gamma>|^2$. We suggest also that the probability to unswer by a vote "blank or null" is $1 - \|\ell\|^2$ in this framework.

The interpretation of the projection process in the quantum measurement for such a first tour of election process is quite clear. During the election, *id est* the particular day where the measure process occurs, the elector ℓ is **obliged** to choose at most one candidate γ_0. In consequence, all his political sensibility is socially "reduced" to this particular candidate. We can write:

$$|\ell> = |\gamma_0>$$

to express the wave function collapse. This quantum interpretation of such voting process clearly shows the **violence** of such king of decision making. Of course, no elector has political opinions that are identical to one precise candidate and this measurement process is a true mathematical projection. Nevertheless, the operational social voting process imposes this projection in order to construct a social choice. The disadvantage and dangers of such process have been clearly demonstrated in France during the presidential election process in 2002 (see *e.g.* [wiki]).

4 Range Voting (i): Quantum Approach for Grading Step

The voting process suggested by M. Balinski and R. Laraki [BL07a] is more complex than the one studied in the previous section. The key point in order to overcome the Arrow impossibility theorem is the fact that in this framework the opinion of electors among the candidates are **codified** by society B through a given set of so-called "grades". These grades are *a priori* very similar to the ones given by the scolar system, as integers between 0 and 20 in France with an associated order

$$0 \prec 1 \prec \ldots \prec j \prec j+1 \prec \ldots \prec 19 \prec 20 ,$$

letters from A to F in the United States with an order

$$A \succ B \succ C \succ D \succ E \succ F ,$$

or numbers from 1 to 6 in Germany with the following (mathematically unusual!) order

$$1 \succ 2 \succ 3 \succ 4 \succ 5 \succ 6 .$$

These grades can be also an ordered list of given words

"very good" $\succ$ "good" $\succ$ "not so bad" $\succ$

$\succ$ "passable" $\succ$ "insufficient" $\succ$ "to be rejected"

as proposed by the previous authors [BL07b] in Orsay experiment for French presidential election in 2007. These grades define an elementary **common language** that is supposed to be endowed by all social actors ℓ of society B. In other terms, a common ordered set G of grades ν is supposed to be given:

$$\nu_1 \succ \nu_2 \succ \ldots \succ \nu_K , \quad \nu_j \in G . \tag{8}$$

As a consequence, an ordering of opinions explicitly refer to this particular set of given grades and to an explicit ordering between these grades like in (8). Remind that in Balinski-Laraki process [BL07a], the society B imposes a commun grading referential to all electors.

The ranking process between the candidates proceeds by two steps. First each elector gives a grade to each candidate. Secondly the candidates are arranged in order through "majority ranking". Each elector ℓ has to express an opinion relative to each candidate $\gamma \in \Gamma$ through a grade $g(\gamma, \ell) \in G$. During the day of the election as in [BL07b], each elector grades each candidate. We propose in this section a quantum model for the first step of this processus. This first step is a measure done by society B on each little "atom" ℓ which constitutes it, as suggested by relation (1). Observe now that each candidate γ has a published political program, is giving radio and television interviews, has a blog, *etc.* We introduce a "political Hilbert space" H_P that refer to **all** this set of political information, following modern approaches for Information Retrieval as suggested by K. von Rijsbergen [vR04]. The family G of grades is imposed by the general laws of society B. Nevertheless, the evaluation of the political program of all candidates is done by the elector ℓ himself in such a process! We suggest that each elector ℓ decomposes this Hilbert space H_P into "grading" orthogonal components E_ν^ℓ through his own **internal** process:

$$H_P = \bigoplus_{\nu \in G} E_\nu^\ell, \quad \ell \in \mathrm{B}. \tag{9}$$

The subspace E_ν^ℓ is the eigenspace giving the grade ν relative to the opinion of elector ℓ. If we denote by A^ℓ the quantum self-adjoint operator associated with the grading process done by elector ℓ, we have

$$A^\ell \bullet |\xi> = \nu\, |\xi>, \quad |\xi> \in E_\nu^\ell \subset H_P, \quad \nu \in G. \tag{10}$$

In other words, we introduce the orthogonal projector P_ν^ℓ onto the closed space E_ν^ℓ. Then these projectors commute

$$P_\nu^\ell\, P_{\nu'}^\ell = P_{\nu'}^\ell\, P_\nu^\ell, \quad \nu, \nu' \in G, \ \ell \in \mathrm{B}$$

and generate a decomposition of the identity operator $\mathrm{Id}(H_P)$ in the political Hilbert space H_P:

$$\sum_{\nu \in G} P_\nu^\ell \equiv \mathrm{Id}(H_P), \quad \ell \in \mathrm{B}. \tag{11}$$

On a very concrete point of view, in front of each political idea, each elector has the capability to give an opinion in the language suggested *a priori* by the set G of grades. The examples of such sets given above show also that the way of decomposition of political space H_P through the grades is strongly influenced by the social choice of the family G.

In some sense, *via* a particular choice of grading, the society B imposes some filtering of space H_P of all political data. Note that the precise way this filter is done depends on each citizen ℓ. In this model, society B imposes the set

G of eigenvalues and each elector ℓ fixes the eigenvectors as in (10). After the elector has interpreted the grades ν in his own vocabulary, *id est* once he has decomposed the space H_P into orthogonal components, we suppose that the grading process, *id est* the result of the measure is *a priori* obtained according to the Born rule. Precisely, we introduce the "perception" ρ^ℓ_γ of political opinion of candidate γ by the elector ℓ. Mathematically speaking, the elector ℓ measurates the political ideas of the candidate γ in a quantum way relatively to the Hilbert space H_P. According to Gleason theorem [Gl57], such a quantum probability is defined by a density matrix, *id est* a positive self-adjoint operator of unity-trace that we denotes also by ρ^ℓ_γ :

$$\rho^\ell_\gamma \text{ positive self-adjoint operator } H_P \longrightarrow H_P, \quad \text{tr}\left(\rho^\ell_\gamma\right) = 1.$$

Then, following A. Gleason [Gl57] and K. von Rijsbergen [vR04], the measure μ^ℓ_γ associated with elector ℓ and candidate γ of any closed subspace $E \subset H_P$ is given in all generality according to

$$\mu^\ell_\gamma(E) = \text{tr}\left(\rho^\ell_\gamma\, P_E\right), \quad E \subset H_P, \quad \ell \in B, \tag{12}$$

where P_E is the orthogonal projector onto space E. Consider now the space $E = E^\ell_\nu$ introduced in (9). Then the (real!) number $\mu^\ell_{\gamma,\nu}$ defined by

$$\mu^\ell_{\gamma,\nu} = \mu^\ell_\gamma(E^\ell_\nu) = \text{tr}\left(\rho^\ell_\gamma\, P^\ell_\nu\right) \tag{13}$$

represents the quantum probability for elector ℓ to give the grade ν to candidate γ. Of course, if we insert the identity operator $\text{Id}(H_P)$ decomposed in (11) inside relation (12), we have due to (13)

$$\sum_{\nu \in G} \mu^\ell_{\gamma,\nu} = 1, \quad \ell \in B, \; \gamma \in \Gamma, \tag{14}$$

and the sum of probabilities for all different grades is equal to unity.

Remark that two different ingredients are necessary to determine the previous probability $\mu^\ell_{\gamma,\nu}$ in (13). First the decomposition (9) of the political space through the grades G. As usual in quantum mechanics, no detailed structure of "atom" ℓ is transmitted through the measure process. In this case, the orthogonal decomposition (9) is not known by the society. Second the "perception operator" ρ^ℓ_γ which represents in some sense the particular "political knowledge" that the elector ℓ has constructed for himself about the candidate γ. Remark that no direct interaction between the candidates occurs in the model. According to Condorcet's ideas [1795], each citizen is adult has make his own opinion through his own way of thinking!

5   Range Voting (ii): Majority Ranking

After this first step of grading, the result of the vote of elector ℓ is a list

$$g(\gamma, \ell) \in G, \quad \gamma \in \Gamma, \quad \ell \in B$$

of grades $\nu = g(\gamma, \ell)$ given by elector ℓ to **each** candidate γ. We give in this section the major points introduced By Balinski and Laraki [BL07a] without any modification. After summation, each candidate γ has a certain number $n_\nu^\gamma \in \mathbb{N}$ of opinions transmitted by the electors:

$$n_\nu^\gamma = \text{Card } \{\ell \in B, \ g(\gamma, \ell) = \nu\} \in \mathbb{N}, \quad \gamma \in \Gamma, \quad \nu \in G. \tag{15}$$

The way of ranking such a list

$$n^\gamma \equiv \left(n_{\nu_1}^\gamma, n_{\nu_2}^\gamma, \ldots n_{\nu_K}^\gamma\right) \in \mathbb{N}^K, \quad \gamma \in \Gamma \tag{16}$$

when the grades $\nu \in G$ are arranged in order without ambiguity by (8) can be explicited with the so-called "majority ranking" introduced by Balinski and Laraki [BL07a]. We give here some details of the algorithm, based on a successive extraction of a **median value** from a list as the one described in (16) and refer to [BL07a], [BL07b] and [PB06].

From an algorithmic point of view, the list n^γ can also be written as a list m^γ of grades written in decreasing order to fix the ideas:

$$m^\gamma = \Big(\underbrace{\nu_1, \ldots, \nu_1}_{n_{\nu_1}^\gamma \text{ times}}, \underbrace{\nu_2, \ldots, \nu_2}_{n_{\nu_2}^\gamma \text{ times}}, \ldots, \underbrace{\nu_K, \ldots, \nu_K}_{n_{\nu_K}^\gamma \text{ times}} \Big) \in \mathbb{N}^{|B|} \tag{17}$$

where $|B| = \text{Card}(B)$ is the number of electors. Then a list m_1^γ can be constructed by omitting the grade $\nu_{j_1}^\gamma$ at the **median** position $\frac{|B|}{2}$ inside the list (17). We obtain in this way a new list extracted from (17)

$$m_1^\gamma = \Big(\underbrace{\nu_1, \ldots, \nu_1}_{n_{1,\nu_1}^\gamma \text{ times}}, \underbrace{\nu_2, \ldots, \nu_2}_{n_{1,\nu_2}^\gamma \text{ times}}, \ldots, \underbrace{\nu_K, \ldots, \nu_K}_{n_{1,\nu_K}^\gamma \text{ times}} \Big) \in \mathbb{N}^{|B|-1} \tag{18}$$

and the integers n_{1,ν_i}^γ are equal to the $n_{\nu_i}^\gamma$ except for index j_1 for which we have

$$n_{1,\nu_{j_1}^\gamma}^\gamma = n_{\nu_{j_1}^\gamma}^\gamma - 1.$$

The grade $\nu_{j_1}^\gamma$ is the first "majority grade" of candidate γ in the majority ranking algorithm of Balinski and Laraki. If $\nu_{j_1}^\gamma \succ \nu_{j_1}^{\gamma'}$ then we have the relative final position $\gamma \succ \gamma'$ between the candidates γ and γ'. If $\nu_{j_1}^\gamma = \nu_{j_1}^{\gamma'}$ we apply the same step from (17) to (18) except that we start with the list (18). Doing this, we extract a second grade $\nu_{j_2}^\gamma$ for each candidate γ. If $\nu_{j_2}^\gamma \succ \nu_{j_2}^{\gamma'}$ or $\nu_{j_2}^\gamma \prec \nu_{j_2}^{\gamma'}$, the conclusion is established. Otherwise the process is carried on until the two majority grades at a certain step are distinct.

It is a main contribution of M. Balinski and R. Laraki [BL07a] to extract an intrinsic order

$$\gamma_1 \succ \gamma_2 \succ \ldots \gamma_j \succ \gamma_{j+1} \succ \ldots, \quad \gamma_j \in \Gamma$$

among the candidates Γ from the given double list (16) of integers n^γ. The important social fact is that the overdetermination of a favorite candidate essentially does **not** influence the final majoritary ranking with this grading method! The proof of this important fact is omitted here and we refer to [BL07a]. We could also think that there is a contradiction between this positive result and the Arrow impossibility theorem. In fact, as pointed in [BL07a], the hypotheses of Arrow theorem are qualitative: each elector consider some ordering of the candidates with his own sensibility. As we have intensively explained with the orthogonal decomposition (9), the social choice of a **given** family of grades is essential for the grading step and the majority ranking.

6   Conclusion

The very elaborated process initialized by M. Balinski and R. Laraki [BL07a] for range voting has been studied in this contribution. The second step of "majority ranking" has been described without adding any new idea to this beautiful article. Concerning the first step of the algorithm devoted to the grading of each candidate by each elector with a given list of grades, we have proposed a quantum algorithm essentially based on modern quantum approaches for Information Retrieval presented in K. von Rijsbergen's book [vR04]. First an orthogonal decomposition of the political Hilbert space supposes that each elector has the capability to have a precise opinion for each political subject. Second, following Gleason theorem [Gl57], we have introduced a "perception operator" that describes mathematically the way a given candidate is politically understood by a given elector. In some sense, a psychological model is incorporated with this description.

With these two ingredients, the computation of the probability for an elector to give a particular grade to each candidate can be evaluated as a result of the model. Of course, it is not actually clear which precise practical advantages has this quantum approach in the description of the voting process. Moreover, we want to find in future works some previsions of the quantum model, and try to compare it with the previsions of a classic model.

In this contribution, we have also presented a first quantum model of a classical election. In this framework, the big scale (the society) imposes a direct generalization of the measure process in quantum mechanics. All the characteristics of the mathematical measure operator are controlled by the large scale. We have noticed the violence of the multiscale interaction through such a the measuring process.

Last but not least, this work is motivated by the fractaquatum hypothesis [Du02]. The case of a voting process is an example of measuring process between two different scales in Nature. If we suppose that the general concepts of quantum mechanics have an extension to all "atoms" in Nature, the process of measuring has to be re-visited to all pairs of "atoms" with different scales. This contribution is a small step in this direction!

Acknowledgments

The author thanks the referees who pointed clearly the importance of Information Retrieval framework for this work and proposed a list of very interesting remarks and an important number of which have been incorporated in the present writing!

References

[Ar51]   Arrow, K.J.: Social Choice and Individual Values. J. Wiley and Sons, New York (1951)

[BL07a]   Balinski, M., Laraki, R.: A theory of measuring, electing and ranking. Proceeding of the National Academy of Sciences of the USA 104(21), 8720–8725 (2007)

[BL07b]   Balinski, M., Laraki, R.: Le Jugement Majoritaire : l'Expérience d'Orsay. Commentaire 30(118), 413–420 (summer 2007)

[Bi96]   Bitbol, M.: Mécanique quantique, une introduction philosophique. Champs-Flammarion, Paris (1996)

[1781]   de Borda, J.C.: Mémoire sur les lections au scrutin. In: Histoire de l'Académie Royale des Sciences, Paris (1781)

[BKNE08]   Bruza, P.D., Kitto, K., Nelson, D., McEvoy, C.L.: Entangling Words and Meaning. In: Proceedings of the Second Quantum Interaction Symposium, pp. 118–124. College Publications, Oxford (2008)

[CDL77]   Cohen-Tannoudji, C., Diu, B., Laloë, F.: Mécanique quantique. Hermann, Paris (1977)

[1785]   de Condorcet, N.: Essai sur l'application de l'analyse à la probabilité des décisions rendues à la pluralié des voix. Imprimerie Royale, Paris (1785)

[1795]   de Condorcet, N.: Esquisse d'un tableau historique des progrès de l'esprit humain. In: Daunou, P.C.F., de Condorcet, M.L.S. (eds.). Agasse, Paris (1795)

[DE02]   D'Espagnat, B.: Traité de physique et de philosophie. Fayard, Paris (2002)

[Du02]   Dubois, F.: Hypothèse fractaquantique. In: Res-Systemica, 5th European Congress of System Science, Heraklion, vol. 2 (October 2002)

[Du05]   Dubois, F.: On fractaquantum hypothesis. In: Res-Systemica, 6th European Congress of System Science, Paris, vol. 5 (September 2005)

[Du08a]   Dubois, F.: Could Nature be quantum at all scales? (preprint) (April 2008)

[Du08b]   Dubois, F.: On the measure process between different scales. In: Res-Systemica, 7th European Congress of System Science, Lisboa, vol. 7 (December 2008)

[Ge01]   Geanakoplos, J.: Three brief proofs of Arrow's Impossibility Theorem. Economic Theory 26(1), 211–215 (2005)

[Gl57]   Gleason, A.M.: Measures on the Closed Subspaces of a Hilbert Space. Indiana University Mathematics Journal (Journal of Mathematics and Mechanics) 6, 885–893 (1957)

[KH07]   Khrennikov, A.Y., Haven, E.: The importance of probability interference in social science: rationale and experiment (September 2007), arXiv:0709.2802

[LMS07]   La Mura, P., Swiatczak, L.: Markovian Entanglement Networks. Leizig Graduate School of Management (2007)

[Ma82]   Mandelbrot, B.: The Fractal Geometry of Nature. W. H. Freeman and Co.,
           New York (1982)
[MMS08]   Mugur-Schächter, M.: Infra-mécanique quantique. Quantum Physics (Jan-
           uary 2008), arXiv:0801.1893
[No98]   Nottale, L.: La Relativité dans tous ses états: au delà de l'Espace-Temps.
           Hachette, Paris (1998)
[RS07]   Rivest, R.L., Smith, W.D.: Three Voting Protocols: ThreeBallot, VAV, and
           Twin. In: Proceedings of the Electronic Voting Technology 2007, Boston,
           MA, August 6 (2007)
[Sa97]   Salem, J.: L'Atomisme antique. Démocrite, Epicure, Lucrèce. Hachette,
           Paris (1997)
[Sm07]   Smith, W.D.: Range Voting satisfies properties that no rank-order system
           can (April 2007)
[PB06]   Peynaud, E., Blouin, J.: Le goût du vin. Dunod, Paris (2006)
[vR04]   van Rijsbergen, C.J.: The Geometry of Information Retrieval. Cambridge
           University Press, Cambridge (2004)
[Vi01]   Vitiello, G.: My Double Unveiled. In: Advances in Consciousness Research.
           John Benjamins Publishing Company, Amsterdam (2001)
[wiki]   Wikipedia (2002),
           http://en.wikipedia.org/wiki/French_presidential_election,_2002

Nonseparability of Shared Intentionality

Christian Flender, Kirsty Kitto, and Peter Bruza

Faculty of Science and Technology,
Queensland University of Technology,
Brisbane, Australia
{c.flender,kirsty.kitto,p.bruza}@qut.edu.au

Abstract. According to recent studies in developmental psychology and neuroscience, symbolic language is essentially intersubjective. Empathetically relating to others renders possible the acquisition of linguistic constructs. Intersubjectivity develops in early ontogenetic life when interactions between mother and infant mutually shape their relatedness. Empirical findings suggest that the shared attention and intention involved in those interactions is sustained as it becomes internalized and embodied. Symbolic language is derivative and emerges from shared intentionality. In this paper, we present a formalization of shared intentionality based upon a quantum approach. From a phenomenological viewpoint, we investigate the nonseparable, dynamic and sustainable nature of social cognition and evaluate the appropriateness of quantum interaction for modelling intersubjectivity.

1 Introduction

How do we relate to others and how do social interactions shape our worldview? What are the processes underlying everyday social encounters and how do these processes contribute to enculturation and our use of symbolic language? Questions of this kind have been of considerable interest ever since the social world became the object of investigation. More recently, studies in developmental psychology and cognitive neuroscience have shed new light on how we interact with each other and the world we inhabit. In conjunction with phenomenological descriptions, potential scientific explanations of intersubjectivity emphasize the role of social perception and the importance of sustained interactions between social beings. From the earliest age, human infants relate to their mothers in an embodied and imitative way which lays the foundation for them to grow into social and cultural reality. The distinction of self and other as well as linguistic abilities are claimed to emerge from an ongoing interaction with other humans [1,2,3,4,5,6].

In adulthood, developmental perspectives also provide evidence of how we are able to adopt a detached or third person view of the world. Experiencing concrete categories like *book* or abstract concepts like *democracy* as distinct or independent from us as observers, is like these objects being there for everyone [7,8,9]. If we perceive others, for instance in a conversation, we do this by presupposing our partners to be subjects like us, but in a way that transcends our

P. Bruza et al. (Eds.): QI 2009, LNAI 5494, pp. 211–224, 2009.

subjectivity due to the other being there for, or accessible to, everyone else. It is this presupposed intersubjectivity that facilitates a shared world of objects as the foundation for enculturation including the development of symbolic language.

Language and the sense of self and other develop in early ontogenetic life through an ongoing interaction between mother and infant. Even before the infant develops a deliberate sense of self, there is a bodily pairing between mother and infant that is characterized by an intermodal link between action, in particular motor behaviour, and perception of the mother [10,11]. It is this perceptual, practical and self-engaging sense of other agents which is essentially not separable into ego and alter ego (self and other) and maintained throughout life. There is always some degree of undifferentiated identification; self and other are never fully distinguished [12].

In this paper, we want to make explicit this presuppositional sense of the other by formalizing shared intentionality using notions borrowed from quantum mechanics. Quantum formalisms lend themselves to model interactive, context-dependent and emerging phenomena [13,14,15,16] and could provide considerable help in developing a more precise understanding of intersubjectivity.

We proceed as follows. In the next section, we introduce intersubjectivity by means of two important concepts. Firstly, *social perception* emphasizes the direct or non-inferential character of social interaction. Secondly, the discovery of *mirror neurons* in cognitive neuroscience supports social perception. Mirror neurons provide an intermodal bridge between action and perception. Acting and perceiving someone else performing are two nonseparable and intentional concepts which we will formally introduce in Section 3. We will show that states involved in mother-infant interactions are essentially nonseparable and that this can be represented as a an entangled quantum state. The mutual anticipation of the other's reaction involves nonseparable states of an emerging interaction process. In Section 4, we conclude that entangled states must derive from the nonseparable time evolutions that govern the dynamic co-emergence [17] of shared intentional states and intentional states of mother and infant. Departing from our developed notion of shared intentionality, in Section 5, we discuss intersubjectivity from the perspective of phenomenology. We contrast intersubjectivity as empathy with intersubjectivity as co-subjectivity and integrate both dimensions under the umbrella of shared intentionality. Lastly, we give a conclusion and outlook toward future work.

2   Intersubjectivity

In this section, we introduce intersubjectivity or shared intentionality according to recent discussions in developmental psychology and neuroscience. In sharing intentions, agents perceive each other based upon a multiplicity of states such as emotions or somatic sensations [18]. Mirror neurons support the nonseparability of such states as they link one's own actions with perception of someone else's movements and actions.

2.1 Social Perception

Classical theories of social cognition have been criticized for overintellectualizing social cognition by underestimating perception [10,19,20]. The two dominant theories, Theory theory (TT) and simulation theory (ST), require subjects to add extra inferential mechanisms in order to understand the other. However, experiments reveal infants reacting with facial expressions or patterns of vocalization and gestures to affordances such as movement or sounds without a need for theory or models [19].

Once the child develops and acquires concepts, the initial smartness of early infant perception (that is perception as inherently active, direct and non-inferential) is maintained [10,11,21,22]. I still recognize my friend as being my friend without necessarily attributing him with a certain attitude towards me. Hence, neither the smartness nor the directness of perception is necessarily dependent on perceiving things under concepts or judgement [10]. Put another way, perception itself is conceptual but not in the sense of explicit deliberate judgement [23]. Perception essentially gets shaped or informed by the ongoing interaction between intentionalities [20]. Of course, the role of perception does not rule out deliberation affecting perception, however, in many social encounters the smartness of direct perception mirrors our ability to skilfully interact with others.

Later in ontogenetic life, perception is also informed by the language and concepts we learn. Empirical studies show that children acquire linguistic abilities through language use [3,2,1] and thus through social interaction. Obviously, this presupposes the social community the child is part of. Moreover, the community itself embodies conventionalized conceptual structure derived from language use. Learning words intersubjectively in early ontogenetic life is essentially a process of extracting elements from the larger linguistic construction of adults. After months of gestural and vocal interaction, most Western middle-class children begin producing linguistic utterances in the months following the first birthday [1]. Such first expressions (holophrases), which are mainly declarative statements, imperative requests and interrogative questions, are learned and used in the same intentional context as for the perceptual or non-linguistic intentionalities during the first year of their life. From there, more complex constructions develop. For instance, conceptual integration, the formation of abstract concepts accross episodes or the reflexive adoption of the perspective of the listener. It is the latter which has only recently been taken seriously into account.

2.2 Mirror Neurons

A phenomenology of intersubjectivity reveals that social interaction is often not inferential. Others have argued against explicit deliberate judgement or deductive reasoning as the only form of sense-making [24,25,26]. When we are absorbed in complementary communicative gestures, dance-like behaviours or language games (in a Wittgensteinian sense), there is no need for detached and deliberate reasoning. However, this does not deny complex processes on a subpersonal or neural level. On a subpersonal or unconsciouss level, mirror neurons are activated both when an agent acts and when this agent perceives someone else acting [27]. Mirror neurons integrate both motor and perceptual properties, i.e. they tightly couple observed behaviour with one's own performance. For instance, in monkeys cells discharge when an experimenter places food in front of the monkey and when the monkey reaches for it [28]. More recently, mirror neurons have received much scientific interest since they do not just mirror observed physical behaviour, but purposeful action. Hence, they support understanding others to be like oneself. Furthermore, it was shown that mirror neurons are not just receptive to the teleological structure of an action, but also to the style or manner of the action in which the goal is achieved [29]. Although mirror neurons were first discovered in the brain area F5 of the ventral premotor cortex of monkeys, there is strong evidence for their existence in humans [28,29,30]. Even more striking is that area F5 in monkeys is homologous to Broca's area in humans, a part of the brain associated with linguistic ability and verbal communication [29,31].

In mother-infant interactions, it has been argued that the mirror system is a necessary though not sufficient condition for imitation [30]. From an early age on children perceive intentions of their mother. In their ability to imitate behaviour, an ability in which humans differ from monkeys and other primates [19], children directly perceive the other without consciously inferring what the mother's goals are. Direct perception in social interaction is facilitated through an intermodal bridge between observing the other and enacting what the other does. Perception of others and motor action are nonseparably mapped through the underlying mirror system [30,10].

3   Sharing Intentions

Infants responsiveness to facial expressions, gestures or sounds and their ability to imitate is not based upon inference, analogy or simulation. It is the direct perception of others and the intentional potentialities of their own body that facilitates understanding others as animate beings like themselves [11]. Mirror neurons are supportive of this view since they provide an intermodal and nonseparable link between action and perception. In the following, we develop a formal model of the concepts introduced in the previous section. We start with the notion of intentionality following the phenomenological tradition of Edmund Husserl [32,9,33].

3.1   Intentionality

According to Husserl, human experience is intentional, i.e. it aims toward something beyond itself. This means that every experienced phenomenon is about, or of something, i.e. it is *directed*. For instance, the infant's perception of its mother or the mother's imagination of her child's well being are intentionally directed experiences. Moreover, intending something is always accompanied by certain flavours such as bodily sensations or moods but also communal norms, conventions and historical traditions. For instance, the infant might be hungry while perceiving its mother or the mother might be depressed while imagining her child being sick. Those constituents of experience are not directed and thus *open*. According to Husserl, many of such presupposed meanings are tacitly taken over from our culture and belong to the so-called lifeworld (Lebenswelt).

Directedness. Intentional experiences are directed though not necessarily toward an object that is explicitly distinguished. For instance, the mother can directly perceive her infant as being her child or she can deliberately reason about the amount of time the child kept sleeping last night. No matter what degree of awareness, directed experience is inherently temporal and has a correlational structure. This structure inseparably connects intentional act (noesis) and sense or appearance of an object (noema). Noemata correspond to all anticipations we have about, or of, an object. For instance, in perceiving her infant the mother presupposes it to be her child. She does not explicitly judge but rather implicitly anticipates the senses which constitute the relationship to her child. Noesis is the way or mode in which this anticipation takes place or unfolds. It discloses the noemata of an object in time and so gives meaning to it. For instance, if the mother sees her child crying, the mode, or way she anticipates her child changes toward a stronger bodily tension [4].

Let $B = \{|0\rangle, |1\rangle\}$ be the orthonormal basis of an intentional object. The orthonormal basis generates a vector space. Each vector represents a possible noema, sense or meaning in that space and can be written as a linear combination of orthogonal vectors. Let an intentional object be:

$$|p\rangle = a_1|0\rangle + a_2|1\rangle \tag{1}$$

where $|a_1|^2 + |a_2|^2 = 1$. Noemata correspond to all anticipated senses of an intentional object while being inseparably connected to the way an object is disclosed. For instance, consider an infant's awareness of its mother as a superposition or anticipation of senses. The basis vector $|0\rangle$ stands for the primordial sense 'do not expect reaction', whereas $|1\rangle$ represents 'expect reaction'. As discussed in the previous section, perceiving others is not believing, thinking or reasoning. It is more to be drawn to do something, a tension that the body aims to reduce and where we do not pay attention to the anticipation itself [11,21]. Let the probability of noesis to disclose noemata be defined as:

$$P(|0\rangle) = |a_1|^2 \text{ and } P(|1\rangle) = |a_2|^2 \tag{2}$$

Noeses represent the degree of bodily tension or anticipation. For instance, the infant can have a strong or weak bodily tension towards its mother. Acts exhibit a certain probability P and unavoidably disturb the intentional object leaving it in a state $|0\rangle$ or $|1\rangle$ determined by the outcome.

Openness. Open intentionality corresponds to Husserl's lifeworld (Lebenswelt) [9]. The lifeworld is always and already pregiven or presupposed and presents directed experience in a certain light. For instance, bodily sensations such as pain, moods such as happiness, or absorbed skillful activities such as driving and dancing are open, prepredicative and complement directed activities. They only take on an object-directed structure in moments of breakdowns, e.g. I attend to my hurting knee or the strange engine sound. Hence, open intentionality is subpersonal but can potentially be brought to awareness. It forms the horizon or ground of all our activities. Husserl considers intersubjectivity as crucial for the generation and transformation of this presupposed horizon [9]. In the next section we will formalize what happens if an infant's directedness towards its mother and vice versa becomes embodied or internalized and so a part of the lifeworld. Anticipations can become entangled and thus they can not be separated into directed intentionalities of either mother or infant.

3.2   Shared Intentionality

If mother and infant interact their intentionalities are directed towards each other. Vector $|p_1\rangle$ represents the mother's awareness of her infant. $|p_2\rangle$ represents the infant's awareness of its mother.

$$|p_1\rangle = a_1|0\rangle + a_2|1\rangle \text{ and } |p_2\rangle = b_1|0\rangle + b_2|1\rangle \qquad (3)$$

where $|a_1|^2 + |a_2|^2 = 1$ and $|b_1|^2 + |b_2|^2 = 1$. Vectors $|p_1\rangle$ and $|p_2\rangle$ have a tension towards each other. To make this clear, we define shared intentionality as a larger combined vector space. Here, $S = B_M \otimes B_I = \{|00\rangle, |01\rangle, |10\rangle, |11\rangle\}$ is the basis of mother and infant. In this combined space, shared senses $|\psi\rangle$ can be generated.

$$\begin{aligned} |\psi\rangle &= |p_1\rangle \otimes |p_2\rangle \qquad (4)\\ &= (a_1|0\rangle + a_2|1\rangle) \otimes (b_1|0\rangle + b_2|1\rangle)\\ &= a_1b_1|00\rangle + a_2b_1|10\rangle + a_1b_2|01\rangle + a_2b_2|11\rangle \end{aligned}$$

where $|a_1b_1|^2 + |a_2b_1|^2 + |a_1b_2|^2 + |a_2b_2|^2 = 1$. However, experiments have shown that an infant's anticipation of its mother is not the crucial point [5]. Infant and mother rely on each other to behave responsively in order to sustain their involvement in interaction [34,6]. Two-month-old infants are able to sustain interaction with their mothers via a live double video link. However, when they are shown recordings of their mothers, they do not coordinate with the unresponsive recording but become distressed and removed. More important is the ongoing anticipation of the mother. The infant performs some actions and anticipates the mother's reactions reflecting the infant's actions and vice versa, i.e. the mother's anticipation of the infant's reactions. Sustained social interactions can only be established when these anticipations are mutual and dynamic [35]. Therefore, intentionalities of mother and infant can not be separated and should be represented as an entangled or nonseparable state.

$$|\psi\rangle = x|00\rangle + y|11\rangle \qquad (5)$$

If $|\psi\rangle$ is entangled, there is no $|p_1\rangle$ and $|p_2\rangle$ such that $|\psi\rangle = |p_1\rangle \otimes |p_2\rangle$. Since $|\psi\rangle$ is not separable into directed intentionalities of mother and infant, it requires both having access to it and thus $|\psi\rangle$ is open. Anticipations of mother and infant (4) evolve towards undirected intersubjectivity (5) which becomes part of the presupposed lifeworld of mother and infant. Crucial is the ongoing interaction that sustains mother's and infant's directedness toward each other. This process does not only emerge from the mutual anticipation of mother and infant, it also shapes the lifeworld of both as shared intentionality becomes internalized and embodied. Hence, individual intentionalities can not be defined independently of the emerging interaction process and the entangled social system can not be defined independently of the single agents. Shared intentional states (5) and directed intentional states of the individuals (1) co-emerge by co-enacting each other [20,22]. Therefore, senses or appearances of an agent, so far represented as state vectors, depend on shared intentional states as much as shared appearances rely on senses of the individual agents. To make this more precise, we need to have a look at shared intentionality evolving in time. Let the time evolution of mother's and infant's directed intentionality be defined as:

$$U_M = e^{iA_1 t} \text{ and } U_I = e^{iA_2 t} \tag{6}$$

The time evolution consists of the exponentiation operation e, a self-adjoint operator A and a time parameter t. Similar to intentional objects (4), time evolutions of two or more agents can be combined by using the tensor product. The only difference is that combined evolutions are products of tensor-product evolutions and not superpositions of tensor-product evolutions [17].

$$\begin{aligned} U_{M+I} &= e^{iA_{1+2}t} \\ &= e^{i(a_1 b_1 \rho_1 + a_2 b_1 \rho_2 + a_1 b_2 \rho_3 + a_2 b_2 \rho_4)t} \\ &= e^{ia_1 b_1 \rho_1 t} e^{ia_2 b_1 \rho_2 t} e^{ia_1 b_2 \rho_3 t} e^{ia_2 b_2 \rho_4 t} \end{aligned} \tag{7}$$

Since our claim is that agents evolve from the background of an always and already pregiven intersubjective lifeworld, states or senses of intentional objects are now represented as density operators ρ; senses are always and already intersubjective and thus combined[1]. The density operators ρ_{1-4} represent the basis vectors in S and thus the state vectors or senses of B_M and B_I. The evolution equation can be written as a product because the members of S, i.e. the basis states of mother and infant, are mutually commuting. From a mathematical point of view U_{M+I} derives from the basis states in S. However, experiments reveal that mother-infant interactions are mutual and dynamic. Therefore, time evolutions U_M and U_I of the density operators representing mother and infant can not be characterized independently of U_{M+I}. For instance, consider the mother's awareness of her infant while perceiving her child as a density operator[2]:

---

[1] In this way we can represent both experiences being intersubjectively open and thus nonseparable (e.g. perceiving someone as being there for, or accessible to, me and others) and intentional objects appearing as distinct and thus separable (e.g. perceiving someone as being transcendent and thus distinct from me and others) [8].

[2] Note that senses, now represented as density operators, derive from the dynamic co-emergence [17]. Senses are transcendental or inexhaustible [36,22].

$$\rho = Tr(\rho(t)) \tag{8}$$

$$= a_1(t)|0\rangle\langle 0| + a_2(t)|1\rangle\langle 1|$$

where $a_1(t) = |\langle 0(t)0|\psi\rangle|^2$ and $a_2(t) = |\langle 1(t)1|\psi\rangle|^2$. If, from her first-person perspective, the mother does not expect her infant to react at time t, then $|0(t)\rangle = U_M(t)|0\rangle$ and if she expects a reaction, then $|1(t)\rangle = U_M(t)|1\rangle$. Although social perceptions are intersubjective and thus open, agents are not always involved in sustained face-to-face interactions. If the mother perceives her child sleeping, her directed experience does not rely on the infant reacting to her perceptions. In this case, U_M and U_I are separated and perceiving her child sleeping depends on the mother's private experience at each instant of time ($a_1(t)$ and $a_2(t)$). Nevertheless, her experience is intersubjectively constituted and thus there is a suppositional reference to nonseparable states $|\psi\rangle$. If the unitary evolution of mother and infant is separable, ψ is presupposed and not needed at each instant of time to describe their combined evolutions; there is no mutual face-to-face encounter. However, as experiments have shown, if mother-infant expectations are mutual and dynamic, $U_{M+I}(t)$ (7) is nonseparable, i.e. it can not be factorized into the tensor product $U_M(t) \otimes U_I(t)$. Consequently, nonseparable states dynamically emerge from mutual anticipations and can not be defined independently of mother and infant. Hence, expectation values of senses to be disclosed from the first-person perspective of mother and infant also rely on shared senses at each instant of time. Therefore, $a_1(t) = |\langle 00|\psi(t)\rangle|^2$ and $a_2(t) = |\langle 11|\psi(t)\rangle|^2$. Compared to shared intentional states (5), the ongoing interaction between mother and infant, i.e. the nonseparable time evolution, provides a higher degree of nonseparability since the nonseparable state $|\psi\rangle$ is needed at each instant of time to describe the evolution of mother's and infant's directed intentionality towards each other. Shared intentional states emerge from social interactions but also submerge or modulate individual agents. Subjectivities of both mother and infant move quite literally as a whole. In the remainder we will consider the possibility of this nonseparability of shared intentionality being not only relevant to face-to-face encounters but also to intentionality-to-lifeworld relations in general.

4   Discussion

In this section we discuss shared intentionality from a phenomenological perspective. Intersubjectivity or shared intentionality is considered as a constitutive aspect of phenomenal experience. An agent's experience is analysed in terms of conditions of possibility for manifestation. In other words, we examine the role of intersubjectivity as a condition for animate beings and inanimate objects to become manifest in experience. We divide our discussion of intersubjectivity in two parts. Firstly, we look at face-to-face encounters like the mother-infant interaction presented in the previous section (empathy). Secondly, we examine the disclosure or manifestation of intentional objects precisely as being there for, or accessible, to others (co-subjectivity). We will argue that both dimensions of intersubjectivity can be understood under the umbrella of shared intentionality as developed in the previous section.

4.1 Empathy

One of the core problems of phenomenological intersubjectivity is the question of how we can have access to other minds [37]. In particular, one of the most intriguing questions is the relation between empathy, the experience of otherness, and our existence in a common or shared world. We shall focus on empathy first. Empathy is a form of social perception enacted in face-to-face encounters and directed toward the experience of the other. Contrary to the argument from analogy[3], empathy is inherently active, direct and non-inferential (cf. section 2). Therefore, perceiving others is not attributing internal mental states inferred from observed external behaviour rather behaviour as perceived is expressive. According to Scheler (1954), expressive behaviour is neither perceived as a mere body nor as a hidden psyche but as a unified whole [12]. Furthermore, perceiving others as animate beings is different from perceiving physical objects. The other is given as bodily present or as a lived body (Leib) and not just as a transcendent object (Körper).

As defined in section 3, empathy is an intentional act that is directed towards the other's lived experiences. In this dynamic process, subjectivity of the other is disclosed from the second-person perspective. The second-person perspective is one's own open lived experience directed toward the open and directed lived experience of the other. Obviously, this nonseparability of first-person perspective and second-person view is reminiscent of the nonseparability of shared intentionality as introduced in the previous section. Crucially, empathy is not a multistage process where one observes mere external behaviour (behaviourism) and then adopts a theoretical stance to infer or compute the internal mental state of the other (cognitivism). As Heiddegger points out, grasping mental states of others is the exception rather than the rule. Under normal circumstances we understand each other well enough through our shared engagement in the common world [38]. This preflecitve otherness (alterity) accompanying everyday social encounters is often called primary intersubjectivity [39].

Moreover, social interaction extends toward secondary intersubjectivity [40] or the ability to share attention and intention. Here, interacting partners do not only relate to each other but refer to objects and events around them. In such triadic situations, agents learn to understand other's intention by means of other's expressive and contextualized behaviour. For instance, gaze-monitoring indicates that agents seek to verify the attention of the other towards the same thing, e.g. a hammer lying on a table, as well as to validate whether their intention is understood and thus shared. Hence, intersubjectivity in social perceptions is not always exclusively directed towards others but often mediated through the

---

[3] According to the argument from analogy, I infer by analogy that observed behaviour of foreign bodies is associated with experiences similar to those I have myself.

pragmatic circumstances of our encounters. For instance, when perceiving a hammer and nails, I see those tools as affordances or possible uses which were essentially learned in an intersubjective and pragmatic context. Likewise, perceiving other agents is pragmatic and context-dependent. Here, affordances are possible intentions associated with the perception of the other embedded or situated in pragmatic contexual situations, e.g. I perceive my friend as someone who is an expert with tools like hammers and nails.

From a developmental perspective, Tomasello (1999) has proposed that we gradually develop our understanding of others starting from (1) animate beings (from birth onwards) over (2) intentional agents (9-12 months) to (3) mental agents (4-5 years) [2]. In the first stage, children solely empathize by perceiving expressive behaviour and so they can distinguish animate from inanimate beings. Approaching their first year of life, expressive behaviour is increasingly experienced as goal-directed and context-dependent. Phenomena such as gaze-following, joint attention, shared engagement and imitative learning are indicators for children being able to see others as intentional agents. Apparently, these stages correspond to primary intersubjectivity and secondary intersubjectivity. The third stage essentially derives from social interactions in the previous stages. To understand others as mental agents requires to understand that others have beliefs and thoughts differing from one's own thoughts and beliefs. Children need to engage in discourses in which diverging perspectives emerge, e.g. disagreements, misunderstandings or requests for clarification. Importantly, understanding others as mental agents requires primary and secondary intersubjectivity upon which narrative competency and skilful practical reasoning develops [41]. However, narrative and practical reasoning skills do not involve reference to unobservable, abstract and general entities as postulated by some kind of theory of mind, e.g. TT. Rather such skills are grounded in observable events that take place in the world. The concrete and particular context is of primary importance for the determination of meaning [42].

In summary, children flourish starting with undifferentiated self-other relations or empathetic face-to-face encounters. From there, they engage in triadic communications laying the groundwork for experiencing others as situated and intentional beings. Eventually, children derive belief systems from language-use. Obviously, this presupposes the social community the child is a part of. Moreover, the community itself embodies conventionalized conceptual structure derived from language use. Therefore, even beyond our development in childhood, there is an always and already pregiven co-subjectivity, an anonymous being-with-others, that is characteristic of intersubjectivity though different from empathy.

4.2 Co-subjectivity

Empathy can be considered as a subtle form of intersubjectivity according to which an agent directly perceives expressive behaviour of other animate beings. However, the other's expressive behaviour is never isolated or separated. There is always and already a shared or common ground that influences how behaviour is perceived. Therefore, empathetically relating to others never requires a theoretical understanding, or model, of someone else's mind rather behaviour is expressive in the sense that a common lifeworld affords me to perceive certain shared aspects tuning the expressiveness of other's behaviour. As Husserl points out, I have been together with others for as long as I can remember, and my understanding and interpretation are, therefore, structured in accordance with intersubjectively handed-down forms of apperception [33]. Obviously, drawing from communal norms, social conventions and historical traditions does not only structure how we empathize and perceive other social beings but shapes our intentional relation to the world, including inanimate things and abstract objects alike. For instance, when visually perceiving an object, e.g. a glass of red wine, the impoverished sense data on my retina does not cause me to internally reconstruct or represent the external object. The object is actively explored as a function of my body movement, a prereflective sensorimotor grasp, where qualities like shape and color of the glass are directly perceived rather than internally represented. We visually experience the world to be rich in detail not because we must represent all that detail inside our heads at any given moment, but because we have constant access to the presence and detail of the world, and we know how to make use of this access [23]. In this way profiles hidden from view, e.g. the backside of the glass, are brought to awareness. Phenomenologically speaking, the glass is precisely given in experiences as being there for or accessible to others. As Zahavi (2006) puts it, subjects intentionally direct toward objects whose giveness in experience bears witness to their openness for other subjects [37]. Husserl was quite clear about the intersubjective or co-subjective nature of experience. According to his analysis, everything objective that stands before me in experience and primarily in perception has an apperceptive horizon of possible experiences. Every appearance that I have is from the very beginning a part of an open endless, but not explicitly realized totality of possible appearances of the same, and the subjectivity belonging to this experience is open intersubjectivity [9]. Hence, appearances or senses are transcendental or inexhaustible (cf. equation 8). Consequently, even prior to empathetic interactions with other social beings, intersubjectivity is already present as co-subjectivity.

What is the relation between empathy and co-subjectivity? According to our model as developed in the previous section, empathy and co-subjectivity are not separable. Empathy essentially refers to the co-directed intentionality that entangles with the pregiven co-subjectivity of the lifeworld. The most obvious form of empathetic relations is the sustained involvement in mutual anticipations. In our example, mother and infant empathize by continuously anticipating the other's directed awareness. The nonseparable time evolution that governs the co-emergence of the interaction and modulation process literally moves as a whole when the mother's co-subjectivity (i.e. her intersubjectively handed-down forms of apperception) entangles with the subjectivity of the child. Presumably, shared intentionality is an important ontogenetic and phylogenetic process that shapes human consciousness. Neither empathy nor co-subjectivity derive from one another, rather empathy and co-subjectivity go hand-in-hand, they dynamically co-emerge in time.

5 Conclusion and Future Work

Intersubjectivity is a necessary condition for self and language to unfold. It develops in early ontogenetic life when interactions between mother and infant mutually shape their relatedness. We presented a model of shared intentionality that represents intersubjectivity as nonseparable or entangled states of an ongoing process of mutual anticipation. Entanglement is a phenomenon that only arises in quantum mechanics. Hence, a quantum model is necessary. It is the sustained involvement in interaction that leads to entangled and shared intentional states. These states are not separable and thus they are not reducible to directed intentionalities of either mother or infant. Moreover, it is the ongoing interaction between mother and infant that is neither reducible to subjectivity of mother nor to subjectivity of infant.

Generally, empathic skills facilitate the understanding of expressive behaviour in others. In addition, empathic skills lay the groundwork for sharing attention and intention and imitative learning, as well as linguistic abilities such as narrative and practical reasoning competencies. Once internalized and embodied, the shared intentional states involved in all of these social skills become essentially part of the lifeworld. The already and always pregiven lifeworld presupposes how objects including other subjects are disclosed in experience and in this way it presupposes noemata co-constituted by other subjects.

Despite this not being obvious from a naïve realist point of view, closer phenomenological examinations reveal experience including perception to be essentially intersubjective. From a first-person perspective, all observed phenomena, whether physical or not, are private. However, there are objects and events to which there is public access. These public phenomena are not transcendent in any subject-independent sense. They are objective precisely in the sense of open intersubjectivity as introduced in this paper. Essentially, there is no necessity for maintaining the dualism of subjectivity and objectivity since intersubjectivity or shared intentionality mediates the latter and includes the former [8]. This understanding of shared intentionality is further strengthened by recent findings in developmental psychology and neuroscience.

Departing from our proposed notion of shared intentionality and its nonseparable nature, future work will investigate how individual intentionalities are transformed and even generated in social interactions. Furthermore, one of the most interesting questions is how such an account deals with failures in everyday social encounters. If breakdowns are mediated by the interaction process itself what are the consequences for understanding miscommunication?

Acknowledgements. This project was supported in part by the Australian Research Council Discovery grant DP0773341.

References

1. Tomasello, M.: Acquiring linguistic constructions. Handbook of Child Psychology 1, 1–48 (2006)
2. Tomasello, M.: Constructing a Language: A Usage-Based Theory of Language Acquisition. Harvard University Press, Cambridge (2003)
3. Tomasello, M.: The Cultural Origins of Human Cognition. Harvard University Press, Cambridge (1999)
4. Gibbs, R.: Intention as emergent products of social interactions. In: Intentions and Intentionality: Foundations of Social Cognition, pp. 105–124. MIT Press, Cambridge (2001)
5. Murray, L., Trevarthen, C.: Emotional Regulation of Interactions between Two-month-olds and their Mothers. In: Social Perception in Infants, pp. 177–197. Ablex, Norwood (1985)
6. Trevarthen, C.: The Self Born in Intersubjectivity: The Psychology of an Infant Communicating. In: The Perceived Self: Ecological and Interpersonal Sources of Self-knowledge, pp. 121–173. Cambridge University Press, Cambridge (1993)
7. Zahavi, D.: Beyond Empathy: Phenomenological Approaches to Intersubjectivity. Journal of Consciousness Studies 8, 151–168 (2001)
8. Velmans, M.: Intersubjective Science. Journal of Consciousness Studies 6, 299–306 (1999)
9. Husserl, E.: Husserliana 14: Zur Phänomenologie der Intersubjektivität. Texte aus dem Nachlass. In: Zweiter Teil: 1921-1928. Martinus Nijhoff, Den Haag (1973)
10. Gallagher, S.: Direct Perception in the Intersubjective Context. In: Consciousness and Cognition (2008)
11. Gallagher, S.: How the Body Shapes the Mind. Oxford University Press, Oxford (2005)
12. Scheler, M.: The Nature of Sympathy. Yale, New Haven (original work published 1923), Translated by Heath, P. (1954)
13. Bruza, P.D., Cole, R.: Quantum logic of semantic space: An exploratory investigation of context effects in practical reasoning. In: Artemov, S., Barringer, H., d'Avila Garcez, A., Woods, J. (eds.) We Will Show Them: Essays in Honour of Dov Gabbay, pp. 339–361. College Publications (2005)
14. Busemeyer, J.R., Wang, Z., Townsend, J.T.: Quantum dynamics of human decision making. Journal of Mathematical Psychology 50, 220–242 (2006)
15. Gabora, L.: The cultural evolution of socially situated cognition. Cognitive Systems Research 9(1-2), 104–113 (2008)
16. Gabora, L., Rosch, E., Aerts, D.: Toward an ecological theory of concepts. Ecological Psychology 20(1), 84–116 (2008)
17. Kronz, F., Tiehen, J.: Emergence and Quantum Mechanics. Philosophy of Science 69, 324–347 (2002)
18. Gallese, V.: The Shared Manifold Hypothesis: From Mirror Neurons to Empathy. Journal of Consciousness Studies 8, 33–50 (2001)
19. Hurley, S.: The Shared Circuits Model (SCM): How Control, Mirroring, and Simulation Can Enable Imitation, Deliberation and Mindreading. Behavioral and Brain Sciences 31, 1–58 (2008)
20. Jaegher, H.D., Paolo, E.A.D.: Participatory Sense-making: An Enactive Approach to Social Cognition. Phenomenology and the Cognitive Sciences 6(4), 485–507 (2007)

21. Gallagher, S., Meltzoff, A.: The Earliest Sense of Self and Others: Merleau-Ponty and Recent Developmental Studies. Philosophical Psychology 9, 211–233 (1996)
22. Thompson, E.: Mind in Life - Biology, Phenomenology and the Sciences of Mind. Harvard University Press, Cambridge (2007)
23. Noë, A.: Action in Perception. MIT Press, Cambridge (2004)
24. Bruza, P., Widdows, D., Woods, J.: A Quantum Logic of Down Below. In: Handbook of Quantum Logic. Elsevier/North-Holland, Amsterdam (2006)
25. Gabbay, D., Woods, J.: The Reach of Abduction: Insight and Trial. Elsevier, Amsterdam (2005)
26. Gärdenfors, P.: Conceptual Spaces: The Geometry of Thought. MIT Press, Cambridge (2000)
27. Rizzolatti, G., Fadiga, L., Gallese, V., Fogassi, L.: Premotor cortex and the recognition of motor actions. Cognitive Brain Research 3, 131–141 (1996)
28. Fogassi, L., Gallese, V.: The Neural Correlates of Action Understanding in Non-human Primates. In: Mirror Neurons and the Evolution of Brain and Language, pp. 13–35. John Benjamins, Amsterdam (2002)
29. Rizzolatti, G., Craighero, L., Fadiga, L.: The Mirror System in Humans. In: Mirror Neurons and the Evolution of Brain and Language, pp. 37–62. John Benjamins, Amsterdam (2002)
30. Wohlschläger, A., Bekkering, H.: The Role of Objects in Imitation. In: Mirror Neurons and the Evolution of Brain and Language, pp. 101–113. John Benjamins, Amsterdam (2002)
31. Gallese, V., Lakoff, G.: The Brain's Concepts: The Role of the Sensory-motor System in Conceptual Knowledge. Cognitive Neuropsychology 22(3/4), 455–479 (2005)
32. Husserl, E.: Husserliana 13: Zur Phänomenologie der Intersubjektivität. Texte aus dem Nachlass. In: Erster Teil, pp. 1905–1920. Martinus Nijhoff, Den Haag (1973)
33. Husserl, E.: Husserliana 15: Zur Phänomenologie der Intersubjektivität. Texte aus dem Nachlass. In: Dritter Teil: 1929-1935. Martinus Nijhoff, Den Haag (1973)
34. Nadel, J., Carchon, I., Kervella, C., Marcelli, D., Reserbat-Plantey, D.: Expectancies for Social Contingency in Two-month-olds. Developmental Science 2, 164–173 (1999)
35. Auvray, M., Lenay, C., Stewart, J.: Perceptual Interactions in a Minimalist Virtual Environment. New Ideas in Psychology (2008)
36. Husserl, E.: The Crisis of European Sciences and Transcendental Phenomenology: An Introduction to Phenomenological Philosophy. Translated by Carr. D. Northwestern University Press (1970)
37. Zahavi, D.: Subjectivity and Selfhood. MIT Press, Cambridge (2006)
38. Heidegger, M.: History of the Concept of Time: Prolegomena. Indiana University Press (1985)
39. Trevarthen, C.: Communication and cooperation in early infancy: A description of primary intersubjectivity. In: Before Speech: The Beginning of Interpersonal Communication, pp. 321–347. Cambridge University Press, Cambridge (1979)
40. Trevarthen, C., Hubley, P.: Secondary intersubjectivity: Confidence, confiding and acts of meaning in the first year. In: Action, Gesture and Symbol: The Emergence of Language, pp. 183–229. Academic Press, London (1978)
41. Hutto, D., Ratcliffe, M. (eds.): Folk Psychology Re-Assessed. Springer Publishers, Dordrecht (2007)
42. Bruner, J.: Actual Minds, Possible Worlds. Harvard University Press (1986)

Semantic Spaces: Measuring the Distance between Different Subspaces

Guido Zuccon, Leif A. Azzopardi, and C.J. van Rijsbergen

Dept. of Computing Science, University of Glasgow,
Glasgow, United Kingdom
{guido,leif,keith}@dcs.gla.ac.uk

Abstract. Semantic Space models, which provide a numerical representation of words' meaning extracted from corpus of documents, have been formalized in terms of Hermitian operators over real valued Hilbert spaces by Bruza et al. [1]. The collapse of a word into a particular meaning has been investigated applying the notion of quantum collapse of superpositional states [2]. While the semantic association between words in a Semantic Space can be computed by means of the Minkowski distance [3] or the cosine of the angle between the vector representation of each pair of words, a new procedure is needed in order to establish relations between two or more Semantic Spaces. We address the question: how can the distance between different[1] Semantic Spaces be computed? By representing each Semantic Space as a subspace of a more general Hilbert space, the relationship between Semantic Spaces can be computed by means of the subspace distance. Such distance needs to take into account the difference in the dimensions between subspaces. The availability of a distance for comparing different Semantic Subspaces would enable to achieve a deeper understanding about the geometry of Semantic Spaces which would possibly translate into better effectiveness in Information Retrieval tasks.

1 Introduction

Semantic Space techniques map words in a high dimensional vector space [4]. The map is usually built by computing lexical co–occurrences between words appearing in the same context where each vector is assigned to a word and represents the co–occurrences between the word and others. In this work, we consider a particular instance of a Semantic Space, the Hyperspace Analogue to Language (HAL). The HAL space is created through the co–occurrence statistics within a corpus of documents. This space has been used as a representation model of semantic memory [5] and has been shown to be compatible with human reasoning in cognitive science [6]. Within the area of Information Retrieval, HAL has been used to perform information inference for query expansion [7].

---

[1] We refer to different Spaces, not different instances of the same space, i. e. the same space rescaled.

P. Bruza et al. (Eds.): QI 2009, LNAI 5494, pp. 225–236, 2009.

In Semantic Spaces (like HAL) words (or concepts [8]) are represented by points in a high dimensional vector space: their position in the space is related to their meaning and inter–relationships. The former can be inferred by examining the components of the high dimensional vector associated with a word, while the latter can be exploited by a similarity measurement between word vectors. For example, in [3] the authors propose adopting the **Minkowski distance**, defined with respect to two vectors $\mathbf{u_i}$ and $\mathbf{v_i}$ of the Semantic Space as:

$$d_M = \sqrt[r]{\sum (|\mathbf{u_i} - \mathbf{v_i}|)^r} \tag{1}$$

Comparing the word vectors is one way to derive meaning from the Semantic Space. An alternative is to compare subspaces of documents, or sets of documents. While Semantic Spaces provide a representation of knowledge generated from a sample of text, a problem arises when we consider two or more Semantic Spaces that have been generated from independent samples of text. Specifically, how do we compare one Semantic Space with another? Once again, the simplest solution is to consider the distance between the representation of the same word vector in the two Semantic Spaces, using for example the Minkowski distance. However, this naïve treatment may be inappropriate, because different words used in the same sense will not be compared. For example, *cat* and *kitten* are semantically related in the context of the concept feline, and thus we would expect them to share the same vector representation. However, when computing the distance in a naïve way we do not take into account such relationships. Then, if in document d_1 we refer to the concept of feline with the common word *cat*, while in d_2 we refer to the close concept but using the term *kitten*, we might not capture the semantic relationship between the two documents. To avoid such problem, we propose to compute the distance between Semantic Spaces not relying on word–representation similarity, but on the more general subspace distance. The subspace correspondent to a document or to a set of documents conveys the meaning expressed by the text traces; comparing subspaces then would provide a distance based on the meaning/topic area associated *to the set* as opposed to the word level.

The paper continues as follow. In Section 2 we illustrate a formalization of Semantic Spaces in terms of Quantum Theory (QT) as it has been introduced in [2]. Moreover, we briefly present how to derive a numeric representation of a Semantic Space from a corpus of documents. In Section 3, several measures to compute the distance between subspaces are illustrated, guiding the reader to the definition of a metric which allows comparisons between Semantic Subspaces. Section 4 illustrates and discusses the preliminary experiments using subspace distance. The paper concludes providing a discussion of the distance between Semantic Subspaces, stating the objects of future investigations (Section 5).

2   Semantic Spaces: A Hilbert Space Representation

In the following, the formalization of Semantic Space in terms of Hilbert spaces [2] is presented. Consider a n–dimensional (real valued) Hilbert space

H, in which the inner product is represented by the Euclidean scalar product. In the following we limit our focus at real valued Hilbert spaces, discarding the analysis of complex valued spaces. Such a limitation is driven by the fact that the spaces are built from statistical data from texts, which uses only real values. Nevertheless, it is clear that complex numbers plays an important role in the description of states of a QT systems [9]. Each dimension of the Hilbert space H corresponds to a word in the vocabulary of a corpus of documents. The global Semantic Space, i.e. the Semantic Space derived considering the whole corpus, is denoted by $\hat{S}$. The Semantic Space derived from document d of the considered corpus is represented by S_d. Similarly the Semantic Space associated to a word w belonging to the vocabulary V of the corpus is denoted by S_w. It is clear that $\hat{S}$ is a subspace of the Hilbert space H since its vectors are instances of the vectors in H; in particular, $\hat{S}$ is a n—dimensional subspace. Similarly S_d is a m–dimensional subspace of H. Note that the subspace relationship $S_d \subseteq \hat{S}$ always holds.

We briefly illustrate the procedure to form the high dimensional matrix which corresponds to the HAL representation of the corpus of documents[2]. A window of text is passed over each document in the collection in order to capture co–occurrences of words. The length of the window is set to l: a typical value of l is 10; different values capture different levels of relationship between words. Words that co–occur into a window do so with a strength inversely proportional to the distance between the two co–occurring words. A thorough study which investigates the most effective function for encoding the inverse proportional weighting can be found in [10, Chapter 8.5]. By sliding the window over the whole collection and recording the co–occurrence values, a co–occurrence matrix A can be created. Since in our approach, as well as in [1,2,7], we are not interested in the order of the co–occurrences, in contrast with the work of Gärdenfors [8], therefore we can compute a symmetric matrix by means of $\hat{S} = A + A^T$, and then normalise the columns.

A symmetric matrix obtained by the illustrated procedure is associated to each subspace and is denoted with the same symbol assigned to the subspace: it is clear from the use if it refers to the subspace itself or to its symmetric HAL matrix. Note that subspace S_d can be defined as the range, or the complement of the range, of matrix S_d. The symmetric matrices $\hat{S}$ and each S_d, S_w are Hermitian linear operators. The following relations between the previous linear operators hold:

$$\hat{S} = \sum_{d \in C} S_d, \tag{2}$$

$$\hat{S} = \sum_{w \in V} S_w \tag{3}$$

where C is a corpus of documents. In the rest of this paper the focus will be on subspaces referring to document or set of documents.

---

[2] The interested reader should refer to [3] for a complete investigation of the procedure.

3   A Distance Measure between (HAL) Spaces

We aim to define a distance measure between Semantic Spaces, in order to be able to geometrically compare Semantic Spaces generated by different sources of evidence, i.e. compare subspaces formed with different subsets of documents.

Consider the general case of comparing the subspaces S_a and S_b derived by different sets of documents (a more particular case is when the set D associated to S_d contains only one document). We can associate to each subspace a $n \times n$ projector operator P. Then the **inner product** between two subspaces of H is the trace inner product for projection matrices:

$$\langle S_a, S_b \rangle = tr(P_a^* P_b) = tr(P_a P_b) \tag{4}$$

The appropriate candidate as distance between Semantic Subspaces has to satisfy several characteristics. Firstly, it would be desirable that the measure turns to be a metric. The inner product between two subspaces is not a metric: the inner product of P_a with itself is maximal rather then minimal. Nonetheless, it represents a measure of the *similarity* between the two subspaces: it is matter of fact that the measure proposed at the end of this Section employs the inner product between projectors of subspaces. An additional constraint to the measure has to be added. When comparing Semantic Subspaces, obtained for example from two documents, it is not guaranteed that they have the same number of dimensions, on the contrary it is frequently the case that the basis for such subspaces differ remarkably. Thus, a right candidate to measure the distance between two Semantic Subspaces should be able to capture differences in the dimensions of the basis of the Semantic Subspaces. The angle between the vectors of the subspaces is a key factor not only for the inner product between projectors, but for a whole family of measures based on the **principal (or minimal) angles**.

Definition 1. *For nonzero subspaces S_a and $S_b \subseteq S$, the principal angle between S_a and S_b is defined as the number $0 \leq \theta \leq \frac{\pi}{2}$ that satisfies*

$$\cos \theta = \max_{a \in S_a, b \in S_b, \|a\|=\|b\|=1} a^T b \tag{5}$$

The principal angle θ is 0 if and only if $S_a \cap S_b \neq \mathbf{0}$, while $\theta = \frac{\pi}{2}$ if and only if $S_a \perp S_b$. It is worthwhile to reformulate definition 1 in terms of projectors; this leads to the following theorem (where the proof is shown in [11]).

Theorem 1. *If P_a and P_b are the orthogonal projectors onto S_a and S_b respectively, then*

$$\cos \theta = \|P_a P_b\| = \|P_b P_a\| \tag{6}$$

These principle angles are related to the eigenvalues of $P_a P_b$: in fact, the first m (where m is the minimum between the subspace dimensions of S_a and S_b) eigenvalues of $P_a P_b$ are $\cos^2 \theta_1, \ldots, \cos^2 \theta_m$. We are however interested in comparing subspaces which have different dimensions, i.e. they do not have the same basis dimension. Unfortunately, the behaviour of a measure based on the principle angles is quite controversial if the subspaces have a different dimension. In fact, the principal angles are defined just for the minimum between the subspace dimensions: thus the measure does not take into consideration all the dimensions of both subspaces. For example, consider two subspaces: S_a of dimension p, S_b of dimension r such that $r \geq p$. Subspace S_b is built such that its first p basis vectors are the same of S_a, while the other $r - p$ basis vectors are arbitrarily constructed. Consider the **geodesic distance** [12] as measure based on principal angles; the measure is defined by:

Definition 2. *Let S_a and S_b be two subspaces and $\theta_1, \ldots, \theta_m$ be the m principal angles between S_a and S_b (where m is the dimension of the smallest subspace). The geodesic distance between S_a and S_b is*

$$d_g(S_a, S_b) = \sqrt{\theta_1^2 + \ldots + \theta_m^2} \tag{7}$$

If S_a and S_b are constructed as illustrated before, then the geodesic distance between S_a and itself will be 0 since each θ_i is zero implying that $S_a \equiv S_a$. However, when measuring the distance between S_a and S_b based on principal angles, we find that all the p angles that are computed are equal to the null angle 0, since S_b shares p basis vectors with S_a. Thus, the measure does not take into account the $r - p$ basis vectors of S_b that are not shared with S_a.

A distance measure based on the principal angles between subspaces, such as the geodesic distance, is then significant if and only if the subspaces have the same dimensions: this is unlikely to be the case when comparing different Semantic Spaces. We refer to such problem as the Zero Distance Problem: the bigger the difference in the number of dimensions of the two subspaces, the greater the extent of the problem since the number of discarded dimensions in the computation of the distance grows. Measures based on the principal angles, such as the geodesic distance, are generally affected by the Zero Distance Problem: the solution to the problem passes through the **chordal distance** [13], a monotonic function of the inner product.

Definition 3. *The chordal Grassmannian distance between two subspaces S_a and S_b is given by means of the associated projectors P_a and P_b by*

$$d_c(P_a, P_b) = \sqrt{m - tr(P_a P_b)} \tag{8}$$

As for the previous measures, also in the case of the chordal distance a difference in the dimensionality of the subspaces S_a and S_b is only partially taken into account: in fact the product $P_a P_b$ depends on the degree of association (or similarity) between the two subspaces, comparing each dimension, but the number m in equation 8 refers to the dimension of the subspaces S_a and S_b, thus needing to be of the same dimension. Anyway, the chordal distance opens the path to the definition of a distance that is not restricted by the dimensionality of the subspaces to be compared. The first step is to introduce the Hausdorff distance which measures the distance between two compact subsets of the space. In our case, we consider the L_2–Hausdorff distance between a vector u_i and a subspace V which is expressed by $d_H(u_i, V) = min||u_i - v||$, where $v \in V$ and $||.||$ is the Euclidean norm. We now have to consider the subspace distance between subspaces of the same dimension proposed in [14].

Definition 4. *The subspace distance $d_s(S_a, S_b)$ for two p–dimensional subspaces S_a and S_b is defined as*

$$d_s(S_a, S_b) = \sqrt{\sum_{i=1}^{p} d_H^2(\mathbf{u_i}, S_b)} \qquad (9)$$

where

1. $\mathbf{u_1}, \ldots, \mathbf{u_i}, \ldots, \mathbf{u_p}$ *is an orthonormal basis for S_a, and*
2. $d_H(\mathbf{u_i}, S_b)$ *is the Hausdorff distance from the end point of the basis vector $\mathbf{u_i}$ to subspace S_b.*

Such distance has been extended to the case where the subspaces have different dimensions. Let $\mathbf{v_1}, \ldots, \mathbf{v_i}, \ldots, \mathbf{v_r}$ be an orthonormal basis for S_b, then

Definition 5. *The subspace distance $d_s(S_a, S_b)$ between the p–dimensional subspaces S_a and the r–dimensional subspace S_b is defined as*

$$d_s(S_a, S_b) = \sqrt{\max(p, r) - \sum_{i=1}^{p}\sum_{j=1}^{r}(\mathbf{u_i}^T\mathbf{v_j})^2} \qquad (10)$$

The introduced distance has several properties. In primis, it is invariant to the choice of the orthonormal basis for the subspaces S_a and S_b. Furthermore, it is symmetric and not negative, in particular $d_s(S_a, S_b) = 0$ if and only if $S_a \equiv S_b$. The upper bound for the subspace distance is given by $d_s(S_a, S_b) \leq \sqrt{\max(p, r)}$ and corresponds to the orthogonality condition $S_a \perp S_b$. Finally, as proved in [15], the subspace distance satisfies the triangle inequality, and thus it is a proper metric defined on subspaces.

The next step is to express the subspace distance in terms of projector operators, and thus finding a relationship with the chordal distance. As demonstrated in [16], equation 10 can be re-written as:

$$d_s(S_a, S_b) = \sqrt{\frac{1}{2}tr\left[(\Lambda_p - \Lambda_r)^2 + (S_aS_a^T - S_bS_b^T)^2\right]} \qquad (11)$$

where $\Lambda_i = diag(1, \ldots, 1, 0, \ldots, 0)$ is a diagonal matrix with i 1's and $n - i$ 0's elements and S_a, S_b are the symmetric HAL matrices associated to the corresponding subspaces. With some algebraic calculation and since the matrix products $S_aS_a^T$ and $S_bS_b^T$ are the projectors P_a and P_b respectively, the subspace distance can be stated as:

$$d_s(S_a, S_b) = \sqrt{\max(p, r) - tr(P_aP_b)} \qquad (12)$$

The proposed subspace distance might be employed to compute the distance between Semantic Subspaces, aiming to obtain a more precise measurement of separation than using a naïve distance based on comparison between single word representations, e.g. the Minkowski distance.

Comparing equation 12 and 8, both formulating the chordal distance between two subspaces, it appears clear the strong relationship between the two distances, differing in taking into account the maximum of the subspace dimensions.

Each rank d projector represent a basis of a Hilbert subspace and can be regarded as a d–(hyper)plane: this provides an embedding of the Grassmannian of d–plane into a flat vector space. Thus, the rank d projector will sit on a sphere in this flat space, more precisely it will be point on the surface of a sphere, and its Euclidian distance provides us with a chordal distance between projectors. The chordal distance has been successfully used to study the packing problems for n–planes, where the aim is to find a set of hyper-planes such that the minimum distance between each pair of planes in the set is as large as possible [17]. Since the chordal distance provides a natural measure of the distance between bases of the same rank in a Hilbert spaces, it has been used to detect Mutually Unbiased Bases (MUB) [18], i.e. bases which spans planes totally orthogonal between them. This condition is reached when the chordal distance between the two bases is maximum.

Previous research in QT focused on the derivation of a suitable measure to judge the distance between quantum states of different preparations. Such a measure can be used to characterize the degree of distinguishability between states (and related preparations). In fact, because of the statistical error introduced when measuring frequencies of possible outcomes for a finite ensemble of identically prepared systems, it is generally difficult to distinguish between preparations that slightly differ [19]. The measure thus is used to judge the degree of separation between states. This is the underlying idea of the *statistical distance* between quantum preparations presented in [19], and is determined entirely by statistical fluctuations. However, it turns out that the statistical distance provides an identical result to the measure of the angle between rays in a Hilbert space associated with the pure quantum states of the preparations. Computing the distance between Semantic Space representations of a word (in terms of HAL subspaces) is similar to measuring the angle between the representative rays spanned by the word in its Hilbert space representations. Another distance that is related to the evaluation of the distinguishability of two quantum states is the so called Bures distance [20]. It measures the distance of the associated density operators ρ_1 and ρ_2 by the formula $d_B(\rho_1, \rho_2) = \sqrt{2}[1 - tr((\rho_1^{1/2}\rho_2\rho_1^{1/2})^{1/2})]^{1/2}$. The Bures distance has been interpreted as a generalization of transition probabilities to mixed states [21].

4   Pilot Experiment

We conducted a pilot investigation in order to examine how well subspace distance performs. In particular, we experimentally demonstrate that related documents are at a closer subspace distance between each other than not related ones. As baseline for the comparison we employed the Minkowski distance with $r = 2$ (Euclidian distance) between Semantic Space representations of words. In the following we describe the details of the experiment, discussing how Semantic Subspaces have been generated and in which terms we compare the subspace distance against the baseline.

We employ a standard IR collection, namely WSJ 87–92, as source of documents used to generate the Semantic Subspaces. This collection has more than 170 thousand newspaper articles, containing over 226 thousand unique terms. For the purpose of our pilot study we consider two subsets of document, R and N. Set R contains only documents that have been judged relevant (by human assessors) to a query q, while set N containing those documents judged as irrelevant to the same query, each query belonging to one of the TREC 1 topics. All the documents have been processed through stop-word removal and stemming.

Two methods for generating the Semantic Spaces have been employed, both inspired by the HAL paradigm. In both methods a window of text is passed over the text. The window size is 11, 5 words on the left and 5 on the right of the target word. We adopted an inverse proportional function to score the strength of co-occurrences with the target word, i.e. closer the co-occurent term is to the target word and higher is the score attributed to the pair. The only difference between the two methods is represented by the text over which the window is passed. The first method, which sticks to the definition of the generating procedure for HAL, passes the window over all the text contained in a document. On the contrary, the second method, which is partially inspired by [2] and [22], passes the window over traces of text extracted from the document. Such traces are extracted considering windows centered on target words. For each TREC 1 topic, target words are extracted from the description of the topic itself. As well as the documents, also the target words are pre-processed by matching against a stop-word list and by stemming. From now on, we refer to this Semantic Subspace generation method as *HAL traces*.

In tables 1 and 2, we report the preliminary results obtained by our study using topic 51 of the WSJ 87–92 TREC collection. The values presented in table 1 contains the average distance values obtained employing *HAL traces*, while table 2 refers to the average distances calculated using the classic HAL representation. In both tables, the Euclidean distance has been calculated as the square root of the squared difference between selected word representations associated with two HAL spaces. The values obtained where then averaged among the documents contained in the set and reported into the tables. Instead, for what concern the subspace distance, the reported values refer to the average over the correspondent set of documents of the following formula:

$$sim_s(S_a, S_b) = 1 - \frac{\sqrt{\max(p, r) - \sum_{i=1}^{p} \sum_{j=1}^{r} (\mathbf{u_i}^T \mathbf{v_j})^2}}{\sqrt{\max(p, r)}} \qquad (13)$$

which expresses the similarity (driven by the subspace distance) between subspaces S_a and S_b. A value of this similarity close to 0 means that the two

Table 1. Average distance between sets of relevant documents (R) and not relevant documents (N) obtained by the subspace distance (*Subspace* in the table) and the Euclidean distance (*Euclidean*) computed over subspaces generated by the *HAL traces* paradigm

	R		N	
	Subspace	Euclidean	Subspace	Euclidean
R	0.0376 ± 0.0116	9.3910 ± 4.6994	0.0182 ± 0.0072	6.7059 ± 5.5936
N	0.0182 ± 0.0072	6.7059 ± 5.5936	0.0386 ± 0.0093	3.3816 ± 2.0667

Table 2. Average distance between sets of relevant documents (R) and not relevant documents (N) obtained by the subspace distance (*Subspace* in the table) and the Euclidean distance (*Euclidean*) computed over subspaces generated by the traditional *HAL* paradigm

	R		N	
	Subspace	Euclidean	Subspace	Euclidean
R	0.1504 ± 0.0142	19.8121 ± 4.9445	0.0124 ± 0.0068	5.7710 ± 5.1289
N	0.0124 ± 0.0068	5.7710 ± 5.1289	0.1181 ± 0.0173	3.5376 ± 2.4407

subspaces are almost orthogonal, with $sim_s(S_a, S_b) = 0$ representing the case $S_a \perp S_b$, while a value close to 1 represents high degree of similarity. Thus, the two distances are not directly comparable. However, it is possible to understand the behavior of the two measure in discriminating between relevant and not relevant documents.

4.1   Discussion of the Preliminary Results

The results of the preliminary experiments reported in this paper refer to topic 51 of TREC 1. The results show that the subspace distance is able to discriminate between subspaces associated with relevant documents and the ones generated from non relevant documents. In fact, in accordance with the values reported for the subspace distance, the degree of Semantic similarity between the non relevant set of documents (labeled N) and the relevant set (labeled R) is lower (0.0182 for *HAL traces*, 0.0124 for *classic HAL*) than the similarity among occurrences of relevant documents (*HAL traces*: 0.0376, *classic HAL*: 0.1504) or not relevant documents (*HAL traces*: 0.0386, *classic HAL*: 0.1181). The same result is not achieved by the Euclidean distance. For Semantic Subspaces generated by *HAL traces* and by the traditional approach, the Euclidian distance between subspaces belonging to R is higher than the accumulated average distance between R subspaces and N ones.

From the tables is possible to evince that the subspace distance tends to flatten the distance among subspaces to the range $[0.9, 1.0]$, while the Euclidean distance is able to provide a greater range of values, making easy to detect significant differences between subspaces.

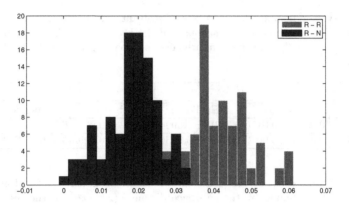

(a) Distribution of frequencies for the subspace distance

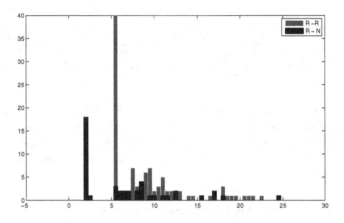

(b) Distribution of frequencies for the Euclidean distance

Fig. 1. Frequencies distribution of pairwise subspace distances (a) and Euclidean distances (b) between subspaces belonging to the set of relevant documents (R) and the non relevant (N) for topic 51. The subspace generation paradigm adopted is *HAL traces*.

Fig. 1 illustrates the frequencies distribution of pairwise distance values (obtained by the subspace distance (a) the Euclidean distance (b)) between Semantic Subspaces generated using the paradigm *HAL traces*, although rather similar figures are obtained when considering subspaces generated by the standard HAL derivation. The figures can be interpreted as follow. Subspaces associated to relevant documents (R) are on average at a closer subspace distance to each other than to non relevant documents (N) (see Fig. 1 (a)). using the Euclidean distance the separation between R and N is not as distinct. This suggests that the subspace distance will be more effective in discriminating relevant documents from non relevant.

5   Conclusion and Future Work

In this work a distance based on the chordal distance has been introduced in order to compare Semantic Subspaces constructed from subsets of a document corpus. Our approach allows to compare directly two sets of documents though their subspace distance, whereas [3] only deals with comparing a word and its meaning. Geometrically, this corresponds in considering the projection of a subspace into another, rather than the intersection between two subspaces.

Future work will be directed towards applying the proposed measure in a number of retrieval applications in order to determine its effectiveness.

Acknowledgement

This work was supported in part by the EPSRC Renaissance project (grant number EP/F014384/1) and by the EPSRC grants EP/E002145/1 and EP/F014708/1. The authors are thankful to prof. D. Song for the useful discussions and the guidance provided to make the first steps into the investigation of Semantic Subspaces, and to Alvaro Huertas-Rosero for the valuable suggestions. Moreover, the authors are grateful to the IRF[3] (Information Retrieval Facility) for the resources and support provided for carrying out the preliminary experiments.

References

1. Bruza, P.D., Cole, R.J.: Quantum Logic of Semantic Space: An Exploratory Investigation of Context Effects in Practical Reasoning. In: We Will Show Them: Essay in Honour of Dov Gabbay, vol. 1, pp. 339–361. College Publications (2005)
2. Bruza, P.D., Woods, J.: Quantum Collapse in Semantic Space: Interpreting Natural Language Argumentation. In: Proceedings of the 2nd QI Symposium, pp. 141–147 (2008)
3. Lund, K., Burgess, C.: Producing High-dimensional Semantic Spaces from Lexical Co-occurrence. Behavior Research Methods 28(2), 203–208 (1996)
4. Osgood, C., Suci, G., Tannenbaum, P., Date, P.: The Measurement of Meaning. University of Illinois Press, US (1957)
5. Burgess, C., Livesay, K., Lund, K.: Explorations in Context Space: Words, Sentences, Discourse. Discourse Processes 25(2,3), 211–257 (1998)
6. Landauer, T.K., Foltz, P.W., Laham, D.: An Introduction to Latent Semantic Analysis. Discourse Processes 25(2,3), 259–284 (1998)
7. Song, D., Bruza, P.D.: Discovering Information Flow Using High Dimensional Conceptual Space. In: Proceedings of the 24th ACM SIGIR, pp. 327–333 (2001)
8. Gärdenfors, P.: Conceptual Spaces: The Geometry of Thought. MIT Press, US (2000)
9. Bowman, G.E.: Essential Quantum Mechanics.. Oxford University Press, UK (2008)
10. Sahlgren, M.: The Word-Space Model. Ph.D thesis. Stockholm University (2006)

---

[3] http://www.ir-facility.org/

11. Ipsen, I.C.F., Meyer, C.D.: The Angle Between Complementary Subspaces. American Mathematical Monthly 102(10), 904–914 (1995)
12. Wong, Y.C.: Differential Geometry of Grassmann Manifolds. In: Proceedings of the National Academy of Science, vol. 57, pp. 589–594 (1967)
13. Bengtsson, I., Bruzda, W., Ericsson, A., Larsson, J.A., Tadej, W., Zyczkowski, K.: Mubs and Hadamards of Order Six (2006), ArXiv Quantum Physics e-prints
14. Wang, L., Wang, X., Feng, J.: Subspace Distance Analysis with Application to Adaptive Bayesian Algorithm for Face Recognition. Pat. Rec. 39(3), 456–464 (2006)
15. Sun, X., Wang, L., Feng, J.: Further Results on the Subspace Distance. Pat. Rec. 40(1), 328–329 (2007)
16. Sun, X., Cheng, Q.: On subspace distance. In: Campilho, A., Kamel, M.S. (eds.) ICIAR 2006. LNCS, vol. 4142, pp. 81–89. Springer, Heidelberg (2006)
17. Conway, J.H., Hardin, R.H., Sloane, N.J.A.: Packing Lines, Planes, etc.: Packings in Grassmannian Spaces. Experimental Mathematics 5(2), 139–159 (1996)
18. Bengtsson, I., Bruzda, W., Ericsson, A., Larsson, J.A., Tadej, W., Życzkowski, K.: Mutually Unbiased Bases and Hadamard Matrices of Order Six. Journal of Mathematical Physics 48(5) (2007)
19. Wootters, W.K.: Statistical Distance and Hilbert Space. Phys. Rev. D 23(2), 357–362 (1981)
20. Braunstein, S.L., Caves, C.M.: Statistical Distance and the Geometry of Quantum States. Phys. Rev. Lett. 72(22), 3439–3443 (1994)
21. Uhlmann, A.: The "Transition Probability" in the State Space of a *-algebra. Reports on Mathematical Physics 9, 273–279 (1976)
22. Huertas-Rosero, A.F., Azzopardi, L.A., van Rijsbergen, C.J.: Characterising through Erasing: A theoretical framework for representing documents inspired by quantum theory. In: Proceedings of the 2nd QI Symposium, pp. 160–163 (2008)

Characterizing Pure High-Order Entanglements in Lexical Semantic Spaces via Information Geometry

Yuexian Hou[1] and Dawei Song[2]

[1] School of Computer Sci. & Tech., Tianjin University, China
yxhou@tju.edu.cn
[2] School of Computing, The Robert Gordon University, United Kingdom
d.song@rgu.ac.uk

Abstract. An emerging topic in Quantuam Interaction is the use of lexical semantic spaces, as Hilbert spaces, to capture the meaning of words. There has been some initial evidence that the phenomenon of quantum entanglement exists in a semantic space and can potentially play a crucial role in determining the embeded semantics. In this paper, we propose to consider pure high-order entanglements that cannot be reduced to the compositional effect of lower-order ones, as an indicator of high-level semantic entities. To characterize the intrinsic order of entanglements and distinguish pure high-order entanglements from lower-order ones, we develop a set of methods in the framework of Information Geometry. Based on the developed methods, we propose an expanded vector space model that involves context-sensitive high-order information and aims at characterizing high-level retrieval contexts. Some initial ideas on applying the proposed methods in query expansion and text classification are also presented.

Keywords: Information geometry, Pure high-order entanglement, Semantic emergence, Extended vector model.

1 Introduction

An emerging line of research in Quantum Interaction (QI) is on capturing the meaning of words based on lexical semantic spaces (as Hilbert spaces) [13][14][17]. The intuition is that humans encountering a new concept often derive its meaning via the accumulative experience of contexts in which the concept appears. Therefore, the meaning of a word can be captured by examining its co-occurrence patterns with other words in the language use (e.g., a corpus of texts). A typical semantic space model is the Hyperspace Analogue to Language (HAL) [15]. The semantic space models have demonstrated a cognitive compatibility with human information processing [15][16].

More formally, in this paper, we generalize a semantic space to a Hilbert space induced by a set of words, in which all possible combinations of the words form the basis vectors. For example, given a word set W≡{Napoleon, invasion, Spain}, we have eight basis vectors $|000\rangle, |001\rangle, ..., |111\rangle$, where the basis vector $|001\rangle$ stands for the occurrence of 'Napoleon' and the absence of 'invasion' and 'Spain'. A pure state of

P. Bruza et al. (Eds.): QI 2009, LNAI 5494, pp. 237–250, 2009.

this semantic space can be written as $a_{000}|000\rangle+...+a_{111}|111\rangle$, where the linear combination coefficients a_{ijk} meet the normalization condition $|a_{000}|^2+...+|a_{111}|^2=1$. In quantum mechanics, the squared norm of a linear combination coefficient is considered as the probability of the corresponding basis event observed. According to this interpretation, it is clear that we can readily recover the marginal probability of any word occurrence from a given pure state of the semantic space. For example, the marginal probability of the occurrence of the first word (Napoleon) can be given by $|a_{100}|^2+|a_{101}|^2+|a_{110}|^2+|a_{111}|^2$. In this sense, a pure state of the semantic space gives a more comprehensive description of the space than the conventional Vector Space Model (VSM or VM for short) [1].

A kind of quantum states of particular importance is the entangled state [2], in which the quantum states of two or more objects are dependent on each other so that one object can no longer be adequately described without a full mention of its counterparts. Technically, several objects are entangled if the state of the compositional system cannot be expressed as the tensor of individual systems' states. Recent study by Bruza el. al. revealed some initial evidence that the phenomenon of entanglement also exist in semantic spaces [12]. Then, the next fundamental research question arise: how to characterize and utilize the entanglements in semantic spaces? This is the very aim of this paper. We are particularly interested in the pure high-order[1] entanglements in semantic spaces, i.e., high-order entanglements that cannot be reduced to the compositional effect of lower-order interactions, which often indicate the emergence of high-level semantic entities.

For illustration, let us consider the example semantic space shown earlier. Given a pure state of this semantic space[2],

$$|\psi\rangle=\sqrt{0.3296}|000\rangle+\sqrt{0.0002}|001\rangle+\sqrt{0.0900}|010\rangle+\sqrt{0.0001}|011\rangle$$

$$+\sqrt{0.3000}|100\rangle+\sqrt{0.0001}|101\rangle+\sqrt{0.2000}|110\rangle+\sqrt{0.0800}|111\rangle$$

it is easy to check that $|\psi\rangle$ cannot be expressed as the tensor of the pure state of its 1-order subsystems, i.e.,

$$|\psi\rangle\neq(x_0|0\rangle+x_1|1\rangle)\otimes(y_0|0\rangle+y_1|1\rangle)\otimes(z_0|0\rangle+z_1|1\rangle)$$

for arbitrary x_0, x_1, y_0, y_1, z_0 and z_1 meeting $|x_0|^2+|x_1|^2=1$, $|y_0|^2+|y_1|^2=1$ and $|z_0|^2+|z_1|^2=1$. Hence, we conclude that $|\psi\rangle$ is an entangled state.

In this paper, we focus on the pure high-order entanglement, i.e., the high-order entanglement that cannot be expressed as the tensor of any lower-order systems that might be entangled too. For example, it is easy to check that the above $|\psi\rangle$ cannot be expressed as the tensor of state vectors of any two subsystems, e.g.,

$$|\psi\rangle=(u_0|0\rangle+u_1|1\rangle)\otimes(v_{00}|00\rangle+v_{01}|01\rangle+v_{10}|10\rangle+v_{11}|11\rangle),$$

---

[1] In this paper, the "high-order entanglement" corresponds to the "multipartite entanglement" in Quantum Mechanics.

[2] Note that the coefficients of $|\psi\rangle$ meet the normalization condition, i.e., 0.3296+0.0002+ 0.0900+0.0001+0.3000+0.0001+0.2000+0.0800=1.

where $|u_0|^2+|u_1|^2=1$, $|v_{00}|^2+|v_{01}|^2+|v_{10}|^2+|v_{11}|^2=1$. In this case, we conclude that $|\psi\rangle$ has a pure 3-order entanglement.

The purpose of this paper is to characterize pure high-order entanglements and investigate their semantic implications in semantic spaces. To this end, there are two fundamental issues. One is how to measure entanglements, and the other is how to distinguish pure high-order entanglements from the compositional effect of lower-order entanglements so that we can illuminate the semantic implication of pure high-order entanglements by a computational method.

For the first issue, there are several well-known statistics measuring 2-order entanglement of a pure state, e.g., Von Neumann entropy [2], relative entropy of entanglement [3], robustness of entanglement [3] and squashed entanglement [3]. The measurement of high-order entanglement is more complicated. Some measures on high-order entanglement are derived by a direct generalization or a simple combination of 2-order measures, e.g., relative entropy of entanglement [3], robustness of entanglement [3] and global entanglement [3]. In addition, there are high-order entanglement measures that do not inherently depend on 2-order measures, e.g., Tangle [3] and Schmidt measure [3]. Although they are useful in general, most of the above statistics have some limitations in certain contexts. For example, many of these statistics cannot effectively distinguish pure high-order entanglements from the lower-order ones. Although, based on the above statistics, a rather satisfactory understanding has been achieved in the bipartite case, there is a certain degree of consensus that there is no universal way to define pure high-order entanglement, even in the simplest case of pure states [18, 19]. The existing pure high-order entanglement statistic often has to depend on some strong presupposition, e.g., symmetric Gaussianity [20].

The second issue requires a method which can not only measure pure high-order entanglements but also easily find or construct surrogate states so that we can investigate their semantic implications exclusively. Here, a surrogate state refers to the state that shares the same (k-i)-order entanglements, where 0<i<k, with the original state but does not have pure k-order entanglement. Hence, by comparing the manifestation of the original state and the surrogate state in a proper context, e.g., information retrieval, we can evaluate the semantic implication of the pure k-order entanglement. In our opinion, the pure k-order entanglements are an important indicator of specific semantic entities.

In this paper, we propose the use of Information Geometry (IG) [4][5] to characterize the pure high-order entanglements. IG provides useful tools and concepts for this purpose, including the orthogonality of coordinate parameters and the Pythagoras relation in the KL-divergence [6][7]. For example, based on parametric orthogonality, we can give a set of statistics and methods for analyzing word occurrence patterns by decomposing the word entanglements into various orders. As a result, pure 2-order, 3-order, and higher-order entanglements are singled out.

It should be emphasized that, owing to the lack of a proper quantum statistic, the proposed IG method in this paper is classical in itself. The usefulness of IG method in a quantum framework roots on the following observation: In a post-measurement configuration, the entanglement degenerates into the statistical dependence between the measurement results. Specifically, it can be shown that several objects are entangled only if the corresponding random variables denoting the measurement results of these objects are statistically dependent on each other (see Subsection 2.1 for details).

Since the occurrence and co-occurrence patterns of words can be naturally explained as the measurement results of semantic spaces, we believe that the proposed IG method is sufficient for our purpose.

2   Preliminaries of Information Geometry

Information Geometry (IG) represents probabilistic distributions as parametric co-ordinate systems, and hence could establish a connection between the properties of statistical distributions and some well-known notions in differential geometry to capture statistical dependencies from a geometric point of view. In this section, we will first discuss the connections between quantum entanglement and Information Geometry (IG), and then give a brief introduction to some relevant concepts and theorems from IG. Note that most theorems presented in Subsections 2.2, 2.3 and 2.4 have been formally proved or implied by the pioneering work in IG, e.g., the work by Rao [8], Jeffreys [9] and Sun-Ichi Amari [4][5]. Here, we restate and inter-pret them for our purpose in the context of semantic spaces, as their original expres-sions are heavily dependent on the notions and symbols of differential geometry and are thus not easy to follow for readers without a strong mathematical background.

2.1   On the Connection of Quantum Entanglements and Classical Dependences

Although IG is expressed in a classical framework of probability theory and originally aims at characterizing classical interactions[3], it can be naturally applied in the quan-tum framework because of the intrinsic connection between quantum entanglements and statistical dependences. For illustration, let $|\psi\rangle$ be a pure state of a two-qubit system A. Then $|\psi\rangle$ can determine a joint distribution on the basis events of A. For instance, if $|\psi\rangle = a_{00}|00\rangle + a_{01}|01\rangle + a_{10}|10\rangle + a_{11}|11\rangle$, then $|\psi\rangle$ determines a joint distribution: $P_{|\psi\rangle} = \left\{ p_{00} = |a_{00}|^2, p_{01} = |a_{01}|^2, p_{10} = |a_{10}|^2, p_{11} = |a_{11}|^2 \right\}$. Let $X_{|\psi\rangle}$ be the (classical) random vari-able obeying the joint distribution $P_{|\psi\rangle}$. We call $X_{|\psi\rangle}$ the denotative random variable induced from $|\psi\rangle$, and denote the value of $X_{|\psi\rangle}$ by x_ψ. For example, if $|\psi\rangle = |10\rangle$, then $x_\psi = 10$. The following proposition, which can be generalized to general cases of multi-compositional systems, illuminates the equivalence between entanglements and statis-tical dependences in the post-measurement configuration.

Proposition 1. Let $|\psi\rangle$ be a pure state of a quantum system A, $\{B, C\}$ be a bipartition of A such that $A = B \otimes C$, and $|u\rangle$ and $|v\rangle$ be the pure states of B and C respectively. Then, $|\psi\rangle = |u\rangle \otimes |v\rangle$ iff $\Pr\left(X_{|\psi\rangle} = x_u \circ x_v\right) = \Pr\left(X_{|u\rangle} = x_u\right) \cdot \Pr\left(X_{|v\rangle} = x_v\right)$ where $X_{|\psi\rangle}$, $X_{|u\rangle}$ and $X_{|v\rangle}$ are denotative random variables induced from $|\psi\rangle$, $|u\rangle$ and $|v\rangle$ respectively, and $\circ$ stands for the conjunction of x_u and x_v, e.g., if $x_u = 01, x_v = 10$, then $x_u \circ x_v = 0110$.

---

[3] In this paper, we use the term 'interaction' or 'dependence' to be the classical counterpart of the quantum entanglement. The connection between these notions is shown in Proposition 1.

Proof: Let $|\psi\rangle = a_{0...0}|0...0\rangle + ... + a_{1...1}|1...1\rangle$ is a state vector of 2^n-dimensional Hilbert space A, $|u\rangle = b_{0...0}|0...0\rangle + ... + b_{1...1}|1...1\rangle$ is a state vector of $A's$ 2^k-dimensional subspace B and $v = c_{0...0}|0...0\rangle + ... + c_{1...1}|1...1\rangle$ is a state vector of $A's$ 2^l-dimensional subspace $A-B$, where $n = k + l$.

If $|\psi\rangle = |u\rangle \otimes |v\rangle$, i.e.,

$$a_{0...0}|0...0\rangle + ... + a_{1...1}|1...1\rangle = (b_{0...0}|0...0\rangle + ... + b_{1...1}|1...1\rangle) \otimes (c_{0...0}|0...0\rangle + ... + c_{1...1}|1...1\rangle)$$

it turns out that $a_{x_1...x_n} = b_{x_1...x_k} \cdot c_{x_{k+1}...x_n}$ for any $x_1,...,x_n \in \{0,1\}$, i.e., the probability of a basis event $|x_1,...,x_n\rangle$ is equal to the product of probabilities of corresponding basis events in subsystems. Sufficiency follows directly from this observation.

Assumes that denotative random variables induced by $|\psi\rangle$, $|u\rangle$ and $|v\rangle$ satisfy

$$\Pr(X_{|\psi\rangle} = x_u \circ x_v) = \Pr(X_{|u\rangle} = x_u) \cdot \Pr(X_{|v\rangle} = x_v) \quad . \qquad \text{Based} \quad \text{on} \quad \text{the} \quad \text{observation} \quad \text{that}$$

$\Pr(X_{|\psi\rangle} = x_u \circ x_v) = |a_{x_u \circ x_v}|^2$, $\Pr(X_{|u\rangle} = x_u) = |b_{x_u}|^2$ and $\Pr(X_{|v\rangle} = x_v) = |c_{x_v}|^2$, it is easy to check the necessity.

The main tenet of IG is that many important structures in probability theory and statistics can be treated as structures in differential geometry by regarding a space of probabilities as a differential manifold endowed with a Riemannian metric and a family of affine connections [4]. In particular, IG provides a novel method to characterize pure high-order interactions among random variables. According to Proposition 1, IG is relevant to the task of entanglement identification in the post-measurement configuration. Note that most current applications of semantic spaces are essentially in the post-measurement configuration. Hence we can directly investigate the entanglement in semantic spaces using IG.

2.2 Statistical Manifold and Orthogonality

We represent a co-occurrence pattern of words by a random vector with binary components so that the joint distribution of co-occurrence can be exactly expanded by a log-linear model [10]. Let $\mathbf{X} \equiv [X_1, X_2,..., X_n]^T$, $X_i \in \{0,1\}$ be a $n \times 1$ random vector and let $p \equiv p(\mathbf{x})$, $\mathbf{x} \equiv [x_1, x_2,..., x_n]^T$, $\mathbf{x}_i \in \{0,1\}$ be its joint probability distribution. Each X_i indicates that the ith word is present ($X_i = 1$) or absent ($X_i = 0$).

Each distribution $p(\mathbf{x})$ is defined by 2^n probabilities:

$$p_{i_1...i_n} \equiv \Pr\{X_1 = i_1,..., X_n = i_n\} > 0, i_k \in \{0,1\}, 1 \le k \le n, \sum_{i_1,...,i_n} p_{i_1...i_n} = 1$$

Hence, the set of all distributions forms a (2^n-1)-dimensional manifold $\mathbf{S}_n$, where the subscript n of $\mathbf{S}$ denotes the number of random variables. Note that we require $p(\mathbf{x}) > 0$ for all $\mathbf{x}$ since the case of a various support set[4] of $p(\mathbf{x})$ poses rather significant difficulties for analysis. This requirement can be met by any common statistical smoothing method, e.g., Good-Turing estimator [11]. A direct coordinate system of $\mathbf{S}_n$ can be constructed by any 2^n-1 terms among $p(\mathbf{x})$. We refer to this coordinate system as *p-coordinates*.

---

[4] In mathematics, the support of a function is the set of points where the function is not zero, or the closure of that set. Here, the support set refers to the set of terms with nonzero probabilities.

Another coordinate system of S_n is given by the expectation parameters:

$$\eta_i = E[x_i], i = 1,\ldots,n; \eta_{ij} = E[x_i x_j]\ i < j; \cdots; \eta_{12\cdots n} = E[x_1 \cdots x_n] \tag{1}$$

which have also 2^n-1 components. This coordinate system is called η-*coordinates*.

On the other hand, $p(\mathbf{x})$ can be expanded by

$$\log p(\mathbf{x}) = \sum_i \theta_i x_i + \sum_{i<j} \theta_{ij} x_i x_j + \cdots + \theta_{1\cdots n} x_1 \cdots x_n - \psi \tag{2}$$

where ψ is the normalization term corresponding to $\psi \equiv \log p(\mathbf{0})$. It is easy to check that the formula (2) is an exact expansion since all x_is are binary. In addition, if $\mathbf{x}=[0,\ldots,0]^T$, we have $\log p(\mathbf{x})=\log p(\mathbf{0})$. All θ_{ijk}s together have 2^n-1 components and form the so-called θ-*coordinates*.

To characterize pure high-order interactions, we first introduce Riemannian metric tensor which is derived from the Fisher information and orthogonality. We will first give their mathematical definitions in general and then illuminate their meaning in a specific context.

Definition 1 (Fisher Information and Riemannian metric tensor). Given a probability distributions $p(\mathbf{x};\xi)$ parameterized by $\xi \equiv [\xi_1,\ldots,\xi_n]^T \in \Xi$, the **Fisher information** of two coordinate parameters ξ_i and ξ_j is defined by

$$g_{ij}(\xi) \equiv E\left[(\partial/\partial\xi_i)l(\mathbf{x};\xi) \cdot (\partial/\partial\xi_j)l(\mathbf{x};\xi)\right] \tag{3}$$

where $l(\mathbf{x};\xi) \equiv \log p(\mathbf{x};\xi)$ and $E[\bullet]$ denotes the expectation with respect to $p(\mathbf{x};\xi)$. If Fisher information matrix $G(\xi) \equiv (g_{ij}(\xi))$ is nondegenerate for any $\xi \in \Xi$, the parameterized family $S \equiv \{p(\mathbf{x};\xi)\}$ is a Riemannian manifold, and $G(\xi)$ is a **Riemannian metric tensor**.

Definition 2 (Orthogonality). Two coordinate parameters ξ_i and ξ_j are **orthogonal** if the Fisher information of ξ_i and ξ_j vanishes for any $\xi \in \Xi$, i.e.,

$$E\left[(\partial/\partial\xi_i)l(\mathbf{x};\xi) \cdot (\partial/\partial\xi_j)l(\mathbf{x};\xi)\right] = 0 \tag{4}$$

We explain the meaning of Definition 2 by a 3-word example. Using three binary variables X_1, X_2 and X_3 to denote the occurrence of the word w_1, w_2 and w_3 respectively, the joint distribution of X_1, X_2 and X_3 is given by $p(\mathbf{x}) \equiv p_{ijk} = \Pr\{x_1=i, x_2=j, x_3=k\}>0$, $i, j, k \in \{0,1\}$, where $\mathbf{x}=[x_1, x_2, x_3]^T$. It is clear that we need seven free parameters to characterize a distribution because of the constraint $\sum_{ijk} p_{ijk}=1$. Hence, the p-coordinates (Note that the p-coordinates is not unique), η-coordinates and θ-coordinates of this system can be given by:

$$\mathbf{p} \equiv [p_{001}, p_{010}, p_{011}, p_{100}, p_{101}, p_{110}, p_{111}]^T, \mathbf{\eta} \equiv [\eta_1, \eta_2, \eta_3, \eta_{12}, \eta_{13}, \eta_{23}, \eta_{123}]^T, \mathbf{\theta} \equiv [\theta_1, \theta_2, \theta_3, \theta_{12}, \theta_{13}, \theta_{23}, \theta_{123}]^T.$$

Given any p-coordinates of a distribution, the computation of η-coordinates is direct, and the θ-coordinates can be obtained by formula (2). For example, it is easy to check that $\theta_1 \equiv \log(p_{100}/p_{000}), \theta_{12} \equiv \log(p_{110}p_{000}/p_{100}p_{010}), \theta_{123} \equiv \log(p_{111}p_{100}p_{010}p_{001}/p_{110}p_{101}p_{011}p_{000})$ etc. The components of η-coordinates, except the unary marginals, can reflect interactions of words. For example, η_{12} measures the co-occurrence between w_1 and w_2 in the sense that the larger η_{12} is, the more frequent the co-occurrence between w_1 and w_2 is.

The effect of an interaction can be evaluated with respect to a likelihood or log-likelihood function. To be specific, given a η-coordinates $\boldsymbol{\eta}$, η_{12} is natural measure of the interaction between w_1 and w_2. An increment $\Delta\eta_{12}$ of η_{12} will result in increments of log-likelihood function at different **xs**. It is convenient to write these increments in the vector form $\Delta\mathbf{l}(\Delta\eta_{12})\equiv[\Delta l_{000}(\Delta\eta_{12}),...,\Delta l_{111}(\Delta\eta_{12})]^T$, where $\Delta l_{ijk}(\Delta\eta_{12}) \equiv l([i, j, k]^T, \boldsymbol{\eta}')-l([i, j, k]^T, \boldsymbol{\eta})$, i, j, $k \in \{0,1\}$, and $\boldsymbol{\eta}'$ is the same as $\boldsymbol{\eta}$ except that the parameter η_{12} becomes $\eta_{12}+\Delta\eta_{12}$. A natural intuition is that, if another component ξ of $\boldsymbol{\eta}$ is irrelevant to the interaction between word w_1 and w_2, then the vector $\Delta\mathbf{l}(\Delta\xi)$ should be orthogonal to the vector $\Delta\mathbf{l}(\Delta\eta_{12})$. It is easy to check that the parameter orthogonality given in Definition 2 is only a weighted generalization of the orthogonality between the above incremental vectors of the log-likelihood function, and hence shares the essentially identical meaning with the original one. It turns out that we have an intuitive reason to consider a parameter ξ independent of all 2-order interactions if ξ is orthogonal to all η_{ij}s. More technically, this is summarized in Theorem 1.

Theorem 1. Given a coordinate system $\xi\equiv[\xi_1...\xi_n]^T$, if ξ_i is orthogonal to ξ_j, then the Maximum Likelihood Estimation (MLE) of ξ_i is independent of the value of ξ_j.

Theorem 1 technically confirms our intuition on the independence between parameters. It guarantees a nice property of orthogonal parameters, which remarkably simplifies some common procedures of hypothesis test relevant to our purpose. We will revisit this issue in later.

According to the above discussion, it is natural to require that any measure reflecting pure k-order interactions should be orthogonal to all parameters reflecting lower-order interactions. The requirement cannot be met by η-coordinates or θ-coordinates alone. For example, there might often be the dependence between η_{123} and η_{12}. Hence η_{123} can not reflect the pure 3-order interaction. On the other hand, Information geometry assures that the η-coordinates and θ-coordinates are dually orthogonal coordinates.

Theorem 2. Let the η-coordinates and θ-coordinates of S_n be $\boldsymbol{\eta}\equiv[\boldsymbol{\eta}_1,...,\boldsymbol{\eta}_n]^T$ and $\boldsymbol{\theta}\equiv[\boldsymbol{\theta}_1,..., \boldsymbol{\theta}_n]^T$ respectively, where $\boldsymbol{\theta}_1\equiv[\theta_1,..., \theta_n]^T$, $\boldsymbol{\theta}_2\equiv[\theta_{12}, \theta_{13},..., \theta_{(n-1)n}]^T$ and so on, and let $\boldsymbol{\eta}_{k-}\equiv[\boldsymbol{\eta}_1,...,\boldsymbol{\eta}_k]^T$ and $\boldsymbol{\theta}_{k+}\equiv[\boldsymbol{\theta}_{k+1},..., \boldsymbol{\theta}_n]^T$, then in the k-cut mixed coordinate $\zeta_k\equiv[\boldsymbol{\eta}_{k-},\boldsymbol{\theta}_{k+}]$, any θ parameter is orthogonal to all η parameters, and vice versa.

Hence, we can construct the mixed-coordinates, e.g., $\zeta_2\equiv[\eta_1, \eta_2, \eta_3, \eta_{12}, \eta_{13}, \eta_{23}, \theta_{123}]^T$, such that θ_{123} is orthogonal to all η_i and η_{ij}. It can also be shown that θ_{123} is orthogonal to other common interaction measures, e.g., $\text{cov}_{ij}\equiv\eta_{ij}-\eta_i\eta_j$ and the correlation coefficient ρ_{ij}. Furthermore, it is easy to check that, if we generalize the definition of cov and ρ to the high-order case, e.g., $\text{cov}_{ijk}\equiv\eta_{ijk}-\eta_i\eta_j\eta_k$, the above claim still holds accordingly. Another important observation is that the independence of $X_1,...,X_k$ implies $\theta_{1...k}=0^5$. Hence, $\theta_{1...k}$ is a relevant measure of pure k-order interactions. By now, we are able to construct the proper coordinate system aiming at measuring pure

---

[5] We should not require that the converse proposition holds, since $\theta_{1...k}=0$ does not entail the independence of $X_1, X_2,...,X_k$ if there are lower-order dependences among them.

high-order interactions. In practice, the measuring procedure of pure high-order interactions can follow two threads: one is directly parametric estimation of mixed coordinates; the other is computing the KL-divergence between the original state and the surrogate state using the Pythagoras relation entailed by the dual orthogonality of mixed coordinates.

2.3 Parametric Estimation of Mixed Coordinates

It is natural to investigate the pure k-order interaction in the (k-1)-cut mixed coordinate $\zeta_{k-1} \equiv [\eta_{(k-1)\cdot}, \theta_{1\ldots k}]^{T}$ of S_k, since the dual orthogonality gives a simple form of the Fisher information metric, and hence simplifies the estimation procedure of $\theta_{1\ldots k}$.

Given $[\eta_{(k-1)\cdot}, \theta_{1\ldots k}]^{T}$, let us consider a standard procedure of hypothesis test concerning the null hypothesis H_0: $\theta_{1\ldots k} = \theta^{(0)}{}_{1\ldots k}$ against H_1: $\theta_{1\ldots k} \neq \theta^{(0)}{}_{1\ldots k}$. Let the log likelihood of models H_0 and H_1 be

$$l_0 = \max_{\eta_{(k-1)\cdot}} \log p\left(x_1, \ldots, x_N; \eta_{(k-1)\cdot}, \theta^{(0)}_{1\ldots k}\right), \quad l_1 = \max_{\eta_{(k-1)\cdot}, \theta_{1\ldots k}} \log p\left(x_1, \ldots, x_N; \eta_{(k-1)\cdot}, \theta_{1\ldots k}\right)$$

where N is the number of observations.

The likelihood ratio test uses the test statistic $\lambda \equiv 2\log(l_1/l_0)$. It can be shown that $\lambda \sim \chi^2(1)$, where the degree of freedom in Chi-squared distribution is determined by the difference of the free parameter number between l_0 and l_1. Since the distribution of test statistics is known, we can obtain the estimated value of $\theta_{1\ldots k}$. However, the free parameters of l_1 and l_0 are often considerably huge. As a consequence, the computational cost might be prohibitive for the coordinates without dual orthogonality. In the mixed coordinates with dual orthogonality, the likelihood maximization with respect to $\eta_{(k-1)\cdot}$ and $\theta_{1\ldots k}$ can be performed independently, and hence we have

$$l_0 = \log p\left(x_1, \ldots, x_N; \hat{\eta}_{(k-1)\cdot}, \theta^{(0)}_{1\ldots k}\right), \quad l_1 = \max_{\theta_{1\ldots k}} \log p\left(x_1, \ldots, x_N; \hat{\eta}_{(k-1)\cdot}, \theta_{1\ldots k}\right)$$

where $\hat{\eta}_{(k-1)}$ can be estimated independently and kept unchangeable for both l_1 and l_0. Hence, the parametric space is remarkably reduced and the likelihood ratio test becomes feasible.

2.4 Kullback-Leibler Divergence and Pythagoras Relation

The properties of dual orthogonal coordinates entail the generalized Pythagoras theorem, which gives a decomposition of the Kullback-Leibler divergence (KL- divergence for short) such that we can examine different contributions in the discrepancy of two probability distributions, or contributions of different ordered interactions of words.

The KL-divergence between two probabilities $p(x)$ and $q(x)$ is defined by $D[p:q] \equiv \sum_x p(x)\log[p(x)/q(x)]$. Given a distribution $p \in S_k$, let p_m be the distribution that is the closest to p and without pure k-order interactions, We then have $p_m = \arg\min_{q \in E_{(k-1)+}(0)} D[p:q]$, where $E_{(i-1)+}(0)$ is the set of all distributions having no k-order interactions, i.e., $\theta_{1\ldots k}=0$. We refer to p_m as the m-projection of p to $E_{(i-1)+}(0)$. Let the mixed coordinates of p be $[\eta_{(k-1)\cdot}, \theta_{1\ldots k}]^{T}$, then the coordinates of p_m is $[\eta_{(k-1)\cdot}, 0]^{T}$.

An important result of Information Geometry guarantees that KL-divergence can been approximated subject to the Riemannian metric tensor derived from Fisher information:

$$ds^2 = \sum_{i,j} g_{ij}(\xi) d\xi_i d\xi_j = 2D\left[p(\mathbf{x};\xi) : p(\mathbf{x};\xi + d\xi) \right] \qquad (5)$$

This approximation would remarkably simplify the computation of KL-divergence between a distribution p and its m-projection p_m. To explain the Pythagoras relation, we need the following definitions:

Definition 3. A coordinate curve is called an *e-geodesic* if it is given by a linear function $\theta(t)=\mathbf{t}\mathbf{a}+\mathbf{b}$ in the θ-coordinates, where **a** and **b** are constant vector. A coordinate curve is called a *m-geodesic* if it is given by a linear function $\eta(t)=\mathbf{t}\mathbf{a}+\mathbf{b}$ in the η-coordinates, where **a** and **b** are constant vectors.

Theorem 3 (Pythagoras relation). Let p, q and r be three distributions. If the m-geodesic connecting p and q is orthogonal at q to the e-geodesic connecting q and r, then we have D[p:r] = D[p:q] + D[q:r].

Based on Theorem 3, given any p_0 with the coordinate $[\boldsymbol{\eta}'_{(k-1)\text{-}},0]^T$, we have D[p:$p_0$]= D[p:$p_m$]+ D[$p_m$:$p_0$]. The first decomposing term of KL-divergence, i.e., D[p:p_m] offers us another relevant statistic to quantitatively evaluate the level of high-order interactions. Note that the D[p:p_m] can be computed by formula (5).

3 Characterizing High-Order Entanglements in Semantic Spaces

3.1 On Semantic Implications of Pure High-Order Interactions

In this section, we illustrate by two artificial examples the semantic implication of pure high-order entanglements in semantic spaces. Our fundamental idea is: If a set of words as a whole has a significant interaction that cannot be reduced to the compositional effect of lower-order interactions, then this pure high-order interaction implies the emergence of some semantic entity.

Example 1. Given a corpus related to the history of French wars, a word set {w_1=revolution, w_2=Waterloo, w_3=Napoleon} and their occurrence/co-occurrence probabilities:

$$\eta_1 \equiv \frac{\#chunk_1}{\#chunk} = 0.43001, \quad \eta_2 \equiv \frac{\#chunk_2}{\#chunk} = 0.40011, \quad \eta_3 \equiv \frac{\#chunk_3}{\#chunk} = 0.67000$$

$$\eta_{12} \equiv \frac{\#chunk_{12}}{\#chunk} = 0.18001 \quad \eta_{13} \equiv \frac{\#chunk_{13}}{\#chunk} = 0.40000, \quad \eta_{23} \equiv \frac{\#chunk_{23}}{\#chunk} = 0.44000$$

$$\eta_{123} \equiv \frac{\#chunk_{123}}{\#chunk} = 0.18000$$

where η_i is the marginal occurrence probability of w_i's in all chunks (a chunk is a unit fragment of text, e.g., within a window, a paragraph, a section or a document.), η_{ij} is the co-occurrence probability of w_i and w_j, η_{123} is the joint co-occurrence probability of w_1, w_2 and w_3, '#chunk' is the total number of chunks, #chunk$_i$ is the number of chunks in which w_i occurs, #chunk$_{ij}$ is the number of chunks in which w_i and w_j co-occur simultaneously and so on.

In example o1, there is a correlation between the occurrences of 'revolution' and 'Napoleon' since the early life of Napoleon is closely related to the France revolution. There is also a correlation between the occurrences of 'Waterloo' and 'Napoleon' since the Waterloo battle ended the myth of Napoleon. Because both 'revolution' and 'Waterloo' are correlated with 'Napoleon', there is also a significant interaction among these three words. We consider the interaction of these three words *significant* if $\eta_{123} > \eta_1\eta_2\eta_3$, e.g., the joint occurrence probability is significantly greater than the product of marginal occurrence probabilities.

It is clear that the set {revolution, Napoleon, Waterloo} cannot be naturally mapped to a realistic event or a specifically semantic entity even if there is an obvious interaction among these three words. One may argue that the whole of these three words is still meaningful since both 'revolution' and 'Waterloo' are related to 'Napoleon', and hence the combination of 'revolution' and 'Waterloo' offers a more complete picture on 'Napoleon'. However, this 3-word correlation is not a *pure* 3-order correlation. Specifically, let us assume that we have already known there were two significant 2-word correlations, i.e., the correlation between 'revolution' and 'Napoleon' and the correlation between 'Waterloo' and 'Napoleon', then it is natural to consider that 'Napoleon' is related to 'revolution' and 'Waterloo' even if we have no any knowledge on the 3-word interaction. It turns out that the extra knowledge on the existence of a 3-word interaction offers nothing new for us. The above insight is confirmed by the observation of $\eta_{123} \approx \eta_{13}\eta_{23}$, which implies that the obvious interaction of w_1, w_2 and w_3 can be explained by a coincidence of two pairwise events. Consequently, in many applications, e.g., query expansion in information retrieval, the 3-order correlation between {revolution, Napoleon, Waterloo} may not bring much added value then the consideration of the individual 2-order correlations, i.e., between 'revolution' and 'Napoleon' and between 'Waterloo' and 'Napoleon'.

Example 2. Given the same corpus, another word set {w_3=Napoleon, w_4= invasion, w_5= Spain} and the corresponding occurrence/co-occurrence probabilities:

$$\eta_3 \equiv \frac{\#chunk_3}{\#chunk} = 0.5801, \quad \eta_4 \equiv \frac{\#chunk_4}{\#chunk} = 0.3701, \quad \eta_5 \equiv \frac{\#chunk_5}{\#chunk} = 0.0804$$

$$\eta_{34} \equiv \frac{\#chunk_{34}}{\#chunk} = 0.2800, \quad \eta_{35} \equiv \frac{\#chunk_{35}}{\#chunk} = 0.0801, \quad \eta_{45} \equiv \frac{\#chunk_{45}}{\#chunk} = 0.0801$$

$$\eta_{345} \equiv \frac{\#chunk_{345}}{\#chunk} = 0.0800$$

The high-order interaction that makes better sense semantically is the pure high-order interaction. In 3-word cases, roughly speaking, a pure 3-order interaction should meet the condition $\eta_{123} > \eta_{12}\eta_{23}$, $\eta_{123} > \eta_{13}\eta_{23}$, $\eta_{123} > \eta_{12}\eta_{13}$, $\eta_{123} > \eta_1\eta_2\eta_3$, $\eta_{123} > \eta_1\eta_{23}$, $\eta_{123} > \eta_2\eta_{13}$ and $\eta_{123} > \eta_3\eta_{12}$, i.e., the joint probability indicating a pure high-order interaction should be greater than any possible compositional effect of lower-order correlations. In Example 2, since Napoleon launched a series of famous invasions, there is a high correlation between 'Napoleon' and 'invasion'. On the other hand, since Spain is not very important during Napoleon's life except for a short period during Spain war, there is only a relatively low correlation between 'Napoleon' and 'Spain'. However, η_{345} is approximately equal to η_{35} and η_{45} since Napoleon's invasion to Spain is the most important event relating Napoleon to Spain. Hence we have $\eta_{345} > \eta_{34}\eta_{35}$.

Furthermore, it is easy to check that we have $\eta_{345}>\eta_{34}\eta_{45}$, $\eta_{345}>\eta_{35}\eta_{45}$, $\eta_{345}>\eta_{3}\eta_{4}\eta_{5}$, $\eta_{345}>\eta_{3}\eta_{45}$ and so on. Therefore, η_{345} is significant greater than any possibly compositional effect of lower-order interactions. Hence, we can conclude that there exists a pure 3-order interaction of w_3, w_4 and w_5, which cannot be explained by a coincidence of lower-order events and implies an emergence of a semantic entity corresponding to the event of Napoleon's invasion to Spain.

It should be noted that we can also define η_i to be consistent with the conventional vector model if the η_i is computed with respect to the chunk of a word. In this case, all the above discussions are essentially similar subject to a minor modification.

The above discussion seems to imply a method identifying pure high-order interaction, i.e., by checking whether η_{345}-$\eta_{3}\eta_{4}\eta_{5}$, η_{345}-$\eta_{35}\eta_{45}$, η_{345}-$\eta_{34}\eta_{45}$, and so on, are greater than zero. However, this naïve method is in general difficult to be applied. For illustration, let us consider the task of identifying k-order pure interactions by an exhaustive search. First, we have to check whether the k-order interaction is significant than any possible bipartition coincidence. Hence we need to compare $\sum_{i=0}^{k}C_{k}^{i}=2^{k}$ configurations. Second, we have to check whether the k-order interaction is significant than any possible tri-partition coincidence. It turns out that we also have to check all possible l-partitions ($l \le k$). In summary, the number of configurations that we need to check is given by the Bell number B_k. Recall that the exponential generating function for Bell numbers is $\sum_{n=0}^{\infty}\frac{B_n}{n!}z^n=e^{e^z-1}$, it is, in general, prohibitively complex.

Furthermore, the difficulty of an exhaustive search strategy also lies in its intrinsic unstableness in practice, especially for small corpus since we can only control the search procedure by a set of ad-hoc thresholds, which is lack of theoretical guarantees. On the other hand, by IG method, the measure of any k-order pure interaction can be given by a closed-form formula. In addition, we can perform some rigorously-established estimation procedure, e.g., the likelihood ratio test introduced in Subsection 2.2, to quantitatively determine how significant our decision is.

3.2 Characterizing Pure High-Order Interactions by Information Geometry

Information Geometry offers a promising method to estimate pure high-order interactions. The likelihood ratio test described in Subsection 2.3 can be directly applied to estimate the statistic $\theta_{1...k}$ which measures pure k-order interactions. Moreover, as described in Subsection 2.4, we can measure the level of high-order interactions by decomposing the KL-divergence with respect to a proper m-projection. As a demonstration, we can directly derive θ-parameters from the p-coordinates as shown in the following.

In Example 1, the η-coordinates is given. It is easy to obtain the p-coordinates from η-coordinates by solving a simple linear system. According to p-coordinates, its θ-parameters are θ_{12}=-0.0004, θ_{13}=5.3932, θ_{23}=11.264, θ_{123}=-3.4584. The negative value of θ_{123} indicates that, although η_{123} is large in absolute value, there is no pure 3-order interaction among the corresponding words. Moreover, the interaction level among w_1, w_2 and w_3 is lower than the compositional effect of lower-order interactions. In Example 2, the θ-parameter are θ_{34}=0.8926, θ_{35}=0.5991, θ_{45}=0.6049 and θ_{345}=6.4852.

The positive value of θ_{345} indicates that, although η_{345} is small in the absolute value, there is still a significant pure 3-order interaction among w_3, w_4 and w_5.

3.3 An Extended Vector Model with Pure High-Order Interactions

To investigate semantic implications of high-order interactions, we extend the conventional vector model so that it can incorporate high-order interactions. Traditionally, the marginal distribution of words has acted as the language model in IR (Information Retrieval), MT (Machine Translation) and NLP (Natural Language Processing) because a general higher-order model is often computationally expensive even in the 2-order case. However, in many practical applications, it is unnecessary to construct a general high-order model involving all high-order interactions. On the other hand, it is often sufficient to comprise only a small proportion of high-order interactions in a context-sensitive way, e.g., the pure high-order interaction corresponding to some specific subject. This idea is formalized in the following.

Definition 4 (Vector Model). Given a word set $\{w_1,\ldots,w_n\}$ derived from a corpus C, a text's (corpus's) Vector Model (VM) with respect to $\{w_1,\ldots,w_n\}$ is the marginal distribution $[p_1,\ldots,p_n]^T$ of this text (corpus), where p_i is the marginal probability of w_i.

Definition 5 (Extended Vector Model). Given a word set $\{w_1,\ldots,w_n\}$ derived from a corpus, a text's (corpus's) Extended Vector Model (EVM) is composed by the marginal distribution and some statistics measuring the pure high-order interaction, and has the following form: $\left[p_1,\ldots,p_n,\theta_{i_1\ldots i_{k_1}},\theta_{j_1\ldots j_{k_2}},\ldots\right]^T$ or $\left[p_1,\ldots,p_n,D_{i_1\ldots i_{k_1}},D_{j_1\ldots j_{k_2}},\ldots\right]^T$, where $\theta_{i_1\ldots i_{k_1}}$ is the θ parameter subject to the joint distribution of $\left\{w_{i_1},\ldots,w_{i_{k_1}}\right\}$, $D_{i_1\ldots i_{k_1}}$ is the KL-divergence between p and p_m subject to $\left\{w_{i_1},\ldots,w_{i_{k_1}}\right\}$ (see Subsection 2.4), $p_1,\ldots,p_n$ is the marginal probability of $w_1,\ldots,w_n$.

3.4 Practical Applications in Text Classification and Query Expansion

The remaining issue is to determine what θ or D should be included in an EVM. This issue can only be clarified in specific application backgrounds. We give two examples to explain this issue.

In the task of supervised text classification, it is useful to extract a set of words for each class representing the class subject so that the classification model can be designed accordingly. These sets of theme words can be obtained, in principle, by finding out the word set having significantly pure high-order interactions with respect to the joint distribution of the corresponding class. This finding procedure can be efficient by the aid of prior knowledge. For example, if a few initial theme words are given, it is natural to only search possible pure high-order interactions involving some of prior theme words. Even if there is no prior knowledge on class' subjects, the pure high-order interactions relevant to a specific class can be found by checking, e.g., the mutual information between high-order interactions and class labels. Another method evaluating the relevance between pure k-order interactions and class subjects is to compare the class label of the original state and the surrogate state (see Section 1) with vanishing pure k-order interactions. The surrogate states can be obtained by direct searching over the corpus or manual construction. In the latter case, the ficti-tious class label of a surrogate state is determined by the classification model trained with respect to the EVM involving the pure k-order interactions.

In query expansion tasks, it is desirable to mine the pure high-order interactions in-volving some of query words so that the marginal language model can be expanded accordingly. We suggest that the pure high-order interaction involving query words would be an indication of relevance of the query theme. The following is a brief algo-rithmic framework:

1. Collect top ranked initial retrieval results into a set S_I
2. Search word subsets involving some query words and other words, and compute the pure high-order interactions.
3. Construct S_I's EVM by incorporating the pure high-order interactions mined in step 2.
4. Get new search results based on the derived EVM. There can be a number of ways to do that, for example, by using the EVM as a relevance language model to filter or re-rank S_I or to expand the initial query using words with pure high-order interactions with query words; etc.

4   Conclusions and Further Work

Pure high-order entanglements in lexical semantic spaces indicate the emergence of high-level semantic entities. To characterize the intrinsic order of entanglements and distinguish pure high-order entanglements from lower-order ones, we develop a set of methods in the framework of Information Geometry. Based on the developed method, we present an expanded vector space model that involves context-sensitive high-order information and aims at characterizing high-level context. Several examples with specific application backgrounds, e.g., query expansion and text classification, are discussed, and an algorithmic framework incorporating our method in query expan-sion are proposed. The further work is to carry out practical experiments and develop more efficient algorithms to implement the proposed framework. To this end, some nice properties of pure high-order correlations, e.g., sub-inheritance, can be used to improve the computational efficiency.

Acknowledgements

This research is funded in part by the UK's Engineering and Physical Sciences Re-search Council, grant number EP/F014708/1 and Natural Science Foundation of China (Grant 60603027).

References

1. Salton, Gerard, Buckley, C.: Term-weighting approaches in automatic text retrieval. In-formation Processing & Management 24(5), 513–523 (1988)
2. Nielsen, M., Chuang, I.: Quantum Computation and Quantum Information. Cambridge University Press, Cambridge (2000)

3. Horodecki, R., Horodecki, P., Horodecki, M., Horodecki, K.: Quantum entanglement, http://arxiv.org/abs/quant-ph/0702225
4. Amari, S., et al.: Method of Information Geometry. AMS series. Oxford University Press, Oxford (2000)
5. Amari, S.: Differential Geometrical Method in Statistics. Springer, Heidelberg (1985)
6. Amari, S.: Information Geometry on Hierarchy of Probability Distributions. IEEE Transactions on Information Theory 47, 1701–1711 (2001)
7. Nakahara, H., Amari, S.: Information-Geometric Measure for Neural Spikes. Neural Computation 14, 2269–2316 (2002)
8. Rao, C.R.: Information and accuracy attainable in the estimation of statistical parameters. Bull. Calcutta. Math. Soc. 37 (1945)
9. Jeffreys, H.: An invariant form for the prior probability in estimation problems. Proc. Roy. Soc., A 196 (1946)
10. Christensen, R.: Log-Linear Models and Logistic Regression. Springer, Heidelberg (1997)
11. Orlitsky, A., Santhanam, N., Zhang, J.: Always Good-Turing: Asymptotically optimal probability estimation. Science 302
12. Bruza, P.D., Kitto, K., Nelson, D., McEvoy, C.L.: Entangling Words and Meaning. In: Proceedings of the Second Quantum Interaction Symposium, pp. 118–124. College Publications (2008)
13. Bruza, P.D., Woods, J.: Quantum Collapse in Semantic space: Interpreting Natural Language Argumentation. In: Proceedings of the Second Quantum Interaction Symposium, pp. 141–147. College Publications (2008)
14. Bruza, P.D., Cole, R.J.: Quantum Logic of Semantic Space: An Exploratory Investigation of Context Effects in Practical Reasoning. In: Artemov, S., Barringer, H., d'Avila Garcez, A.S., Lamb, L.C., Woods, J. (eds.) We Will Show Them: Essays in Honour of Dov Gabbay. College Publications (2005)
15. Burgess, C., Livesay, L., Lund, K.: Explorations in Context Space: Words, Sentences, Discourse. In: Foltz, P.W. (ed.) Quantitative Approaches to Semantic Knowledge Representation, Discourse Processes, vol. 25(2, 3), pp. 179–210 (1998)
16. Landauer, T., Dumais, S.: A Solution to Plato's problem: The latent semantic analysis theory of acquisition, induction, and representation of knowledge. Psychological Review 104(2), 211–240 (1997)
17. Widdows, D.: The Geometry of Meaning. Center for the Study of Language and Information/SRI (2004)
18. Plenio, M.B., Virmani, S.: Quant. Inf. Comp. 7, 1 (2007)
19. Eisert, J., Gross, D.: Lectures on Quantum Information. In: Bruß, D., Leuchs, G. (eds.). Wiley-VCH, Weinheim (2007)
20. Adesso, G., Illuminati1, F.: Genuine multipartite entanglement of symmetric Gaussian states: Strong monogamy, unitary localization, scaling behavior, and molecular sharing structure, October 11 (2008), arXiv:0805.2942v2 [quant-ph]

Semantic Vector Combinations and the Synoptic Gospels

Dominic Widdows[1] and Trevor Cohen[2]

[1] Google, Inc.
widdows@google.com
[2] Arizona State University
trevor.cohen@asu.edu

Abstract. This paper applies some recent methods involving semantic vectors and their combination operations to some very traditional questions, including the discovery of similarities and differences between the four Gospels, relationships between individuals, and the identification of geopolitical regions and leaders in the ancient world. In the process, we employ several methods from linear algebra and vector space models, some of which are of particular importance in quantum mechanics and quantum logic.

Our conclusions are in general positive: the vector methods do a good job of capturing well-known facts about the Bible, its authors, and relationships between people and places mentioned in the Bible. On the more specific topic of quantum as opposed to other approaches, our conclusions are more mixed: on the whole, we do not find evidence for preferring vector methods that are directly associated with quantum mechanics over vector methods developed independently of quantum mechanics. We suggest that this argues for synthesis rather than division between classical and quantum models for information processing.

1   Introduction

Semantic vector approaches have been used with considerable research success in recent years. Applications have included information retrieval, automatic word sense discrimination, ontology acquisition, and the creation of practical aids to document annotation and translation.

During the recent period in which these tools have been developed, most empirical research in computational linguistics has been devoted to large and rapidly growing corpora. This is for very good reasons. Many current information needs are greatest when dealing with the recent explosion in the scale of available information. The rapidity with which information sources such as the World Wide Web have developed has forced the adoption of new information search and exploration strategies, some not previously possible or necessary.

At the same time, much cultural and literary scholarship focusses (appropriately) on comparatively small and well organized corpora — studying (for example) works that have long been established as scriptures and classics. Resources in the form of concordances, cross references, and commentaries, have

P. Bruza et al. (Eds.): QI 2009, LNAI 5494, pp. 251–265, 2009.

been readily available in paper form for many of the scriptures and classics for some centuries, and these information modalities are very much the prototypes for today's electronic indexes, hyperlinks, and commenting, tagging, annotation and collaborative filtering systems.

This paper tries to take a step that may be considered retrograde, or at least retrospective: to see what recent advances in empirical semantic vector analysis may have to say on some simple issues in literary scholarship, particularly Biblical scholarship. Naturally, our goal is not to discover something as yet unseen in a field which has had many careful research lifetimes already devoted to it: rather, it is to see if a very simple mathematical machine can retrieve any comparable results, and to see if this sheds any useful light on techniques of automatic information analysis more generally. In the process, we hope to demonstrate and test some recent developments in semantic vector methodology, particularly with regard to semantic combination and composition operations.

It is hoped that this latter aspect of the work presented will be of particular interest to the quantum interaction community: specifically because some of the vector combination techniques relate directly to operations used in quantum mechanics (in particular eigenvalue decomposition) and quantum logic (particularly the non-distributive disjunction). At the same time, other techniques in vector mathematics including permutation and clustering are also useful in semantic analysis. If vector operations can be largely categorized as "quantum" or "non-quantum", there seems to be no experimental reason at this stage for preferring the "quantum" over the "non-quantum" vector operations. This may help to inform the investigation of questions about the developing focus of "quantum interaction" as an area derived from quantum physics, or an area evolving at least somewhat independently, and about how this field should be characterized.

2   Methods Used

The semantic vector methods used in this paper are descendants of the vector model for information retrieval, and the subsequent development of latent semantic analysis, which compresses the sparse information in the vector model's term by document matrix into a more condensed, lower-dimensional representation. The relationship between these structures and the quantum logic of Birkhoff and von Neumann [1] has been further recognized in recent years (see particularly [2,3]).

Particular methods used from these models include:

- Vector sum for composition, from the earliest vector model search engines [4].
- Singular Value Decomposition for more compressed semantic representation, from Latent Semantic Analysis [5].
- The use of subspaces as another more generalizing model for disjunction [2,6,3].
- The use of orthogonality to model complementation and negation [3, Ch. 7].

Other methods from the wider literature that are used particularly include:

- Clustering for finding more stable units (c.f., 'quanta') among observed results, as developed for word sense discrimination [7].
- Visualization of groups of word vectors using principal component plotting [8].
- The recent permutation based construction of semantic vectors [9].
- Pathfinder link analysis, a graph construction and visualization method [10].

The data used in our experiments is principally the King James Bible, that is, the translation into English of Jewish (Hebrew language) and Christian (Greek language) scriptures, authorised under King James (VI of Scotland, I of England), dated to 1611.

Nearly all of the software and corpora used in this paper is freely available and relatively easy to use. The corpus is from Project Gutenberg (www.gutenberg. org). Free software components are from Apache Lucene (lucene.apache.org), the Semantic Vectors project (semanticvectors.googlecode.com), and the Java Matrix Package (math.nist.gov/javanumerics/jama) used for singular value decomposition.

3   Semantic Vectors and the Synoptic Gospels

This section describes our single most deliberate experiment: testing to see if vector analysis discerns the similarity of the Synoptic Gospels. Since at least the second century AD, the Christian writings gathered in the New Testament have included four canonical accounts of the activities of Jesus of Nazareth (ca. 5BC - 30AD), and these writings, called the Gospels, have since the earliest times been attributed to authors called Matthew, Mark, Luke, and John. A basic tenet of New Testament scholarship is that Matthew, Mark and Luke are closely related, with much material drawn from one another or at least from common sources. For this reason, these three Gospels are referred to as the Synoptic (Greek, "joined eye") Gospels.

3.1   Vector Sum Similarity

In this experiment, we set out to discover whether a semantic vector model built from the text of the King James Bible shared the view that Matthew, Mark and Luke are similar and John is the odd one out. Semantic vectors for terms (frequency > 10, stopwords removed) were produced using random projection (reduced dimension = 200) on the Lucene term-by-document matrix, as implemented in the SemanticVectors package [11]. Random projection is a computationally efficient variant of Latent Semantic Analysis: instead of computing exactly orthogonal latent axes using Singular Value Decomposition, latent axes are chosen randomly, based on the mathematical property that randomly chosen axes can be demonstrated to be nearly orthogonal in a suitably quantifiable sense [12].

Table 1. Cosine similarities between the Gospels, whole Bible model

	Matthew	Mark	Luke	John
Matthew	1	0.995	0.998	0.990
Mark		1	0.996	0.987
Luke			1	0.989
John				1

Table 2. Cosine similarities between the Gospels, Gospels only model

	Matthew	Mark	Luke	John
Matthew	1	0.990	0.994	0.969
Mark		1	0.991	0.968
Luke			1	0.969
John				1

Document vectors for each chapter were produced using the (normalized) weighted vector sum of term vectors, and combined vectors for each of the four Gospels were computed as a normalized vector sum of the document vectors representing the chapters of each book. (This latter sum is implemented on the fly using a useful regular expression matching query builder applied to the filesystem paths: this technique can be easily used for other potentially interesting aggregate queries, such as producing query terms combining many morphological variants of the same root.) Pairwise similarities between the four resulting vectors were computed, and are shown in Table 1 and Table 2. The first table shows similarities in a model computed using the entire King James Bible, the second one shows similarities in a much smaller model computed using only the Gospel texts themselves.

Two things are immediately apparent. Firstly, the similarities are on the whole very high. Often nearest neighbour similarities in such models range from 0.3 to 0.7 (see the Tables later in this paper for a sample of reasonably typical values), so any cosine similarity greater than 0.9 is very high. It appears that the commonalities between the Gospels (e.g., use of frequent terms) outweigh their differences by a long way. This may be due to the "bag of words" nature of the creation of document vectors. In bag of words methods, the order of words is not taken in to account — in this case, due to the commutative property of vector addition. Thus if the Gospels share many common words with typical frequencies, they will have similar document vectors. By comparison, average similarities between the Gospels and earlier Old Testament works tend to be in the range of 0.9 to 0.95 (see Table 3). It is reasonable that these are lower similarities, though they are still high, and some statistical analysis of document creation and term reuse may help to account for this.

Secondly, even within these very close results, John is clearly the odd one out, having lower similarities with all of the other Gospels than are found in between the three Synoptic Gospels. This is particularly apparent in the smaller model, though this appears to be partly because the smaller model shows similar

Table 3. Cosine similarities between the Gospels and a sample of Old Testament books

	Matthew	Mark	Luke	John
Exodus	0.945	0.932	0.946	0.921
1 Kings	0.949	0.942	0.956	0.926
Psalms	0.934	0.912	0.934	0.929
Jeremiah	0.950	0.931	0.950	0.934

comparisons but distributed across a wider range of scores. We note in passing that these experiments were repeated several times with different dimensions (ranging from 100 to 1000), with remarkably similar and often exactly the same results.

3.2   Cluster Comparison of Chapters

Another way of analysing similarities and differences between the Gospels is to cluster the individual chapter vectors (instead of summing them into combined book vectors). Clustering chapters provides a much richer qualitative analysis, at a greater computational cost. However, for a dataset the size of the Gospels (89 chapter vectors), this cost is trivial in contemporary terms. The clusters in our experiments are produced using the k-means algorithm: at each stage of the algorithm, each vector is assigned to its nearest cluster centroid, and then the centroids of the clusters are recomputed based on the new assignment. An implementation of this algorithm is included in the SemanticVectors package.

The results with 20 clusters clearly demonstrate the distinctive nature of John's Gospel. The chapters of John's Gospel tend to appear in tight clusters, a majority of whose members are from the same Gospel: on the other hand, if a cluster contains chapters from one of the Synoptic Gospels, it is far more likely to include chapters from others of these Gospels. A simple quantitative measure of the distinct nature of John's Gospel can be obtained using conditional probability: given that one chapter in a cluster is from a particular Gospel, what is the probability that another chapter in the same cluster is from the same Gospel? Typical results obtained in this experiment were:

John: 0.66 Matthew: 0.28 Luke: 0.24 Mark: 0.18.

Note that due to the random initialization of clusters, results from clustering runs are not identical each time. In each of several runs, the score for John was above 0.5, a threshold never breached by any of the other Gospels. This shows that that the chapters in John's Gospel have, on average, stronger mutual similarities than those of the other three Gospels, which are much more easily mixed together.

A further interesting experiment would be to extend the cluster analysis to cover pairwise conditional probabilities, to see if these reflect the known patterns of how the Synoptic Gospels borrowed from each other.

It is interesting to note that authorship, though important, is only one variable that influences similarity in our results. Sometimes describing similar content is more clearly responsible for similarity: for example, the four element cluster {Luke 23, Matthew 27, John 19, Mark 15} appears in several experimental runs, and each of these four chapters contains the author's account of the crucifixion.

We may conclude that, when asked "Which of the Gospels are similar?", the vector model answers "Matthew, Mark and Luke are similar, John is a bit different", but the model is also sensitive to factors other than authorship, that sometimes produce stronger affinities between texts.

4 Visualization of Disjunctions

This section describes experiments in visualizing the effects of different combination operations on search results. Lists of related terms to the query "jesus + abraham + moses" were obtained using three different query building and search ranking methods:

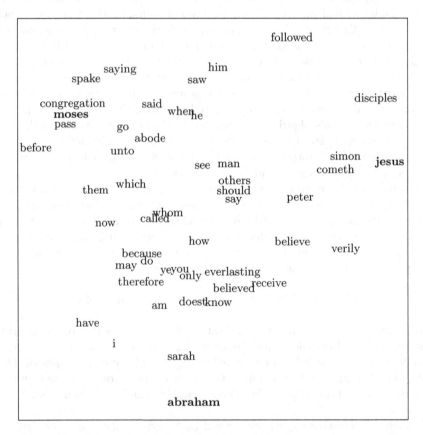

Fig. 1. Neighbours of "Jesus", "Moses" and "Abraham", using Vector Sum

1. Vector sum of the constituent vectors. (Figure 1.)
2. Quantum disjunction of the constituent vectors: that is, results are ranked according to their proximity to the subspace spanned by the query vectors. (Figure 2.)
3. Minimum distance (maximum similarity) to any one of the constituent vectors. (Figure 3.)

The search results are projected down to 2 dimensions by computing the singular value decomposition (using the Jama package) and by plotting the vectors according to the second and third coordinates of their reduced vectors (the first component often mainly says "all the data is somewhere over here in the semantic space" [8]). The plotting itself is performed using a small Java Swing utility from SemanticVectors.

On analysis, the main distinction in the results is between the maximum similarity method and the other two. The maximum similarity method produces, as expected, several results that are similar to just one of the constituents, rather

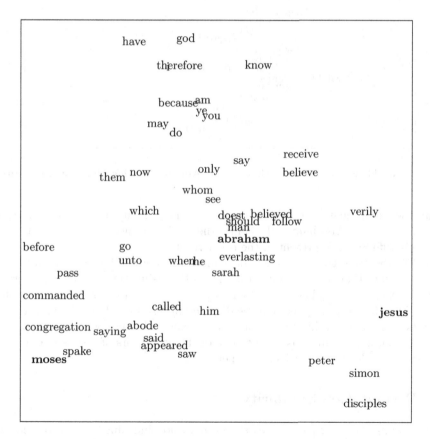

Fig. 2. Neighbours of "Jesus", "Moses" and "Abraham", using Vector Subspace

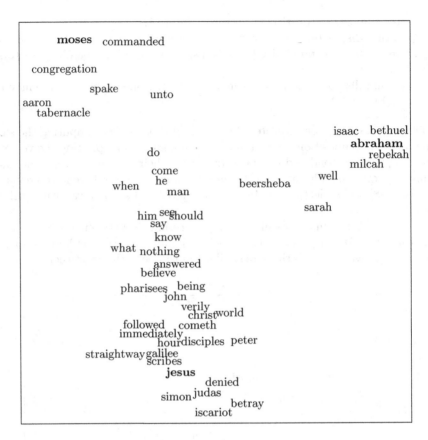

Fig. 3. Neighbours of "Jesus", "Moses" and "Abraham", using Maximum Similarity

than their more general combination. For example, many of the close relatives and associates of Abraham make it into the minimum distance results, whereas only his wife Sarah is present in the other results.

While the other two results sets have much in common, including many more general terms, there is a small suggestion that the disjunction similarity preserves some of the close neighbours as well as the more general terms, for example, Moses' brother Aaron appears in the disjunction results and not the vector sum results. This is something we should have expected, since with the quantum disjunction, if an element is close to one of the generators of a subspace, it will naturally be close to the subspace generated.

5   Permutation Similarity

Our final set of experiments uses the permutation indexing method developed by Sahlgren et al [9], and demonstrates that this method is a powerful enhancement over raw vector similarity at the task of extracting the names of ancient

Table 4. Permutation and similarity results for geopolitical entities

Permutation query "*king of ?*"		Similarity query "*king*"		Similarity query "*assyria*"	
0.728	assyria	1.000	king	1.00	assyria
0.699	babylon	0.441	province	0.653	sennacherib
0.662	syria	0.408	reign	0.628	rabshakeh
0.647	zobah	0.380	had	0.626	hezekiah
0.604	persia	0.378	did	0.575	hoshea
0.532	judah	0.377	came	0.509	amoz

kingdoms from the Bible corpus. In essence, the permutation method works by indexing each term not only as a sum of terms in the surrounding context, but as a *permuted* sum, the permutation in coordinates being governed by the relative positions of the words in question. (A more geometric interpretation can be obtained by noting that many permutations of coordinates are effectively rotations in the semantic space.)

Table 4 shows that, in these cases, results from the permutational query (left hand column) are much more specific in their relationships than those of traditional similarity queries (center and right hand column). In the permutation results, the query "king of ?" finds fillers for the target "?" based on cosine similarity with the permuted vectors for "king" and "of", and picks out purely the names of geopolitical regions in the ancient world. By contrast, if we were to try and construct such a list using traditional cosine similarity, either with a seed example such as "assyria" or one of the same query terms, "king", the results are much less accurate.

Note that in the results presented here, the permutation model was built without removing stopwords (which preserves the integrity of patterns based on exact word order), whereas the similarity results were obtained by removing stopwords as usual. From studying examples, we believe this choice is optimal for each model so makes for a reasonably fair comparison.

As the "king of ?" permutation query illustrates, near neighbours in permutation derived spaces tend to be of the same semantic type (in this case, they are all kingdoms). However, these neighbours need not be thematically related. For example, a query for "adam" in a permutation-based space retrieves the cast of (male) biblical characters in the left-hand column of Table 5. Several of these characters neither appear together in the scripture, nor are they genealogically related.

In contrast, the nearest neighbours of "adam" in a vector space constructed using term-document statistics without regard for word order appear in the right-hand column of Table 5. While these results do include biblical characters (some of Adam's descendants), other elements of the Story of the Fall are also included.

These two types of indexing capture different types of relations between terms. Moreover, it is possible to construct a vector space that combines these relations by using trained (rather than random) term vectors as the basis for

Table 5. Neighbours of "adam" in different semantic spaces

Permutation-based space (word order encoded)	Term-document space (word order ignored)
1.00 adam	1.00 adam
0.676 joseph	0.552 enoch
0.654 saul	0.518 garden
0.641 aaron	0.505 lamech
0.639 noah	0.444 eden
0.638 david	0.407 sixty

Table 6. Combining order-based and order-agnostic vectors

Search for "king ?"		Search for "queen ?"	
Random vector basis	Term vector basis	Random vector basis	Term vector basis
0.756 ahasuerus	0.604 ahasuerus	0.187 desiring	0.332 vashti
0.643 agrippa	0.571 agrippa	0.184 exhort	0.314 ahasuerus
0.493 ahaz	0.465 rehoboam	0.181 whithersoever	0.302 agrippa
0.464 rehoboam	0.451 ahaz	0.172 vashti	0.288 absent
0.401 delighteth	0.450 delighteth	0.168 equity	0.287 darius

a permutation-based space. Each term is then indexed as the coordinate-based permuted sum of a set of meaningful term vectors. This hybrid vector space includes both thematic and order-based associations, supporting a simple sort of inference: queries for "queen ?" retrieve the names of kings as well as queens (see Table 6).

Another way to combine the strengths of these types of indexing procedures is to use the associations generated with one indexing procedure to evaluate relations between nearest neighbours generated in the other. This combination allows for the construction of queries such as "what thing of the same semantic type as 'abraham' is most strongly associated with him" (isaac 0.538).

Figure 4 illustrates the sort of information that can be extracted by combining order-based and order-agnostic representations. The nodes in the network were determined by finding the thirty nearest neighbours of the normalized sum of the vectors for the terms "abraham" and "moses" in in a permutation-based space (d=500, frequently occurring terms included). Nearest-neighbor searches in permutation-based spaces tend to produce results of the same semantic type as the search terms, in this case male biblical characters (aside from the cities Ekron and Hazor). However, these neighbours are not necessarily thematically related: many of these characters are not genealogically related, nor do they appear together in any biblical stories.

In contrast, the links in Figure 4 were determined using an order-agnostic vector space. Initially all nodes were linked according to the cosine similarity between them. The most significant links were identified using Pathfinder network scaling [10], which prunes networks such that no two nodes are linked directly if there is a shorter pathway between them via other nodes. Scaling and

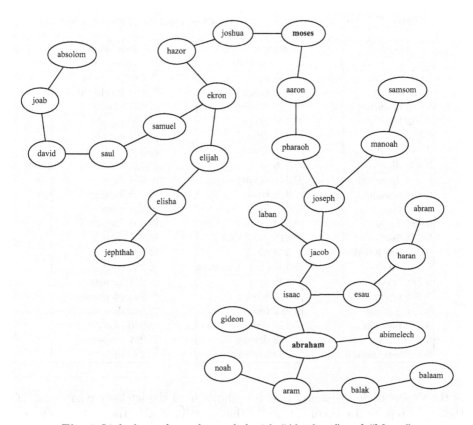

Fig. 4. Linked search results seeded with "Abraham" and "Moses"

visualization were performed with a specially provided version of the Pathfinder software package, presently under development by Roger Schvaneveldt (the diagram has been redrawn by hand in Figure 4 for presentation in print). Pathfinder has preserved several genealogical links, such as the subtree linking Abraham, Isaac, Esau, Jacob and Joseph, and the link between Moses and Aaron. Other personal relationships are also preserved. Elijah is linked to his disciple Elisha, Saul is linked to his successor David, and Absalom is linked to his murderer, Joab. The development of further methods to combine these types of vector spaces is likely to be a fertile area for future research.

The connections from "pharaoh" to the terms "aaron" and "moses" on the one hand and "joseph" on the other are of particular interest as it indicates that the vector representation for the term "pharaoh" refers to at least two distinct individuals. Two different Pharaohs, generations apart from one another, were involved with these different characters. As is the case with ambiguous terms, it is possible to use quantum negation [3, Ch 7] to isolate different senses of a particular vector representation, as illustrated in Table 7. Initially (leftmost column), the vector representation for "pharaoh" is dominated by elements of the biblical story in which Joseph averts famine in Egypt by interpreting the dreams

Table 7. Teasing apart ambiguous pharaoh using quantum negation

pharaoh	pharaoh NOT joseph	pharaoh NOT joseph famine
1.000 pharaoh	0.839 pharaoh	0.783 pharaoh
0.626 egypt	0.501 magicians	0.514 hardened
0.616 favoured	0.492 hardened	0.497 magicians
0.562 kine	0.488 egypt	0.419 egypt
0.543 joseph	0.442 kine	0.362 egyptians
0.543 magicians	0.432 favoured	0.358 enchantments
0.523 ill	0.358 ill	0.333 flies
0.504 dreamed	0.347 egyptians	0.326 frogs
0.499 famine	0.340 river	0.326 kine
0.452 food	0.324 land	0.317 river
0.439 dream	0.324 famine	0.307 favoured
0.435 land	0.315 plenteous	0.305 intreat
0.425 hardened	0.308 plenty	0.290 stretch
0.420 plenty	0.300 enchantments	0.252 rod
0.378 seven	0.296 flies	0.251 locusts
0.368 egyptians	0.290 stretch	0.243 plenteous
0.361 goshen	0.287 frogs	0.242 dream
0.360 plenteous	0.279 seven	0.240 hail
0.340 river	0.273 dream	0.237 houses
0.327 interpreted	0.269 locusts	0.234 ill

of the Pharaoh. Subsequently (second column from the left) the component of the vector representation of "pharaoh" that is orthogonal to the vector representation of "joseph" is isolated and normalized. In this representation, elements of the story of the Exodus from Egypt such as plagues of "flies", "frogs" and "locusts" appear in the list of nearest neighbours. As further elements of Joseph's story are removed (rightmost columns), the terms related to the Exodus improve their rankings in the list of near neighbours.

Other investigations in capturing word-order influences in semantic space models include experiments using tensor products in the SemanticVectors system [13] and convolution products using the BEAGLE system [14]. BEAGLE uses convolution products to obtain representations to encode term position that are close-to-orthogonal to the term vectors from which they are derived. They are also reversible such that this information can be decoded. As shown by Sahlgren et al [9], both of these conditions are also met by permutation of sparse random vectors, though research comparing such approaches is still in its infancy.

The high quality of the permutation results raises the question of how they compare to results obtainable by n-gram modelling [15, Ch 6]. The ability to retrieve the names of kings as well as queens for the query "queen ?" suggest that the vector permutation method generalizes slightly compared with raw n-grams, and perhaps behaves more like a smoothed adaptation of the basic n-gram model. The comparison between n-grams and vector permutations would be fruitful to investigate further, especially since the tradeoffs between exact deduction and intelligent induction are central in discussing the relative usefulness of classical versus quantum logic (see for example [6]).

6   The Relevance of This Work to Quantum Interaction

If our goal was to produce evidence that quantum mechanics and logic provides a correct model for natural language semantics, and classical mechanics and logic provides a flawed model, then it may be argued that these experiments are a failure. We have not (for example) demonstrated that the vectors and similarities used to model natural language violate Bell's inequalities, or that the correct combination techniques for vectors necessarily involve entanglement. While we have used the quantum disjunction and eigenvalue decompositions to good effect, there is as yet no solid ground for always preferring the quantum disjunction to one of the other options, or for viewing the eigenvalue decomposition as the single correct way to obtain distinct meanings corresponding to pure states. Thus far, these appear to be useful tools, and other tools such as clustering and permutation appear to be equally valuable, and sometimes more valuable, in analysing semantic phenomena.

However, we do not believe that this is a failure: it is not our goal to demonstrate that classical is wrong and quantum is right, any more than to demonstrate that quantum is wrong and classical is right. What we believe these experiments demonstrate is that a range of tools, drawn from the same mathematical substratum as those of quantum theory, can be usefully applied to provide relatively simple models of semantic phenomena which, in spite of their simplicity, usefully parallel the findings of human scholars. Natural language (and cognition in general) is often very complex: however, we believe our results demonstrate that some reasonable approximation to this subtlety can be obtained using mathematical tools whose history and development is closely intertwined with the methods of quantum theory. If we accept the loose generalisation that classical mechanics promotes deterministic rationalism and quantum mechanics promotes probabilistic empiricism, then our experiments demonstrate that the quantum family of approaches has much to offer, even in small and tightly encapsulated domains such as the analysis of Biblical texts.

We do not think these experiments promote quantum models as a singularly privileged path forward: rather, we think our work demonstrates that the tension between classical and quantum methods is a useful dialectic that encourages synthesis.

7   Conclusions

We have demonstrated that semantic vector methods, using the same underlying mathematical models as those of quantum theory, produce reasonable results when faced with very traditional literary tasks: in particular, analysing the relationships between the Gospel writers, and identifying geopolitical entities in the ancient world. While it is no surprise that this can be done (none of our findings are new), it is somewhat startling that it can be done based on such simple mathematical assumptions.

As well as the dialectic between classical and quantum approaches to semantic analysis, we believe our work highlights an often underappreciated potential for communication between large scale empirical approaches to analysing information (typified by new fields such as information retrieval and machine learning), and the more traditional literary approach to small scale works that are deemed to be particularly important. New developments in information retrieval and machine learning will hopefully provide tools that promote fresh analysis of important texts: meanwhile, the tradition of literary scholarship may provide deep knowledge, encouraging empirical researchers to ask more significant questions with a richer sense of what sorts of relations may be analyzed.

Acknowledgements

The first author would like to give thanks for years of teaching and influence from Kit Widdows (1946–2007), whose passion for St John's Gospel is manifested in the novel "Fourth Witness" [16].

References

1. Birkhoff, G., von Neumann, J.: The logic of quantum mechanics. Annals of Mathematics 37, 823–843 (1936)
2. van Rijsbergen, C.: The Geometry of Information Retrieval. Cambridge University Press, Cambridge (2004)
3. Widdows, D.: Geometry and Meaning. CSLI publications, Stanford (2004)
4. Salton, G., McGill, M.: Introduction to modern information retrieval. McGraw-Hill, New York (1983)
5. Landauer, T., Dumais, S.: A solution to Plato's problem: The latent semantic analysis theory of acquisition. Psychological Review 104(2), 211–240 (1997)
6. Widdows, D., Higgins, M.: Geometric ordering of concepts, logical disjunction, learning by induction, and spatial indexing. In: Compositional Connectionism in Cognitive Science, Washington, DC. AAAI Fall Symposium Series (October 2004)
7. Schütze, H.: Automatic word sense discrimination. Computational Linguistics 24(1), 97–124 (1998)
8. Widdows, D., Cederberg, S., Dorow, B.: Visualisation techniques for analysing meaning. In: Sojka, P., Kopeček, I., Pala, K. (eds.) TSD 2002. LNCS, vol. 2448, pp. 107–115. Springer, Heidelberg (2002)
9. Sahlgren, M., Holst, A., Kanerva, P.: Permutations as a means to encode order in word space. In: Proceedings of the 30th Annual Meeting of the Cognitive Science Society (CogSci 2008), Washington D.C. (2008)
10. Schvaneveldt, R.W.: Pathfinder associative networks: studies in knowledge organization. Ablex Publishing Corp., Norwood (1990)
11. Widdows, D., Ferraro, K.: Semantic vectors: A scalable open source package and online technology management application. In: Proceedings of the sixth international conference on Language Resources and Evaluation (LREC 2008), Marrakesh, Morroco (2008)

12. Papadimitriou, C.H., Tamaki, H., Raghavan, P., Vempala, S.: Latent semantic indexing: A probabilistic analysis. J. Comput. Syst. Sci. 61(2), 217–235 (2000)
13. Widdows, D.: Semantic vectors products. In: Proceedings of the Second International Symposium on Quantum Interaction, Oxford, UK (2008)
14. Jones, M.N., Mewhort, D.J.K.: Representing word meaning and word information in a composite holographic lexicon. Psych. Review 114(1) (2007)
15. Manning, C.D., Schütze, H.: Foundations of Statistical Natural Language Processing. The MIT Press, Cambridge (1999)
16. Widdows, K.: Fourth Witness. Writersworld Limited (2004)

Eraser Lattices and Semantic Contents:
An Exploration of the Semantic Contents in Order Relations between Erasers

Alvaro F. Huertas-Rosero, Leif A. Azzopardi, and C.J. van Rijsbergen

Dept. of Computing Science, University of Glasgow,
Glasgow, United Kingdom
{alvaro,leif,keith}@dcs.gla.ac.uk

Abstract. A novel way to define Quantum like measurements for text is through transformations called Selective Erasers. When applied to text, an Eraser acts like a filter and preserves part of the information of the document (tokens surrounding a central term) and erases the rest. In this paper, we describe how inclusion relations between Erasers can be used to construct an Eraser Lattice for relevant content. It is posited that given a new piece of text, the application of elements of the Eraser Lattice, will result in the destruction or preservation of the content depending on the relevancy of the document. The paper provides the theoretical derivations required to perform such transformations, along with some example applications, before outlining directions and challenges of future work.

1 Introduction

In [1], *Selective Erasers* were proposed as a means for the representation of text documents in a quantum inspired Information Retrieval System. Selective Erasers provide a scheme for lexical measurements in documents, which is analogous to physical measurements on quantum states. In this way, the representation of the text is only known after measurements have been made, and because the process of measuring may destroy parts of the text, the document is characterised through erasure. A Selective Eraser (or simply *Eraser*) is a transformation $E(t, w)$ which erases every token that does not fall within any window of w positions around an occurrence of term t in a text document. These Erasers act as transformations on documents producing a modified document with some erased tokens, much as projectors act on vectors or other operators. The count of terms after the transformation is analogous to the formal property of *norm*, and can be represented as such. Given the definition of an Eraser, different lexical measurements can be defined based on it, for example:

1. Occurrence of a term t in document D: $|E(t, 0)D|$
2. Frequency of occurrence of a term t in document D: $\frac{|E(t,0)D|}{|D|}$
3. Co-occurrence frequency of terms t_1 and t_2 in document D with a minimum distance w: $\frac{|E(t_2,0)E(t_1,w)D|}{|D|}$

P. Bruza et al. (Eds.): QI 2009, LNAI 5494, pp. 266–275, 2009.

where $|.|$ is a counting operation. While this constitutes a basis of representation of text documents, a method is required in order to harness the analogy, that is, to perform some higher level retrieval operation. In this paper, we extend the formalisation of Selective Erasers to Selective Lattices, which are used to performing ranking or classification based on the "quantum" representation of documents. We posit that it is possible to define a set of compatible Erasers which characterise relevancy, such that the application of these Erasers will either preserve a document or destroy it. If a document is preserved (or largely preserved) then this is indicative of its relevance, while if a document is destroyed (or largely destroyed) then this is indicative of its non-relevance. Specifically, we hypothesise that:

> for a given query, the relations between a set of optimally chosen Erasers will differ significantly in the subset of relevant documents and in the subset of the non-relevant documents.

Thus, we believe that we can characterise the relevancy and non-relevancy through erasure. The intuition is that the usage of language within relevant documents will be similar and that the erasers will preserve this usage, while in non relevant documents the usage of language will be different, even if the same vocabulary is used, and thus be erased.

The remainder of this paper will be as follows: The next section will defined the necessary order relations between erasers, i.e. strict ordering and orthogonality. Section 3, will describe how the Eraser Lattice can be constructed using a partially ordered set, before describing how to use the Eraser Lattice to classify documents as either relevant or non relevant. Then, in Section 5, we perform an empirical study on a standard IR test collection (AP88) where we demonstrate the utility of the method and show how relevance information can be preserved through optimally selected Erasers. Finally, we conclude with a discussion of this work and directions for future work.

2   Erasers and Their Order Relations

As a strategy to catch the context in which words tend to occur, in this work we propose to examine relations between Selective Erasers associated with the occurrence of different terms. Several relations can be defined between Selective Srasers as acting on a certain document, but in this work we focus on two of them, orthogonality and strict ordering (others are also mentioned in appendix A):

- **Orthogonality (Disjointedness):** Two Erasers are orthogonal when there is no common fractions of a document D they both preserve:

$$E_1 \perp_D E_2 \iff \forall D_i \quad |E_1[E_2 D_i]| = 0 \tag{1}$$

- **Strict Ordering (Inclusion):** An order relation exists when one Eraser includes the other, that is, when everything one Eraser preserves in document

D, the other preserves as well. A formal way of stating it for two Erasers E_1 and E_2 is that defined for projectors in [2]:

$$E_1 \geqslant_D E_2 \iff \forall D_i \; E_2[E_1 D_i] = E_2 D_i \tag{2}$$

These relations could also be defined within a subset of the documents, when they hold *for every document in subset s*. This relation within a subset is represented with a subscript on the symbol of the relation. For example, for strict inclusion, it would be

$$E_1 \geqslant_S E_2 \iff \forall D_i \in S \; E_2[E_1 D_i] = E_2 D_i \tag{3}$$

The number of possible Erasers for all the vocabulary in a collection is astronomical, so the practical applicability of this criterion relies on a sensible scheme for selecting Erasers and relations between them. Our approach to that problem is based on the measurement of extremal (maximal or minimal) distances between occurrences of terms.

2.1 Distances between Occurrences and Order Relations

Let us suppose that terms t_1 and t_2 occur in document D $n_1 \neq 0$ times and $n_2 \neq 0$ times respectively. If d_{min} is the minimum number of tokens between neighbour occurrences and $d_{max(t_1,t_2)}$ is the maximum number of tokens between any occurrence of t_1 and the nearest occurrence of t_2, two nontrivial relations can be defined that are fulfilled within this document:

$$E(t_1, d_{min} - \delta_1) \quad \perp_D \quad E(t_2, \delta_1) \tag{4}$$
$$E(t_1, d_{max(t_1,t_2)} + \delta_2) \quad \geqslant_D \quad E(t_2, \delta_2 - \delta_3) \tag{5}$$

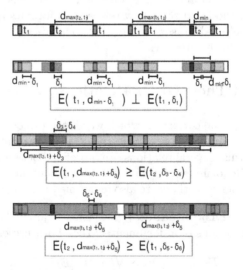

Fig. 1. Relations between Erasers for maximum and minimum distances between occurrences

where δ_1, δ_2 and δ_3 are natural numbers that can vary freely (as long as the width factor remains equal or bigger than zero). Extremal distances show how wide or narrow an Eraser have to be to include or avoid another, as is illustrated in figure 1. The difference between $d_{max(t_1,t_2)}$ and $d_{max(t_2,t_1)}$ is also explicit in the figure.

3   Computation of Eraser Lattices

Any set of Erasers forms, with their order relations, a *Partially Ordered Set* (poset), since relation $\geqslant$ is a proper order relation (reflexive, antisymmetric and transitive). It is not a totally ordered set because there are not order relations between every pair of Erasers [3]. Furthermore, to make it a *lattice* it is necessary to augment it with an *infimum*, a transformation that erases everything, and a *supremum*, a transformation that does not erase anything.

A lattice can be represented by a Hasse diagram, where the infimum is below, the supremum is above, and the elements are in the middle connected with vertical or diagonal lines whenever an order relation holds. In figure 2 the lattice corresponding to the example in figure 1 is depicted.

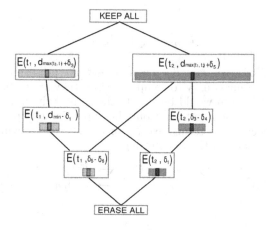

Fig. 2. Hasse diagram representing a lattice

What is *Quantum* About This Scheme? A close inspection of figure 2 can reveal a very interesting feature of the lattices of Erasers: they are non-distributive. Missing crossed relations between the four Erasers below the upper point (supremum) produce 4 possible sub-lattices with the shape of a pentagon called N_5 (for the pentagon, see [3], and for non-distributive logics, [4]) which are known to be a signature of non-boolean lattices. This is not just a mathematical curiosity in the structure of order relations: it means that usual boolean relations are restricted to hold *within* sets of compatible measurements, but they will not hold between elements of two different (incompatible) sets. In Quantum

Theory, sets of compatible observations are related to a particular experimental or operational context; thus in this work, we suggest that sets of compatible lexical measurements could also be related to some kind of context. In particular, we explore the possibility to relate them to a *topical* context, therefore using the contextual nature of Quantum Logics to explore topicality.

Relevancy-Sensitive Eraser Lattices. Relations between Erasers are determined in most cases by the usage of terms. A set of Erasers can be chosen, that when applied to a set of relevant documents, possesses a lattice structure. It is our hypothesis that such lattice would encode semantic information about the topic the documents in the set are relevant to, and it could be used to define a transformation that preserves as much as possible of relevant documents and as least as possible of nonrelevant documents (a *Topical Eraser*).

The first approximation to a Topical Eraser is through a set of orthogonal (disjoint) Erasers. Orthogonality tends to enforce a low window width (Erasers with a width factor of 0 are all orthogonal to each other), so it is desirable to choose them with maximally wide windows, to enhance document preservation in the set of relevant documents. Thus, a set of orthogonal Erasers with *maximum window width* are chosen for the set of relevant documents, and the Topical Eraser can be defined as one that preserves what any of these preserve, that is, their join (union):

$$E_{topic} = \bigcup_i E(t_i, n_i) \qquad (6)$$

The fraction X of a relevant document $D \in topic$ preserved by this Topical Eraser would be extremely easy to compute. Since they are orthogonal, the fraction preserved by the join would be simply the sum of the individual preserved fractions. And this fraction, in turn, would be approximately proportional to the occurrence frequency of the terms, except for border effects (windows truncated by the beginning or end of the document):

$$X_{topic}(D) = \frac{|E_{topic}D|}{|D|} = \frac{|(\bigcup_i E(t_i, n_i))D|}{|D|} = \sum_i \frac{|E(t_i, n_i)D|}{|D|} \approx \sum_i (2n_i + 1)F(t_i) \qquad (7)$$

where $F(t_i)$ is the frequency of occurrence of term t_i (the occurrence divided by the length of the document) and n_i is the window width parameter. This way, we get something like a TF (Term Frequency) scoring with occurrence of terms and weighting factors (widths) tuned with the set of Relevant Documents.

In a non-relevant document, on the other hand, the sum expression would not be valid, since the Erasers would not be necessarily orthogonal, and the fraction preserved by the join would be less than the sum of the fractions preserved by the individual Erasers (since terms in overlaps are not counted twice).

The terms chosen for this set could still occur frequently in nonrelevant documents, producing high *preserved fraction* in nonrelevant documents, and therefore poor sensitivity to context. To avoid this, a further set of Erasers can be used. With the data of *maximal* distances between occurrences, a set of Erasers can be defined such that each one of them *includes* one of the previous Erasers. Since inclusion relation (5) tends to favour Erasers with wide windows, those with *minimal window width* can be chosen to enhance sensitivity. For ED_i being one of the chosen maximal disjoint Erasers, and EI_i the corresponding including Eraser, the condition would be:

$$EI_i = E(t_j, n_j) \quad \text{such that} \quad (EI_i \geqslant ED_i) \wedge (n_{k \neq j} > n_j) \tag{8}$$

On a relevant document, the consecutive application of the including Eraser and the disjoint Eraser would produce the same result than just the application of the disjoint Eraser, but would erase more than the disjoint Eraser in a nonrelevant document, where the inclusion relation (8) does not necessarily hold.

4   Choice of Erasers

Central Terms: To choose the central terms for the Erasers, the ratio between the *average distance to occurrence of other terms* and the *average distance to occurrences of itself* can be used as a criterion to choose terms. Terms would be ranked according to the following quantity:

$$R(t_i) = \begin{cases} \frac{\langle d_{(t_i, t_i)} \rangle}{\langle d_{(t_i, t_{j \neq i})} \rangle} & \text{when present} \\ 0 & \text{when absent} \end{cases} \tag{9}$$

where $\langle \cdot \rangle$ means average and $d_{(t_i, t_j)}$ is the distance between an occurrence of t_i and the nearest occurrence of t_j. If the term is absent in a document, this would count as a 0 in the averaging.

A term that tends to be evenly spaced in the text and occur relatively near to everyone of the others would score high, and one that either occurs very concentrated or does not occur much, will get a low score.

Window Widths: There are two possible criterion that we can use to assign window widths both in disjoint and including Erasers:

1. Maximum *preserved fraction*: This criterion favours maximal window widths for disjoint Erasers
2. Minimum overlap: This criterion favours minimal window widths for including Erasers.

In different documents, the maximum widths compliant to orthogonality condition (4) and the minimum widths compliant to inclusion condition (5) can be different, so we maximise or minimise them, correspondingly, over the whole set of documents. Minimum distance will then be minimum in all the set of documents, and maxima will be also maxima on all the set.

5   A Practical Example in Collection AP88

To check to what extent semantic contents is encoded in the order relations between Erasers, we chose 2 sets of 20 Erasers for the set of relevant documents

Table 1. Erasers chosen for query 82 of AP88

Query 82: Genetic Engineering			
term disjoint	width disjoint	term including	width including
said	13	field	81
field	59	corn	36
new	13	said	25
year	46	genetically	144
s	22	scientists	144
t	33	s	36
used	34	test	81
research	30	t	121
tests	50	aids	81
test	55	disease	100
researchers	31	genetic	100
disease	46	research	100
gene	40	used	81
cancer	37	gene	81
scientists	21	researchers	100
corn	61	vaccine	36
aids	47	cancer	36
genetically	45	new	64
vaccine	46	tests	64
genetic	22	year	121

for different topics in the collection AP88, and for each topic compared the relations holding in the set of *relevant* documents and those holding in a random subset of *nonrelevant* documents.

Results are in table 1 for topic 82. Central terms are clearly related to the topic. The fulfilment of an order relation can be approximately evaluated by comparing the documents acted upon by both Erasers and only one of them, as follows:

$$X(E_1 \geqslant_D E_2) = sim(E_2 E_1 D, E_2 D) \qquad (10)$$

where $sim(A, B)$ is a measurement of the similarity of documents A and B. In figure 3 a part of the lattice for topic 58 is depicted. The most important test for this scheme is the measure of the discrimination between relevant and

Table 2. Topics that were well characterised (easy) and poorly characterised (difficult). The average number of documents and percentages of preservation of different kinds of documents are presented. These values correspond to Topical Erasers with 20 central terms.

Queries	documents	% relevants	% nonrelevants	% nonasessed
easy queries	42.7 ± 26.87	$(72.36 \pm 8.81)\%$	$(30.47 \pm 12.87)\%$	$(9.64 \pm 7.23)\%$
difficult queries	93.69 ± 33.15	$(46.62 \pm 8.33)\%$	$(70.31 \pm 6.06)\%$	$(11.15 \pm 9.48)\%$

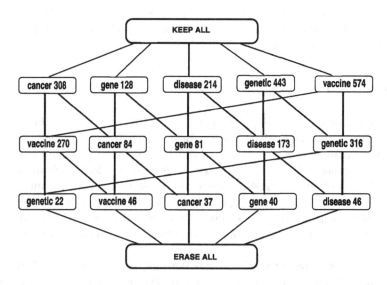

Fig. 3. Order relations between Erasers for topic 82 of TREC-1. Relations are obtained from the list of including Erasers with formulas in appendix B.

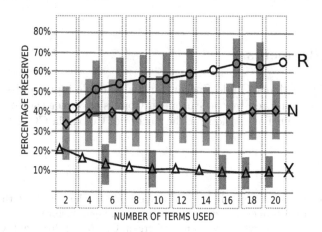

Fig. 4. Average preserved percentage of relevant (R), non-relevant (N) and non-assessed (X) documents for queries 51-100 of TREC-1, for different numbers of central terms

nonrelevant documents. In figure 4 the results are shown for the percentages preserved for queries 51 to 100 of TREC-1 [5] using different numbers of central terms to build the Topical Eraser. Relevant documents were well characterised by the Topical Eraser for most of the queries, but 13 of them had bigger preserved percentage for nonrelevant than for relevant documents. In table 2, results of preserved percentage are shown for the topics (queries) in two groups: the 13 queries for which the results were anomalous (more non-relevant than relevant preserved) and those for which the methodology worked as expected, preserving more relevant than non-relevant.

6   Conclusions

In this paper, we have extended the notion of Selective Erasers to form higher order constructs called selective lattices. Constructing lattices from a set of example relevant documents is a novel way in which to capture the semantic relations within the content. The application of transformations derived from elements of the lattice, will either (mostly) destroy or preserve the relevant information in a new unseen document, and provides a formal mechanism for classifying documents. Examples with sets of documents of a standard IR test collection were used to check the ability of this scheme to capture semantic contents, with positive results.

Future work will be directed towards deriving the optimal set of Selective Erasers, formulating a ranking algorithm and performing a large scale empirical study of one of the first quantum inspired Information Retrieval System. It could also be possible to check situations where incompatible observations exist, like a possible incompatibility between topicality and relevancy mentioned in [4, chapter 6].

Acknowledgements

We would like to thank Guido Zuccon for his valuable input and suggestions. This work was sponsored by the European Comission under the contract FP6-027026 K-Space and Foundation for the Future of Colombia COLFUTURO.

References

1. Huertas-Rosero, A., Azzopardi, L., van Rijsbergen, C.: Characterising through erasing: A theoretical framework for representing documents inspired by quantum theory. In: Bruza, P.D., Lawless, W., van Rijsbergen, C.J. (eds.) Proc. 2nd AAAI Quantum Interaction Symposium, pp. 160–163. College Publications, Oxford (2008)
2. Beltrametti, E.G., Cassinelli, G.: 9. In: The logic of Quantum Mechanics, p. 87. Addison-Wesley, Reading (1981)
3. Burris, S., Sankappanavar, H.P.: A Course on Universal Algebra. Springer, Heidelberg (1981)
4. van Rijsbergen, C.J.: The Geometry of Information Retrieval. Cambridge University Press, Cambridge (2004)
5. Harman, D.K.: Overview of the first text retrieval conference (trec-1). In: Harman, D.K. (ed.) NIST Special Publication 500-207: The First Text REtrieval Conference (TREC-1). NIST Special Publications, National Institute of Standards and Technology, NTIS, vol. 500, pp. 1–20 (1992)

A   More Relations between Erasers

1. Trace Ordering (does not rely on the identity between transformed documents but only between their preserved number of tokens).

$$E_1 \geqslant_{(st,D)} E_2 \iff |E_2 E_1 D| = |E_2 D| \tag{11}$$

2. Weak Trace Ordering (this is rather trivial, and not very useful):

$$E_1 \geqslant_{(wt,D)} E_2 \iff |E_1 D| = |E_2 D| \tag{12}$$

It can be easily shown that the three defined relations form a chain of implication $(3) \Rightarrow (13) \Rightarrow (12)$.

3. Compatibility

$$E_1 \sim_D E_2 \iff [E_2 E_1]D = [E_2 E_1]D \tag{13}$$

This relation implies all the other relations defined in this paper, but is not necessary.

B   Deducing More Relations

Two relations are very useful to deduce more order relations from a list, like that of narrow-window Erasers and wide-window Erasers:

1. Transitivity:

$$(E(A, w_A) \geqslant E(B, w_B)) \wedge (E(B, w_B) \geqslant E(C, w_C)) \Rightarrow (E(A, w_A) \geqslant E(C, w_C)) \tag{14}$$

2. Invariance under simultaneous widening:

$$\forall \alpha > 0, (E(A, w_A) \geqslant E(B, w_B)) \Rightarrow (E(A, (w_A + \alpha)) \geqslant E(B, (w_B + \alpha))) \tag{15}$$

Beyond Ontology in Information Systems

Christian Flender, Kirsty Kitto, and Peter Bruza

Faculty of Science and Technology,
Queensland University of Technology,
Brisbane, Australia
{c.flender,kirsty.kitto,p.bruza}@qut.edu.au

Abstract. Information systems are socio-technical systems. Their design, analysis and implementation requires appropriate languages for representing social and technical concepts. However, many symbolic modelling approaches fall into the trap of underemphasizing social aspects of information systems. This often leads to an inability of ontological models to incorporate effects such as contextual dependence and emergence. Moreover, as designers take the perspective of people living with and alongside the information system to be modelled social interaction becomes a primary concern. Ontologies are too prescriptive and do not account properly for social concepts. Based on State-Context-Property (SCoP) systems we propose a quantum-inspired approach for modelling information systems.

1 Introduction

The idea of capturing and representing real world knowledge in information systems has long been recognized in many fields (see [1] for an overview). However, the problem of solidly grounding such knowledge is widely acknowledged. Philosophical debates on the foundations of knowledge date back to ancient greek thinkers. Since then, many western philosophers have agreed that experience of particulars as it comes moment by moment through the senses is unreliable. Many concluded that only stable, abstract, logical, universal categories can function as objects of reference for the meaning of concepts [2]. Nowadays, classical and rule-based approaches to conceptual modelling still consider such predefinitions. They rely on ontological presuppositions according to what Husserl would have called the natural attitude, a kind of naïve positing of the world as existing independent of the observer [3].

In the light of Husserl's natural attitude, the ontological approach raises ongoing questions of how to adequately represent and transform the cultural world we inhabit. Information systems are cultural artefacts and meant to represent socio-technical phenomena. Conceptual modelling lies at the heart of their design, analysis and implementation. Here, modelling grammars like data-, object-, or process-oriented notations provide concepts for systems analysis and design. However, such modelling notations are limited in their expressiveness. Firstly, modelling emergent properties of composites runs into the problem of ambiguity

P. Bruza et al. (Eds.): QI 2009, LNAI 5494, pp. 276–288, 2009.

so that there is no reasonable justification whether one should represent such properties as associations or objects in their own right [4,5]. Furthermore, conceptual structure changes whenever context causes it to change [6]. In predefining (Cartesian) product spaces entities become inflexible with regard to their actual usage [2]. For instance, the concept Chair has quite different meanings in relation to its usage, e.g. sitting or standing on it. Secondly, formal ontologies tend to claim the existence of things and properties (e.g [7]), however this leads to infinite regress as it might be claimed that facts are meaningless and aggregation of meaningless facts is hopeless unless their actual usage evokes relevance [8], e.g. when disposing of a chair its affordance to sit on becomes meaningless. Even if one aims at framing the scope, in order to distinguish what is relevant, one ends up in infinite regress as many skills are embodied and subpersonal and thus not accessible to deliberate distinctions. These problems have become particularly bothersome in Artificial Intelligence (AI) where their different manifestations are understood under the guise of the commonsense knowledge problem [8,9], the frame problem [10] and the symbol grounding problem [11].

In this paper, we demonstrate that social concepts like Team, Group or Department, can be modelled adequately beyond ontology or symbolic representations. State-Context-Property (SCoP) systems [12,13,14,15] facilitate the specification of interactions between individuals or agents in a quantum-like manner. From the flow of mutual and directed acts states of composite concepts like Team emerge. Such states are neither reducible to the agents from which they emerge nor can they be specified independently from individuals. Generally, states of concepts depend on the intent, or context, from which they arise.

This article is structured as follows: In the next section, problems of symbolic representation are revisited. By means of an example we discuss the limitations of formal ontology and the necessity of accounting for interaction, emergence and context-sensitivity. In Section 3, we argue for the importance of social interactions in socio-technical settings. Humans are intentional beings and relate to each other quite differently from simple message exchange. Understanding human communication requires to adopt different viewpoints, in particular the intentional perspective of others. In Section 4, we formalize this understanding of agent communication and show how to model contextual meaning as well as emerging states of social concepts with SCoP. Lastly, in Section 5, we point out the consequences of our approach, summarize what has been proposed and provide an outlook towards future work.

2   Representational Modelling

In this section we briefly introduce some problems that arise within a representational approach. We will show that emergent properties arising out of contextualized behaviour of interacting entities can not be modelled with traditional modelling approaches. These difficulties have been well known in both AI and the cognitive sciences for many years. Since they arise from the insistence of these fields upon representation they will hold accordingly for information systems building upon formal ontology, which is again a representational approach.

2.1   General Problems of Formal Ontology

Problems of formal ontology or symbolic representations have been extensively discussed in AI and the cognitive sciences, but solutions are yet to be found in the current literature.

1. The commonsense knowledge problem [8,9] recognizes that implicit knowledge is more or less taken for granted but, nevertheless, needs to be made explicit in order to proceed with the representational project. Cumulative aggregations of facts about the world, whether implicit or not, never solved the problem. As Dreyfus (1991) puts it: 'Facts and rules are, by themselves, meaningless. To capture what Heidegger calls significance or involvement, they must be assigned relevance' [8]. However, such an assignment of relevance just adds more meaningless facts, a problem that very quickly leads to infinite regress.
2. The frame problem [10] is characterized by the challenge of limiting the scope of propositions in the light of actions. Propositions represented in symbolic terms must be updated with regard to all relevant effects caused by an action. In order to frame the situation in hand meaningless facts must be ignored. However, deciding relevance through reasoning on facts once again causes the commonsense knowledge problem.
3. The symbol grounding problem [11] adds to the misery of representationalists in asking how the meanings of the meaningless symbol tokens can be grounded in anything but other meaningless symbols.
4. Experiments in cognitive science have shown people generally rate guppy neither as a good example or instance of the concept Fish nor of the concept Pet, but as good example of the combined concept Pet Fish [13,14]. Hence, activiation of Pet or Fish alone does not cause activiation of guppy. According to classical logic and set theory, joint entities are described by means of the product state space, e.g. the Cartesian product space of Pet and Fish. However, the conjunction of both concepts cannot describe the situation wherein novelty (e.g. guppy) is generated.

Thus, there are some quite strong reasons to be wary of representations. In the next section we shall focus upon some of the problems particular to social concepts in information systems that arise within representational approaches.

2.2   Symbolic Representation of Composites

Most approaches to modelling information systems share ontological assumptions which determine the built-in terms offered and therefore their range of applicability [16]. For instance, static languages provide concepts for existing things, their attributes and interrelationships. Dynamic modelling grammars cover temporal aspects like states, state transitions or processes. However, when it comes to emerging or composite concepts they run into trouble. According to Wand et al. (1999) the composite always gains at least one property that did not exist previously [5]. For instance, consider the emergence of a team out

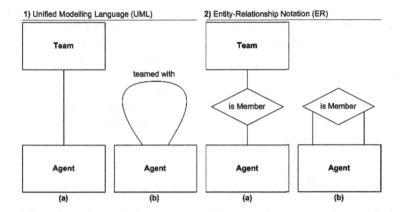

Fig. 1. Representational modelling of socially emergent concepts is ambiguous. Composites like Team can be represented as objects (entities) or associations (relationship classes).

of a number of agents in Figure 1. According to Unified Modelling Language (UML) [17] and Entity-Relationship Notation (ER) [18] composites can be represented as object classes (UML) or entity classes (ER): rectangles in 1a and 2a. Furthermore, combined concepts can be symbolized as associations (UML) or relationship classes (ER): connecting line in 1b and diamond in 2b. Obviously, modelling combined concepts like Team is ambiguous. Generally, if a team had not at least one property like number of agents it would not make sense to model it. But how should one represent that property, as a relationship or an entity in its own right? It was argued that if such emergent properties are represented as relationship classes or associations between entities, the emergent property of the composite cannot be represented [4]. Furthermore, there is no way to represent mutual properties between the composite concept and its components. Features like the number of agents a team needs to exist, the assignment of an agent to a team or the multiple roles an agent can occupy within a team, are hardly represented with associations. In order to dissolve ambiguity, representational approaches to information systems modelling suggest putting social concepts like Team into separate ontological categories, i.e. object classes (UML) or entity classes (ER) [4]. Thus 1a (and 2a respectively) as shown in Figure 1 should be the preferred representation. However, ontological emergence of inherent high-level properties with causal powers is witnessed nowhere [19]. To claim that social composites should be modelled ontologically as entities or objects lacks any scientific evidence. Furthermore, irrespective of representing emergent properties as entities or relationships, there is no chance to account for their dynamic nature. For instance, teams might emerge temporarily and spontaneously, e.g. within a meeting for reasons of group work. Furthermore, a team might modify team members and properties like purpose or duration but still remain the same socially emergent entity with respect to its organisation, i.e. maintain viability as a team. In summary, neither associations (UML) and relationship classes (ER) nor objects (UML) and entities (ER) are capable of representing part-whole relationships of social concepts adequately. Emergent properties are not intrinsic to the composite and mutual properties belonging to both Agent and Team are as much transient as emerging properties. Furthermore, Team is neither separable into agents nor should Agent be specified independently of Team. Social concepts like Team and Agent co-emerge by co-enacting each other [20,21]. In the following section we will discuss how agents relate to each other intentionally and how the viewpoints involved in speech facilitate the emergence of combined social concepts.

3   Shifting Viewpoints

As discussed in the previous section, representational modelling approaches fall short of modelling social concepts adequately. One of their shortcomings is an inability to represent composites like Team which emerge context-dependently in the course of social interactions between agents. In this section, we introduce a natural account of human communication.

3.1   Social Interaction as Message Exchange

Social interactions are inherently dynamic or active. In contrast to static modelling notations like ER or UML class diagrams, dynamic modelling grammars conceptualize interactions between agents in terms of states, state transitions and input-output behaviour (e.g. [22]). For instance, Figure 2 shows the behaviour of two interacting agents in terms of input-output relations or directed edges thus leaving white-box behaviour opaque (left-side of Figure 2). Moreover, internal behaviour is specified in terms of task dependencies between and within agents (right-side of Figure 2). Sending tasks (outgoing edges) trigger receiving tasks (incoming edges) whereas internal events (circles) denote transitions from receiving to sending tasks and vice versa. Agents and input-output symbols (Self, Help) are denoted by nouns, whereas sending and receiving tasks (introduce, request, provide) are labeled as verbs. For instance, Agent 1 sends a request for help (s_3) whereupon Agent 2 receives the request and replies informatively (s_4). Besides sequential behaviour as specified in Figure 2, logical operators facilitate the modelling of more complex control-flow behaviour (cf. Figure 3). However, it is obvious that rule-based behaviour is not the way human agents do naturally

Fig. 2. In social interactions agents actively relate to each other. Agents are conceptualized from an inside and outside view. The latter shows input-output behaviour (left-side), whereas the former mirrors an agent's internal behaviour (right-side).

interact and using language is much more subtle than just following patterns of acts like sending, receiving and processing. To make this clear we need a closer look at the speech acts involved.

3.2   Social Interaction as Shared Intentionality

According to speech act theory [23,24], human agents are conceptualized as inherently active, intentional and embodied beings situated within their background of cultural practices and purposes, that is their contextual situation. Intentional aspects are formalized using the fourfold structure of speech acts. Speech acts are composed of:

1. Utterance acts (the actual embodied expression) like gestures and vocalizations.
2. Illocutionary acts (the social intention) like informing, requesting or sharing attitudes.
3. Propositional acts (the reference to some state of affairs) like pointing or pantomiming.
4. Perlocutionary acts (the expectation of a listener's response) like informing, requesting or sharing attitudes.

Speaking is primarily an intentional act of relating to the world from a first-person point of view. Moreover, agents develop and flourish in social environments in which they share intentional and attentional states with others [25]. During the lifetime of an agent, shared intentionality settles, becomes embodied and shapes a common ground [26]. For instance, this common ground includes the language spoken (e.g. English), communal norms (e.g. legal terms) and social conventions (e.g. welcoming someone to a meeting). It is through such cultural practices and purposes that agents assign meaning to symbols and understand intentional acts of others including the significance of being part of a community of other agents. From the perspective of speech act theory and shared intentionality, Figure 2 can be interpreted quite differently. Instead of sending, receiving and processing symbols, agents interact by actively relating to each other. Agents do not exchange symbols but actively interpret, or bring forth, propositional content based upon a common or shared background. Hence, agents mutually and dynamically relate to each other by actively and directly anticipating each other. They do this through:

1. Illocutionary acts: introduce (s_1 and s_2), request (s_3) and provide (s_4)
2. Propositional acts referring to content: Self, Help
3. Perlocutionary acts: introduce (s_1 and s_2), request (s_3) and provide (s_4)

Illocutionary acts mirror the social intention of an agent, whereas perlocutionary acts represent the anticipation of the other's reaction. Hence, an agent's second-person perspective and first-person view are inseparable. Due to their common ground agents expect one another to (re-)act according to their shared cultural practices and purposes. For instance, as shown in Figure 2, Agent 1 expects

Agent 2 to provide support while asking for help. Similarly, agents presuppose welcoming each other in a particular way, e.g. through hand-shaking, when they meet. In the next section, we will formalize this understanding of human communication using State-Context-Property systems (SCoP) [12,13,14,15]. We will demonstrate that mutual anticipations based upon a common ground lead to emerging states of composite concepts like Team which are neither reducible to states of Agent nor can they be specified independently of Agent.

4   Conceptual Modelling Beyond Ontology

It has been argued that conceptual modelling of information systems should be carried out as the coordination of social intentions [27,22]. Social agents actively coordinate behaviour through intentional acts such as speech acts. Based upon a common background, intentional acts (e.g. requesting, informing or sharing attitudes) constitute the contextual situation that evokes the meaning of an intentional object or concept. In the following we will formalize an understanding of concepts as the coordination of intentional acts and argue for a quantum-like structure in social interactions.

4.1   Example

Two agents introduce each other and talk about a venture they would like to share. In Figure 3, Agent 1 proposes an option, whereas Agent 2 either accepts the proposal or suggests an alternative. Once both have agreed upon a venture a new concept, Team, emerges. Team is neither modelled a priori as an emergent concept having predefined properties and domain nor is it separated from agents. Rather a team is brought forth, or disclosed, by two socially coupled agents in action. Intentional acts s are inseparably entangled with any concept

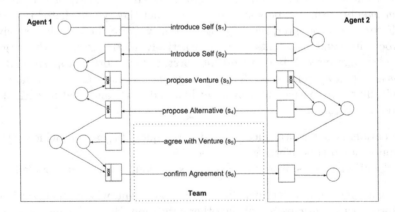

Fig. 3. Two agents talk about a venture they would like to share. Once both have agreed the social concept Team emerges. Team is neither reducible to one or both of the agents nor can it be specified independently of agents.

they evoke including the emerging concept Team. Team is not ontological in the sense of a property existing independently of an observer rather it emerges context-dependently in the course of interactions. Compositions of intentional acts (e.g. introduce, propose, agree etc.) constitute the context which evokes the meaning of an intentional object or concept (e.g. Self, Venture, Team etc.). Each agent is situated within its contextual situation from its first-person point of view. As discussed in the previous section, the first-person perspective of an agent is inseparably entangled with its second-person point of view, i.e. its anticipation of the other's reaction. Obviously, anticipating someone else's reaction to one's own action requires a state of potentiality. Furthermore, such a state of anticipation depends on earlier interaction dynamics, that is on previous illocutionary acts and perlocutionary acts enacted from the respective first-person perspective of the agents involved. To make this more precise, we formalize a particular contextual situation e, shared by two or more agents, as a composition of pathways, trajectories or sequences of intentional acts s^1. Hence, contextual situations e arise from interaction dynamics (cf. Figure 4). It is such a contextual situation e which constitutes the meaning of a propositional content or concept evoked by a given act s. In the following we will show that emerging concepts like Team can be modelled adequately with SCoP. States of social concepts like Team should neither be modelled independently from agents nor are they reducible to one or both of the agents. Agents and Team, and more generally intentional beings and their social understanding are mutually dependent [25].

4.2 State-Context-Property (SCoP) Systems

Let a social interaction be composed of agents and (propositional) objects each represented as State-Context-Property (SCoP) concepts S [12,13,14,15]. A SCoP concept S is a five tuple².

$$S = (\Sigma, M, L, \nu, \mu) \tag{1}$$

Σ is the set of states an agent or object can attain. M is a lattice structure of context elements e. Likewise, L is a lattice structure of properties. Both context elements and properties can be concepts. The function ν returns the probability of a property typical of a concept under a given context. The probability that a contextualized state changes to another state giving rise to a new context is defined in μ. Let us have a closer look at each of the components. The state $p \in \Sigma$ of an agent or referential object is dependent on the context $e \in M$. Since social interactions are essentially composed of agents intending some propositional content, e.g. S^{Agent} intends proposing $S^{Venture}$, intentional acts form part of the context of propositional content. For instance, as soon as one of the agents

---

[1] Theoretically, all conversational pathways could be modelled at once. Practically, this is neither feasible nor desired. We will show that models as simple as Figure 3 already involve shared intentional states which can not be modelled adequately with ontological approaches.

[2] A translation of SCoP concepts into vector spaces can be found in [14].

$$e_1 = s_1$$
$$e_2 = s_1 \wedge s_2$$
$$e_3 = s_1 \wedge s_2 \wedge s_3$$
$$e_4 = s_1 \wedge s_2 \wedge s_3 \wedge s_4$$
$$e_5 = s_1 \wedge s_2 \wedge s_3 \wedge s_5$$
$$e_6 = s_1 \wedge s_2 \wedge s_3 \wedge s_4 \wedge s_6$$
$$e_7 = s_1 \wedge s_2 \wedge s_3 \wedge s_5 \wedge s_6$$

Fig. 4. Social interactions are context-dependent. A contextual situation e is composed of interactive pathways. Interactive pathways are composed of sequences of speech acts s enacted from the respective first-person perspective of the agents involved.

in Figure 3 has agreed on joining a venture (context e where s_5 = true or s_6 = true), states are assigned to S^{Team}. There are three types of states depending on context or intentional acts.

- Theoretical states are unrelated to any context. Before agreement S^{Team} is essentially meaningless and not evoked by any intentional act.
- Eigenstates are actualized. $p \in \Sigma^{Venture}$ = 'proposed' is an eigenstate once the related context (context e where s_3 = true) is given.
- Superposition states afford to get actualized in relation to a specific context. A state $p \in \Sigma^{Team}$ in which S^{Agent} has not yet decided whether to propose an alternative (context e where s_4 = true) or to agree with venture (context e where s_5 = true) is a superposition state.

Agents and objects have properties $a \in L$ dependent on context. For instance, the propositional object S^{Team} may have properties like purpose and duration dependent on a given speech act. If the intention is proposing (s_3), then S^{Team} has not acquired these properties yet. It acquires them with dependence upon state and context. The function ν takes a state, context and feature and returns the applicability of the property. Hence, $(p, e, a) \mapsto \nu(p, e, a)$.

$$\nu : \Sigma \times M \times L \rightarrow [0, 1] \tag{2}$$

For example, $\nu(p, e, a)$ could be the weight of a feature a = 'purpose' of the concept $S^{Venture}$ in a state p = 'confirmed' under the context e where s_6 = true.

Similarly, the function μ describes the transition probability from one state to another under the influence of a particular context. It takes a contextualized state and returns the probability that it changes to another one. Hence, $(f, q, e, p) \mapsto \mu(f, q, e, p)$.

$$\mu : \Sigma \times M \times \Sigma \times M \rightarrow [0, 1] \tag{3}$$

For example, $\mu(f, q, e, p)$ could be the probability that state $p \in \Sigma^{Venture}$ = 'proposed' under the influence of context e changes to the state $q \in \Sigma^{Venture}$ =

'confirmed', giving rise to the new context f. In the following we will have a closer look at context elements $e \in M$.

Contextual structure in social interactions mirrors conversational progress in terms of compositions of intentional acts. M is a complete and orthocomplemented lattice. Lattices are partially-ordered sets. A lattice is orthocomplemented if for each context element e there exists a complement $e^{\perp}$. It is complete if for each subset of M there exists both infimum (greatest lower bound) and supremum (least upper bound). Hence, for all $\{e_i\}_{i \in I}, e_i \in M$, there exists an infimum and supremum. Infimum is the logical conjunction of context elements $e_i \in M$.

$$\forall i \in I : \wedge_{i \in I} e_i \in M \tag{4}$$

The infimum requires a context $f \in M : f \leq e_j$ such that $\forall j \in I \Rightarrow f \leq \wedge_{i \in I} e_i$. For instance, take the situation as shown in Figure 3 where $s_6 = $ true. Here, previous actions led to a state $p \in \Sigma^{Venture} = $ 'confirmed'. Several paths could have led to this contextual situation. For instance, Agent 2 might have had to propose alternatives (s_4) before the agreement could have been established. The infimum is the logical conjunction of such pathways all leading to an eigenstate in the conversation. Note that for any eigenstate under context e where $s_6 = $ true, the *actual* pathway is decided in real time on the fly and thus it is not yet determined. To make this clear we need to define the supremum. In contrast to the infimum, supremum is the logical disjunction of context elements $e_i \in M$.

$$\forall i \in I : \vee_{i \in I} e_i \in M \tag{5}$$

The supremum requires a context $f \in M : f \geq e_j$ such that $\forall j \in I \Rightarrow f \geq \vee_{i \in I} e_i$. For instance, in situations where an agent has to make a decision in relation to another agent who is expecting a response, the logical disjunction of elements e_i denotes a superposition context. Whether an agreement is established without iterative negotiations beforehand (e_7) or with preceding discussions (e_6) emerges on the fly in action.

Zero context elements (0) represent context for which there are no eigenstates, e.g. motor behaviour in relation to some external state of affairs. Unit context elements (1) mirror context for which every state is an eigenstate, e.g. self-movement in relation to speech. For instance, for S^{Team} the subset $\{0, 1, e_6, e_7\} \in P(M^{Team})$ returns the following infimum and supremum.

$$e_6 \wedge e_7 \text{ (Infimum)}$$

$$e_6 \vee e_7 \text{ (Supremum)}$$

Contextual behaviour in social interactions relates infima and suprema to states $p \in \Sigma$. Suprema relate to superposition states. When a superposition state affords to get actualized under a given context e, it changes to an eigenstate in this context. In speech, enacting or intending propositional content actualizes a concept's superposition state. It is defined as as a mapping λ.

$$\lambda : M \to P(\Sigma) \tag{6}$$

The probability function λ takes a context e, maps it to a subset of Σ and returns an eigenstate p under e. Hence, $e \mapsto \lambda(e)$ where $\lambda(e) = \{p|\mu(f, p, e, p) = 1\}$. The probability that a state $p \in \Sigma$ under a superposition context $e \in M$ collapses to an eigenstate under this context is not epistemological, i.e. due to an agent's lack of knowledge. From the background of a given contexual situation, actualizing superposition states while negotiating depends on the agent's freely performed actions.

Generally, if for two contexts e and f, we find that e is stronger or equal than f, then eigenstates under f include eigenstates under e. Hence, $e \leq f \Leftrightarrow \lambda(e) \subset \lambda(f)$. For instance, $e_1 \leq e_2 \Leftrightarrow \lambda(e_1) \subset \lambda(e_2)$. From (4) and (5) follows that interaction dynamics can take different paths and thus conversations are composed of several contextual elements e_i. For infima, a state p under a context e is only an eigenstate, if it is an eigenstate of each of the e_i.

$$\cap_{i \in I} \lambda(e_i) = \lambda(\wedge_{i \in I} e_i) \tag{7}$$

For instance, since $\lambda(e_6) \cap \lambda(e_7) = \lambda(e_6 \wedge e_7)$ it follows that $p \in \Sigma^{Team}$ is an eigenstate of both e_6 and e_7. A state $p \in \Sigma^{Team}$ in which proposals were made iteratively (s_3 after s_4) is an eigenstate. However, this does not hold for the supremum which makes it quantum-like.

$$\cup_{i \in I} \lambda(e_i) \subset \lambda(\vee_{i \in I} e_i) \tag{8}$$

The context M of a concept evolves within $\lambda(M)$ which is a closure space [12]. If we add superposition context to the union of eigenstates then we get the closure space $\lambda(M)$ which makes a SCoP concept (agent or object) topological in the sense that it covers all actual and possible states. Given a superposition context $e \vee f$, we add this context to the set $\lambda(e) \cup \lambda(f)$. Hence, for $e, f \in M : \overline{\lambda(e) \cup \lambda(f)}$ is the closure space of $\lambda(e) \cup \lambda(f)$. For instance, since $\lambda(e_6) \cup \lambda(e_7) \neq \lambda(e_6 \vee e_7)$ (8) it follows that a state $p \in \Sigma^{Team}$ under this context is a superposition state. This is exactly the case if agents have to decide whether they want to proceed negotiating or not. By adding the supremum $e_6 \vee e_7$ to $\lambda(e_6) \cup \lambda(e_7)$ we gain the closure space $\overline{\lambda(e_6) \cup \lambda(e_7)}$ thus giving S^{Team} a topological meaning. When not in its ground state, $p \in \Sigma^{Team}$ is either an eigenstate or superposition state. In the latter case $p \in \lambda(e_6 \vee e_7)$. However, $p \notin \lambda(e_6) \cup \lambda(e_7)$ and thus it is neither an eigenstate of e_6 nor an eigenstate of e_7. Hence, the concept S^{Team} acquires meaning through a common background or shared intentional state $p \in \lambda(e_6 \vee e_7)$ composed of intentional acts or states of both agents. This nonseparability of shared intentional states requires to adopt a second-person viewpoint where both agents have to relate to their common proposition from the perspective of each other. Agent 1 expects Agent 2 to agree while Agent 2 anticipates Agent 1 to confirm venture. Obviously, such mutual expectations are at work during the whole conversation. In our example as shown in Figure 3, it is the situation where agents negotiate (s_3 and s_4) that mutual expectations and role-reversal imitations [28] are explicitly visualized.

5   Conclusion and Future Work

Representational modelling approaches are not capable of modelling social concepts adequately. Putting concepts like Team or Group into ontological categories leads to an inability to incorporate effects such as contextual dependence and emergence. However, as recent studies increasingly reveal, language is context-dependent and emerges from social interactions [25]. Moreover, as information systems are essentially socio-technical systems such effects gain importance. In the end, both social and technical concepts acquire meaning through their actual usage driven by cultural practices and purposes [28].

As a response to the lacks of expressiveness in representational approaches, we presented a novel approach based on State-Context-Property (SCoP) systems. SCoP draws from quantum mechanics and incorporates effects such as context, emergence and interaction. We adopted SCoP to social interactions, in particular to social concepts which acquire relevance dependent on intentional acts or context. Humans are intentional beings and their acquisition and usage of linguistic constructs seems to be driven by something quantum-like. Superposition states involved in mutual expectations are not separable into individual acts from which they emerged. Therefore, in SCoP concepts like Team gain a structure where states involved in social interactions are neither ontological in any agent-independent sense, nor are they reducible to agents from which their significance evolved.

Future work will investigate how social interactions relate to machine communication. Many modelling approaches to information systems development derive from machine languages. For instance, finite-state machines process symbols in order to carry out context-free tasks as defined in a rule-based manner. An useful method for systems analysis and design could support the identification of automation potentials based upon domain models specified with SCoP.

Acknowledgements. This project was supported in part by the Australian Research Council Discovery grant DP0773341.

References

1. Guarino, N.: Formal ontology and information systems. In: Proceedings of FOIS 1998, Trento, Italy, June 6-8, pp. 3–15. IOS Press, Amsterdam (1998)
2. Rosch, E.: Reclaiming concepts. Journal of Consciousness Studies 6(11), 61–78 (1999)
3. Husserl, E.: The Crisis of European Sciences and Transcendental Phenomenology: An Introduction to Phenomenological Philosophy. Translated by Carr., D. Northwestern University Press (1970)
4. Shanks, G., Tansley, E., Weber, R.: Representing composites in conceptual modelling. Communications of the ACM 47(7), 77–80 (2004)
5. Wand, Y., Storey, V., Weber, R.: An ontological analysis of the relationship construct in conceptual modeling. ACM Transactions on Database Systems 24(4), 494–528 (1999)

6. Gabora, L., Rosch, E., Aerts, D.: Toward an ecological theory of concepts. Ecological Psychology 20(1), 84–116 (2008)
7. Bunge, M.: Treatise on Basic Philosophy. Ontology I: The Furniture of the World, vol. 3. Reidel, Boston (1977)
8. Dreyfus, H.L.: What Computers Still Can't Do: A Critique of Artificial Reason. MIT Press, Cambridge (1992)
9. Dreyfus, H.L., Dreyfus, S.: Mind over Machine. Free Press (1986)
10. McCarthy, J., Hayes, P.: Some philosophical problems from the standpoint of artificial intelligence. Machine Intelligence 4, 463–502 (1969)
11. Harnard, S.: The symbol grounding problem. Physica D 42, 335–346 (1990)
12. Aerts, D.: Being and Change: Foundations of a Realistic Operational Formalism. In: Probing the Structure of Quantum Mechanics: Nonlinearity, Nonlocality, Computation, Axiomatics, pp. 71–110. World Scientific, Singapore (2002)
13. Aerts, D., Gabora, L.: A State-Context-Property model of concepts and their combinations I: The structure of the sets of contexts and properties. Kybernetes 34(1,2), 167–191 (2005)
14. Aerts, D., Gabora, L.: A State-Context-Property model of concepts and their combinations II: A Hilbert space representation. Kybernetes 34(1,2), 192–221 (2005)
15. Gabora, L., Aerts, D.: Contextualizing concepts using a mathematical generalization of the quantum formalism. Journal of Experimental and Theoretical Artificial Intelligence 14(4), 327–358 (2002)
16. Mylopoulos, J.: Information modelling in the time of the revolution. Information Systems 23(314), 127–155 (1998)
17. Rumbaugh, J., Jacobson, I., Booch, G.: The Unified Modeling Language Reference Manual. Addison-Wesley, Reading (1999)
18. Chen, P.: The Entity-Relationship Model-Toward a Unified View of Data. ACM Transactions on Database Systems 1, 9–36 (1976)
19. Bitbol, M.: Ontology, matter and emergence. Phenomenology and the Cognitive Sciences 6, 293–307 (2007)
20. Jaegher, H.D., Paolo, E.A.D.: Participatory sense-making: An enactive approach to social cognition. Phenomenology and the Cognitive Sciences 6(4), 485–507 (2007)
21. Thompson, E.: Mind in Life - Biology, Phenomenology and the Sciences of Mind. Harvard University Press (2007)
22. Hettel, T., Flender, C., Barros, A.: Scaling Choreography Modelling for B2B Value-Chain Analysis. In: Dumas, M., Reichert, M., Shan, M.-C. (eds.) BPM 2008. LNCS, vol. 5240. Springer, Heidelberg (2008)
23. Austin, J.L.: How to do things with words. Oxford University Press, Cambridge (1962)
24. Searle, J.R.: Speech acts. Cambridge Univ. Press, Cambridge (1969)
25. Tomasello, M.: Origins of Human Communication. MIT Press, Cambridge (2008)
26. Flender, C., Kitto, K., Bruza, P.D.: Nonseparability of Shared Intentionality. In: Bruza, P., Sofge, D., Lawless, W., van Rijsbergen, K., Klusch, H. (eds.) QI 2009. LNCS (LNAI), vol. 5494, pp. 211–224. Springer, Heidelberg (2009)
27. Winograd, T., Flores, F.: Understanding computers and cognition. Ablex Publishing Corp., Norwood (1986)
28. Tomasello, M.: Constructing a Language: A Usage-Based Theory of Language Acquisition. Harvard University Press, Cambridge (2003)

Structured Information Retrieval and Quantum Theory

Benjamin Piwowarski[1] and Mounia Lalmas[2]

[1] University of Glasgow
benjamin@bpiwowar.net
[2] University of Glasgow
mounia@acm.org

Abstract. Information Retrieval (IR) systems try to identify documents relevant to user queries, which are representations of user information needs. Interaction, context, and document structure are three important and active themes in IR research. We present how we propose to model the task of Structured IR (SIR) based on a QT inspired framework, with a focus on how to exploit user contextual information and user interaction in the search process.

1 Introduction

Information Retrieval (IR) aims at automatically matching a user's query, usually a set of keywords typed by the user, with a set of relevant documents. Structured IR (SIR) breaks away from the traditional retrieval unit of a document as a single large (text) block and aims at returning document fragments (e.g. a chapter, a section, or a paragraph), instead of whole documents in response to a user query. The structure of the document, whether explicitly provided by a mark-up language (e.g. XML) or derived, is exploited to determine these most relevant document fragments. In addition, SIR users may formulate queries with constraints on the content and on the structure of the units to be retrieved. SIR is believed to be of particular benefit for information repositories containing long documents, or documents covering a wide variety of topics (e.g. books, user manuals, legal documents), where the user's effort to locate relevant content within a document can be reduced by directing them to the most relevant parts of the document.

SIR has been extensively experimented within the INEX evaluation forum[1]. Unfortunately, experimental results so far indicate that, contrary to expectation, exploiting the structure in IR has not led to any significant increase in retrieval performance. One reason seems that models developed for SIR have mainly been adaptation of classical IR models. Even within standard IR, incrementally extending the classical IR models (e.g. adding pseudo/implicit relevance feedback and query expansion components) and adjusting parameters have not led to major improvement in retrieval performance [1].

---

[1] http://www.inex.otago.ac.nz/

P. Bruza et al. (Eds.): QI 2009, LNAI 5494, pp. 289–298, 2009.

One reason might be that the clues of relevance go beyond topical relevance (i.e. a document is relevant to a query if it is about the topic of the query), even when considering other aspects of the document content (e.g. the document style). More importantly, the *context* (defining the user information need) and the *interaction* (between the user and the IR system) are two important facets that have to be integrated directly into IR models *and* experiments, rather than being a controlled factor. With respect to SIR, structural context (of a document fragment within a document) and interaction (how the user uses structure to navigate within the document when searching) might play an even more important role, perhaps, than in standard IR.

It has been argued that the Quantum Theory (QT) formalism provides new tools for modeling the context and the interaction in IR. We also postulate that QT will allow the modeling of these between the user and the system, and between structured documents parts in SIR.

The paper follows a constructive approach to investigate the construction of a model for (S)IR based on QT. In Section 2, we discuss related works, first the different QT/IR approaches and then the use of structure in IR. In Section 3, we discuss the factors that should be taken into account, and discuss how a suitable model for SIR can be developed in Section 4.

2   Previous Works

2.1   QT and IR

In this section we focus on previous attempts of building IR models with the QT formalism. We can distinguish three kind of works – not necessarily incompatible, those (1) adapting an IR model to QT, (2) capturing the user-system interaction and (3) trying to define an adequate space for IR.

Adapting IR models to QT is one of the most direct way to incorporate this formalism into IR. It has two potential benefits. First, it shows that the QT formalism is powerful enough to *at least* encompass those models. Secondly, it might provide an insight on how to modify them in order to leverage the QT "added power". In [2], Van Rijsbergen shows that some classical IR models could be easily translated within the QT formalism. For instance, the vector space model can be easily expressed with the document defining the density and the query as an observable (or vice-versa).

Interestingly, Widdows [3] extended the classical IR vector space model with some concepts brought from QT. Two ideas where put forward. First, non-relevance is equated to the orthogonality in a vector space, and this gives a way to represent negation. Second, the disjunction of queries (q_1 or q_2) is modeled by the subspace spanned by the two subspaces associated with the two queries. Although this adaptation leads to nice results in particular with respect to negation, it does not fundamentally depart from classical IR models.

More recently, Guido et al. adapted [4] the logical imaging formalism [5] to the QT formalism. Logical imaging provides a way to compute a conditional probability (or an implication) through a non-uniform redistribution of the probability mass. In the classical case, p($\cdot$|A) implies that the initial mass associated to $\neg A$ is uniformly redistributed to the space of A. Transposed to IR, it allows for the redistribution of a term probability to an associated term belonging to the query (e.g. synonym). The interesting part of this formalism is the use of a kinematic operator that is an alternative to the standard Shrödinger unitary evolution, and that matches nicely the general imaging framework.

It can be interesting to develop from scratch a framework based on QT, hence avoiding to be restricted by previous standard IR approaches. Such approaches have been attempted in particular to model interaction. The latter has an increasing importance in IR research, as shown by the development of new theoretical frameworks for interaction in IR. Within QT, the work of [6, section 3.7.1.] clearly states one possible operational definition of interaction: *"The [IR system] is a type of oracle which detects a user's question with minimal ed required by the user to express the question, and then provides an answer that maximally satisfies the user"* which in the case of an IR system poses the problem as how to select a set of (part of) documents that satisfies *as much as possible* this requirement. In this work, interaction is modeled as a unitary evolution. Operationalising this framework is still an open question even in the case of "simple" flat documents. Nevertheless we support such a view where system and user are separately modeled, which we discuss in Section 3.

A more practical approach to make use of user interaction was proposed by Melucci [7], who makes the assumption that it is possible to define a Hilbert subspace that contains relevant documents, and that this subspace[2] can be built through user interaction. The QT formalism is involved when, given a document d, the model computes the probability that a document d is within the subspace RS by p$(RS|d)$ where RS is the constructed subspace.

Somehow orthogonal to the problem of building a model or a space is the question of how to define a space. Usual vector space in IR are spaces where each dimension is associated to one term. Building a concept space (as opposed to a *simple* term space) might be a way to exploit the geometry of the space. Many works [3,8,9,10] have suggested that QT might be adapted to represent concepts. The reasons are two-fold. First, there are interesting connections between orthogonality in the space and the different senses of a word. Second, contextual information could be captured by building up structures where word senses are entangled. For example, knowing in which sense the term "bat" is used might help to identify in which sense "cricket" is used (either animal or sport). These works have the potential to define suitable conceptual spaces for IR.

Two QT based approaches try to build projectors lattices, which in turn can be used to define a Hilbert space. The first one is related to disambiguating words [11]. A lattice of projectors associated to contexts is built, where atomic contexts (e.g. "the *animal* is a tiger") completely determine the context leaving no other possible interpretation for a word (here *animal*). The other work [12] attempts to build a document space where co-occurrence is a central notion.

---

[2] Melucci names it a context space, but we use this word differently so we do not use his terminology here.

The idea is to characterise texts through *erasing* projectors that only keep words within a given window of a word. While it is not clear how to use these specific frameworks in IR, these approaches are interesting since they define a part of the Hilbert space structure (orthogonality relationships) by the possible observation one could make on the systems represented into that space.

Overall, IR built upon QT foundations is still in an early stage and there are no solid foundations upon which one could develop a sound framework. In section 3 onwards, we discuss some points we deem important for (at least our) future work in that field.

2.2 Structure, Psycholinguistics and IR

It is interesting to consider psycholinguistic studies since they might provide an insight on how humans actually use structure. A good summary of current research can be found in [13]. They report that the facilitatory effect of headings in a text is reflected both in the fixations made during the first-pass reading as well as in the later look-backs directed to the topic sentences. At the finer grained level, sentences in the middle of a paragraph can be understood from the structural context of the previous sentence(s), which is not true of the first sentence of a paragraph; transitions between paragraphs and sections thus require more work from the reader. The effectiveness of headings lies in the fact that they provide a mental frame into which upcoming text information may be integrated. At a coarser level, two common hypotheses on why structure facilitates comprehension are stated: (1) facilitate processing of the text topic structure during reading, and (2) readers use text structure to guide text recall (going back to some parts of the text). It hence appears that structure has semantics that could be exploited in IR, because it provides a good way to organize information.

Within the IR community, the use of structure in IR has been extensively studied and evaluated within INEX. Summarizing, structure in IR has been used as a mean to (1) provide more focused material to the user (e.g. return a section of a chapter instead of the whole chapter), (2) specify user constraints on content and structure (e.g. return sections about wine within a chapter about Chile) and (3) provide structural context to a given document part (e.g. a section about jaguar within a book about cars is not the same as within a book about animals). Note that the latter is one possible use of the psycho-linguistics findings, and one that has been shown to improve significantly the performance of SIR systems. Apart from these achievements, structure has not been shown to enhance traditional IR search. We believe new models that use structural context and interaction could make a difference, since they would complement the lack of explicit information about what the user really wants.

3 Factors to Consider

To consider interaction in SIR using the QT formalism, our QT-based model should be able to respond to the interaction between the user and the (S)IR system, which during a search session may include the queries typed and submitted to the system, the clicks users make on links returned by the system, and if available more fine-grained information such as the seen elements (as obtained through the use of an eye tracking tool, for instance). We stress that *all* interactions, including the interaction with the list of results, have to be taken into account in order to build a fully interactive SIR model.

Our QT-based model should also be able to integrate information related to the context of the information need (e.g. previous searches, time, location). The fact that different document fragments may be deemed relevant for a same set of interactions, indicates that relevance is dependent on the search context. This should be captured by the model. Such situations arise for example when the typed query is ambiguous (e.g. "jaguar" as an animal or a car) or when the expertise level of the users are different.

As pointed out in [2], at least two QT features are particularily important to IR. First, the intertwinement of geometry and probabilities, where two distance-wise close vectors representing system states generate almost the space probability distribution on the Hilbert space, and hence the same probabilities of making a given measurement. An example of the usefulness of this principle, is that close-by documents in a term space would imply close-by probabilities of, say, relevance. The second important feature is that measures made on the system might interact with each other in a non standard way, which might prove useful for interactive IR, where for instance a series of observations on the user might change the user state (if we assume that the user state lies in a Hilbert space).

4 A Framework for SIR Based on QT Formalism

We discuss here which space we could be working with, and how it could be constructed for modeling SIR. We first discuss the choice of the representation, and propose to use an information need space. We then discuss how this representation can be used to model interactive IR, and to which extent document structure can be included in the model.

Among the different spaces we could be working with, various choices are possible, but among the most straightforward choice is the topical space [2] – or its approximation, the term space. In such a space, a document is represented by the terms or concepts it contains. Whether this corresponds in QT to a superposition (i.e. a document is a unique combination of terms) or to a mixture of pure term-states is subject to debate, but in both cases a query (or rather the relevance to a query) is an observable, and one can ask the question: "is this document [system] relevant to this query [observable]?". Another kind of questions that can be asked are "is this document [system] about topic X [observable]?".

While this seems to be an intuitive choice, we argue that from a theoretical point of view it is not a sensible choice if we want to use the QT formalism, since it does not exhibit proper quantum properties and does not seem to be adapted to interactive IR.

To uphold the former statement about quantum properties, let us imagine that we have two observable T_A and T_B associated with the observation "this document is about topic A (resp. B)". It can be argued that the two observables interact since the fact that a document is about one topic might influence the fact that it is about another topic. We could even say if we measure T_A, then T_B and eventually T_A, the first measurement of T_A can be different from the second one because asking if the document is about topic B changed the topicality *as perceived by the user*. However, continuing this series of measurement, that is performing $T_A T_B T_A T_B T_A T_B T_A \ldots$, one would expect that the observed values remained the same for both observables T_A and T_B since no new information is brought. This series of measurements cannot happen within QT if no interaction happens, which in this case stems from the fact that users are expected to learn.

In our opinion, these remarks underline two things. First, document topicality is constructive in the sense that any information adds up to previous ones, and this does not match QT measurement in general, since, while measuring, a part of the information is "destructed". Second, we cannot hope to model directly the user perception of topicality as an observable within a document topicality space, since we believe it is a learning process that saturates (i.e. the opinion of the user does not change with further interaction).

4.1 An Information Need Space

Instead we propose the use of an *information need space* where a state, and more generally a density, corresponds to a user information need. Mixed states could naturally be used to model ambiguous information needs, and context/interaction would provide a way to specify what is the actual information need. The density would be pure when the information need is completely determined, as for example when the model can fully predict what are the relevant documents. For an exploratory search (e.g. "I want to learn about Glasgow"), the need density is mixed, whereas for a navigational search (e.g. "I want the University of Glasgow home page") the need density is pure. The relevance of a document (in IR) or document fragment (in SIR) would then be modeled as an observable. This is different from [7], where relevance is modeled as a yes/no observable within a space where documents are the observed systems, and the corresponding subspace is expanded through user interaction.

We think the information need space can model interactive IR since users change their point of view during a search, and relevance, contrarily to topicality, is expected to evolve within a search session [14]. The mechanisms of this change are yet to be understood, but QT could possibly shed a new light on that matter, since this process is not constructive as the document topicality is – users might change their opinion on what they find relevant.

In more details, the information need space could be a tensor product of smaller spaces, each one related to the different dimensions related to the relevance of an information need. A non-exhaustive list of such dimensions would be the topicality, the style (e.g. review, literature, FAQ, etc.), the position in the structure (e.g. is it a whole book, a section?) and the novelty of the document. Please refer to [15] for a more complete analysis of relevance dimensions.

Without considering context, at the beginning of the search process, the information need space could be seen as a mixed density that corresponds to *all* possible needs, weighed by their probability. What is nice about this is that we could (in theory) provide a list of documents without any interaction and without *any information or interaction* from the user, since it is possible to measure to which extent a document fragment is relevant to an information need density. The context of the search and each interaction would then be extra steps towards the retrieval of relevant information.

Within the various dimensions of relevance, topical relevance is an aspect of the information need that seem to be well adapted to a QT-based model. Let us use an example to illustrate this fact. Consider a user who wants to plan his holidays in Barcelona, and who will be searching for various informations ranging from activities to hotels. Whereas one part of the information need remains untouched (it is about Barcelona in Spain – and not in Venezuela or the Philippines), the other part can drift (from leisure activities to hotels). Interaction through measurement, as described in the next section, would be used to both restrict the subspace to documents about Barcelona in Spain, and to follow the user topical drift from activities to hotels.

4.2   Evolution: Interaction in Information Retrieval?

The evolution of a system is an important topic both in QT and interactive IR. In this section, we study the various forms of evolutions in QT and relate them to our (S)IR.

The first form is measurement. It would account for a partial collapse in the corresponding information need subspaces. An example scenario of interaction would be a user searching for a place to order pizza. At the beginning of the search, the density associated with the information need is not determined and could be a mixture of all possible information needs. The user then types "pizza", which restricts the information need to a given subset of densities and hence to a given subspace of the whole information need space. Knowing that it is 8pm, and that this person is living in a given city would further restrict the density to a smaller subset of densities. More precisely, each new observation (e.g. typed keywords, clicks, time, etc.) would correspond to a possible measurement/projector, and hence to an observable. This integrates nicely within the IR model adapted to QT in [2], since the simplest T would be a projector along the vector representing the keywords defined as in standard IR. In general, the more ambiguous the keywords, the bigger the subspace associated to the projector T.

Note that typed keyword observations can influence more relevance dimensions than the topical one. For instance, if a query contains "review of..." then this is more related to the style of the relevant documents than to their topicality. Linking interactions and measurements would be an iterative process where past interactions could be analysed e.g. in order to compute the exact form of the observables associated to some keywords.

Another possible use of measurement would be to deal with novelty and the related problem of result diversity (that is, how to select a set non redundant pieces of information with respect to a given information need). Documents would be associated to observables within a "knowledge" space for which a user is the system under observation. When a document is read, then the user state would be projected in a subspace that corresponds to a subspace of knowledge where the read document information is known. This process, coupled with the information need specification and drift discussed in 4.1, would be used to build up a list of documents to return to the user.

The second form of evolution would be a unitary one, which describes the evolution of the information need in the absence of interaction. This would be particularly suited to the time observation, since time evolves in a non-interactive manner. Similarly to physics, unitary evolution could also account for the natural evolution of the user's need in the absence of interaction. One possible use would be for example to build up evolution operators using previous user interactions. Again, we can use the holidays in Barcelona example: Users starting to be interested by hotels would turn up to be interested by activities (and vice-versa), leaving the geography-related dimensions untouched.

The third form is through interaction with the environment which in our context is both the user interface and the user memory. This form of evolution should be used when a measurement conducted twice gives two different results. This is the case when, for example, the user interacts with the IR system, and subsequently deems a document to be relevant and latter non relevant, since the user has already read this document. Note that with respect to relevance, if we assume that there is no interaction with the user (i.e. we could use an oracle to tell us that the document is relevant for the current information need state), then we would use standard QT measurement.

To handle the interaction between the user and the SIR system, we would build a user behaviour model. We would define a system space, different from the information need space, where we can represent the current state of the IR system. The state would include information such as which document fragments (or rather hyperlinks to these fragments) are displayed. We would then make the entangled user and system states evolve, taking into account the fact that the user inspects the result list and, in the case of SIR, the behaviour within the document structure, so that to predict which parts of the document collection would be explored by the user. Some part of the interaction would correspond to observations like e.g. when a user clicks on an hyperlink. The result of the interaction ρ can then be measured in this new space, and interaction specific observations like clicks can then be taken into account. The new information need density can be extracted using the partial trace operator, which is useful if we want to reuse this density for new observations and/or predictions.

4.3   Structure

In this section, we discuss how the framework could integrate with structured information.

As discussed in Section 2.2 and in the INEX workshops, structure can help to obtain a better representation of a fragment of text within the document structure whether it be a topical representation, a style or other relevance related dimensions. In our case, to build the topical relevance observable, we could use structure to define the number of dimensions of the associated subspace – ideally, one per topic. An oversimplified example would be to associate each paragraph with a low dimensional subspace of the information need space, and then to build the subspace associated with the section that contains those paragraphs by joining all these subspaces.

Let us note that the bigger (in size) the structural part, the bigger the associated subspace in the information need space, which in turn means that there is a higher chance that a bigger document fragment covers an information need. Consider two document fragments F_1 and F_2, F_2 being included in F_1 (e.g. a paragraph in a section). The projector associated with the relevance of F_1 would "include" (in the sense of inclusion of the projector associated subspace) the subspace associated with the relevance of F_2 (i.e. $F_2 \leq F_1$). Then, if we know that F_2 is relevant to a given query, this would imply that F_1 is also relevant to that query, since the density would be projected into the subspace defined by the projector for F_2, and this subspace is included into the one of the projector for F_1. Deciding which of F_1 or F_2 is better for the user is a matter of user behaviour modeling, as discussed at the end of the previous section.

This nesting property of document fragments also implies that it is not only necessary to find a fragment that covers (*exhaustivity*) the information need, but this fragment has also to be specific to the information need. In order to achieve this, we could build an observable who would measure the percentage of the fragment that deals with the topic of interest. It is relatively easy to build such an observable, since the subspace it spans corresponds to the subspace spanned by the relevance observable associated to the document fragment, but in this case the *specificity* is not a projection observable as the *exhaustivity* is. Both exhaustivity and specificity dimensions are being used in INEX relevance assessments done by human judges, and could be used to compare the output of the algorithms producing the two observables with the values set by the judges.

5   Conclusion

In this paper, we have sketched how contextual and interactive SIR can be modeled borrowing ideas from QT, by defining a space where the user information needs would evolve according to their interactions with the retrieval system. We proposed to use an information need space, as opposed to the standard topical space, as it seems to be better adapted to both IR (allowing interaction) and QT (leveraging a part of the QT framework potential). We briefly described how our information need space, emphasising the fact that it should capture various relevance dimensions beside topical relevance. We then discussed how interaction could be modeled with this representation, and how it would be possible to model the document structure dimension (i.e. what document fragment granularity to return – a paragraph, a section, etc.). While there are still many details to be set in order to get an operational system, we believe this path would allow to capture faithfully the complexity of the search process in SIR.

Acknowledgments. This research was supported by an Engineering and Physical Sciences Research Council grant (Grant Number EP/F015984/2).

References

1. Jones, K.S.: What's the value of trec: is there a gap to jump or a chasm to bridge? SIGIR Forum 40(1), 10–20 (2006)
2. van Rijsbergen, C.J.: The Geometry of Information Retrieval. Cambridge University Press, New York (2004)
3. Widdows, D.: Geometry and Meaning. CLSI Lectures notes, vol. 172. CSLI (2004)
4. Zuccon, G., Azzopardi, L., van Rijsbergen, C.J.: A formalization of logical imaging for information retrieval using quantum theory. In: IEEE proceedings of the 5th International Workshop on Text-based Information Retrieval. IEEE, Los Alamitos (2008)
5. Crestani, F., van Rijsbergen, C.J.: A study of probability kinematics in information retrieval. ACM Trans. Inf. Syst. 16(3), 225–255 (1998)
6. Arafat, S., van Rijsbergen, C.J.: Quantum theory and the nature of search. In: Proceedings of the First Quantum Interaction Symposium, QI 2007 (2007)
7. Melucci, M.: A basis for information retrieval in context. ACM Trans. Inf. Syst. 26(3), 1–41 (2008)
8. Widdows, D.: A mathematical model for context and word-meaning. In: Blackburn, P., Ghidini, C., Turner, R.M., Giunchiglia, F. (eds.) CONTEXT 2003. LNCS, vol. 2680, pp. 369–382. Springer, Heidelberg (2003)
9. Bruza, P., Woods, J.: Quantum collapse in semantic space: Intepreting natural language argumentation. In: [16]
10. Bruza, P.D., Kitto, K., Nelson, D., McEvoy, C.L.: Entangling words and meaning. In: [16], pp. 118–124
11. Aerts, D., Gabora, L.: A theory of concepts and their combinations ii: A hilbert space representation. Kybernetes (2005)
12. Huertas-Rosero, A.F., Azzopardi, L., van Rijsbergen, C.J.: Characterising through erasing: A theoretical framework for representing documents inspired by quantum theory. In: [16]
13. Hyona, J., Lorch, R.F.: Effects of topic headings on text processing: evidence from adult readers' eye fixation patterns. Learning and Instruction 14(2), 131–152 (2004)
14. Xu, Y.: The dynamics of interactive information retrieval behavior, part i: An activity theory perspective. Journal of the American Society for Information Science and Technology 58(7), 958–970 (2007)
15. Saracevic, T.: Relevance: A review of the literature and a framework for thinking on the notion in information science. part ii: nature and manifestations of relevance. J. Am. Soc. Inf. Sci. Technol. 58(13), 1915–1933 (2007)
16. Bruza, P.D., Lawless, W., van Rijsbergen, K., Sofge, D.A., Coecke, B., Clark, S. (eds.) Proceedings of the Second Quantum Interaction Symposium, QI 2008 (2008)

Hilbert Space Models Commodity Exchanges

Paul Cockshott

Dept Computing Science University of Glasgow
wpc@dcs.gla.ac.uk

Abstract. It is argued that the vector space measures used to measure closeness of market prices to predictors for market prices are invalid because of the observed metric of commodity space. An alternative representation in Hilbert space within which such measures do apply is proposed. It is shown that commodity exchanges can be modeled by the application of unitary operators to this space.

1 Linear Price Models

The context of this paper is the empirical testing of linear models of economic activity. Whilst these originated in an informal way in the work of Adam Smith and Quesney, and were partially formalised by Marx in volume 3 of Capital, an adequate formal treatment had to wait for von Neuman[21] and Kantorovich[9]. Both von Neumann and Kantorovich were mathematicians rather than economists. Their contributions to economics were just one part of a variety of research achievements. In both cases this included stints working on early nuclear weapons programs, for the US and USSR[15] respectively. At least in von Neumann's case the connection of his economic work to atomic physics was more than incidental. One of his great achievements was his mathematical formalization of quantum mechanics[22] which unified the matrix mechanics of Heisenberg with the wave mechanics of Schrodinger. His work on quantum mechanics coincided with the first draft of his economic growth model[21] given as a lecture in Princeton in 1932. In both fields he employs vector spaces and matrix operators over vector spaces, complex vector spaces in the quantum mechanical case, and real vector spaces in the growth model. Kurz and Salvadori [11]argue that his growth model has to be seen as a response to the prior work of the mathematician Remak[14], who worked on 'superposed prices'.

Remak then constructs 'superposed prices' for an economic system in stationary conditions in which there are as many single-product processes of production as there are products, and each process or product is represented by a different 'person' or rather activity or industry. The amounts of the different commodities acquired by a person over a certain period of time in exchange for his or her own product are of course the amounts needed as means of production to produce this product and the amounts of consumption goods in support of the person (and

P. Bruza et al. (Eds.): QI 2009, LNAI 5494, pp. 299–307, 2009.

his or her family), given the levels of sustenance. With an appropriate choice of units, the resulting system of 'superposed prices' can be written as

$$p^T = p^T C$$

where C is the augmented matrix of inputs per unit of output, and p is the vector of exchange ratios. Discussing system Remak arrived at the conclusion that there is a solution to it, which is semipositive and unique except for a scale factor. The system refers to a kind of ideal economy with independent producers, no wage labour and hence no profits. However, in Remak's view it can also be interpreted as a socialist economic system [11].

With Remak the mathematical links to the then emerging matrix mechanics are striking - the language of superposition, the use of a unitary matrix operator C analogous to the Hermitian operators in quantum mechanics[1]. Remak shows for the first time how, starting from an *in-natura* description of the conditions of production, one can derive an equilibrium system of prices. This implies that the *in-natura* system contains the information necessary for the prices and that the prices are a projection of the *in-natura* system onto a lower dimensional space[2]. If that is the case, then any calculations that can be done with the information in the reduced system p could in principle be done, by some other algorithmic procedure starting from C. Remak expresses confidence that with the development of electric calculating machines, the required large systems of linear equations will be solvable.

The weakness of Remak's analysis is that it was limited to an economy in steady state. Von Neumann took the analysis on in two distinct ways:

1. He models an economy in growth, not a static economy. He assumes an economy in uniform proportionate growth. He explicitly abjures considering the effects of restricted natural resources or labour supply, assuming instead that the labour supply can be extended to accommodate growth. This is perhaps not unrealistic as a picture of an economy undergoing rapid industrialization (for instance Soviet Russia at the time he was writing).
2. He allows for there to be multiple techniques to produce any given good - Remak only allowed one. These different possible productive techniques use different mixtures of inputs, and only some of them will be viable.

von Neumann again uses the idea of a technology matrix introduced by Remak, but now splits it into two matrices A which represents the goods con-

---

[1] Like the Hermitian operators in quantum mechanics, Remak's production operator is unitary because p is an eigen vector of C and $|p|$ is unchanged under the operation.

[2] Suppose C is an $n \times n$ square matrix, and p an n dimensional vector. By applying Iverson's reshaping[8,7] operator ρ, we can map C to a vector of length n^2 thus $c \leftarrow (n \times n)\rho C$, and we thus see that the price system, having n dimensions involves a massive dimension reduction from the n^2 dimensional vector c.

sumed in production, and B which represents the goods produced. So a_{ij} is the amount of the j th product used in production process i, and b_{ij} the amount of product j produced in process i. This formulation allows for joint production, and he says that the depreciation of capital goods can be modeled in this way, a production process uses up new machines and produces as a side effect older, worn machines. The number of processes does not need to equal the number of distinct product types, so we are not necessarily dealing with square matrices.

Like Remak he assumes that there exists a price vector y but also an intensity vector x which measures the intensity with which any given production process is operated. Later the same formulation was used by Kantorovich. Two remaining variables β and α measure the interest rate and the rate of growth of the economy respectively.

He makes two additional assumptions. First is that there are 'no profits', by which he means that all production processes with positive intensity return exactly the rate of interest. He only counts as profit, earning a return above the rate of interest. This also means that no processes are run at a loss (returning less than β). His second assumption is that any product produced in excessive quantity has a zero price.

He goes on to show that in this system there is an equilibrium state in which there is a unique growth rate $\alpha = \beta$ and definite set of intensities and prices. The intensities and prices are simultaneously determined.

Von Neumann's work was influential in economic theory, spawing a number of similar models, probably the most famous of which was Sraffa's[18].

At the time that von Neumann was writing, despite Remak's optimism, it was not possible to empirically test different linear theories of prices because of the problem of collecting the necessary data, and the problem of solving large matrix equations. From the 1950s onwards though, the empirical problem of obtaining the A and B matrices was reduced by the publication of national input output tables. Since the ready availability of computers emprical testing became possible.

In 1983 Farjoun and Machover published a seminal work applying statistical mechanics to the dynamics of capitalist economies[6]. One of the predictions of their book was that what they called vertically integrated labour coefficients would be good predictors for market prices. One can view their price model as being similar to that of Remak with added thermal noise. Their predictions have largely been born out by subsequent empirical studies [2,19,12,3,13,17,4], though there have been isolated studies questioning this [10,20]. There has been some controversy as to what metric was appropriate for determining the closeness of market prices to integrated labour coefficients. In the recent literature discussing this [10,13,16,20]it has been taken as given that the use of vector space measures is appropriate. For example one measure proposed has been to determine the angle between two price vectors. I wish to point out that this approach is questionable.

2   The Vector Space Problem

Vector spaces are a subclass of metric space. A metric space is characterized by a positive real valued metric function $\delta(p, q)$ giving the distance between two points, p, q. This distance function must satisfy the triangle inequality $\delta(p, q) \leq \delta(p, r) + \delta(q, r)$. In vector spaces this metric takes the form:

$$\delta(\mathbf{p}, \mathbf{q}) = \sqrt{\sum (p_i - q_i)^2} \tag{1}$$

We have argued elsewhere[1] that the metric of commodity space does not take this form. Let us recapitulate the argument.

Conjecture 1. Commodity space is a vector space.

Assume that we have a commodity space made up of two commodities, gold and corn and that 1oz gold exchanges for 100 bushels of corn. We can represent any agent's holding of the two commodities by a 2 dimensional vector $\mathbf{c}$ with c_0 being their gold holding and c_1 being their corn holding. Given the exchange ratio above, we can assume that (1,0) and (0,100) are points of equal worth and assuming that commodity space is a vector space thus

$$\delta((0,0), (1,0)) = \delta((0,0), (0, 100)) \tag{2}$$

This obviously does not meet equation 1 but if we re-normalise the corn axis by dividing by its price in gold, we get a metric

$$\delta_c(\mathbf{p}, \mathbf{q}) = \sqrt{(p_0 - q_0)^2 + (\frac{p_1 - q_1}{100})^2} \tag{3}$$

which meets the equation we want for our two extreme points:

$$\delta_c((0,0), (1,0)) = \delta_c((0,0), (0, 1)) \tag{4}$$

If this is our metric, then we can define a set of commodity holdings that are the same distance from the origin as holding 1oz of gold. Let us term this U the unit circle in commodity space:

$$U = \{a \in U : \delta_c((0,0), a) = 1\} \tag{5}$$

Since these points are equidistant from the origin, where the agent holds nothing, they must be positions of equal worth, and that movements along this path must not alter the net worth of the agent. Let us consider a point on U, where the agent holds $\frac{1}{\sqrt{2}}$oz gold and $\frac{100}{\sqrt{2}}$ bushels of corn.

Would this in reality be a point of equal worth to holding 1 oz of gold?

No, since the agent could trade their $\frac{100}{\sqrt{2}}$ bushels of corn for a further $\frac{1}{\sqrt{2}}$oz gold and end up with $\sqrt{2}$oz> 1oz of gold. Thus there exists a point on U that is not equidistant from the origin, hence equation 3 can not be the form of the metric of commodity space and thus conjecture 1 falls, and commodity space is not a vector space.

3   The Metric of Commodity Space

The metric actually observed in the space of bundles of commodities is:

$$\delta_b(\mathbf{p},\mathbf{q}) = \left| \sum \alpha_i \left[p_i - q_i \right] \right| \tag{6}$$

where $\mathbf{p}$, $\mathbf{q}$ are vectors of commodities, and α_i are relative values. The 'unit circle' in this space actually corresponds to a pair of parallel hyperplanes on above and one below the origin. One such hyperplane is the set of all commodity combinations of positive value 1 and the other, the set of all commodity combinations of value -1. The latter corresponds to agents with negative worth, i.e., net debtors.

Because of its metric, this space is not a vector space and it is questionable whether measures of similarity based on vector space metrics are appropriate for it. However it is possible to posit an underlying linear vector space of which commodity space is a representation.

4   Commodity Amplitude Space

We will now develop the concept of an underlying space, commodity amplitude space, which can model commodity exchanges and the formation of debt. Unlike commodity space itself, this space, is a true vector space whose evolution can be modeled by the application of linear operators. The relationship between commodity amplitude space and observed holdings of commodities by agents is analogous to that between amplitudes and observables in quantum theory.

Let us consider a system of n agents and m commodities, and represent the state of this system at an instance in time by a complex matrix $\mathbf{A}$, where a_{ij} represents the amplitude of agent i in commodity j. The actual value of the holding of commodity j by agent i , we denote by h_{ij} an element of the holding matrix $\mathbf{H}$. This is related to a_{ij} by the equation $a_{ij} = \sqrt{h_{ij}}$.

4.1   Commodity Exchanges

We can represent the process of commodity exchange by the application of rotation operators to $\mathbf{A}$. An agent can change the amplitudes of their holdings of different commodities by a rotation in amplitude space. Thus an initial amplitude of 1 in gold space by an agent can be transformed into an amplitude of 1 in corn space by a rotation of $\frac{\pi}{2}$. Borrowing Dirac notation we can write these as 1—gold¿, and 1—corn¿. A rotation of $\frac{\pi}{4}$ on the other hand would move an agent from a pure state 1—gold¿ to a superposition of states $\frac{1}{\sqrt{2}}$|gold> $+\frac{1}{\sqrt{2}}$|corn¿ . Unlike rotation operators in commodity space this is value conserving since on squaring we find their assets are now $\frac{1}{2}$gold $+ \frac{1}{2}$corn.

The second conservation law that has to be maintained in exchange is conservation of the value of each individual commodity, there must be no more or less of any commodity after the exchange than there was before. This can be modeled by constraining the evolution operators on commodity amplitude space to be such that they simultaneously perform a rotation on rows and columns of the matrix $\mathbf{A}$.

Suppose we start in state:

$$\mathbf{A} = \begin{bmatrix} 1 & 0 \\ 0 & 2 \end{bmatrix}, \mathbf{H} = \begin{bmatrix} 1 & 0 \\ 0 & 4 \end{bmatrix}$$

Where agent zero has 1 of gold and no corn, and agent one has no gold and 4 of corn. We can model the purchase of 1 of corn by agent zero from agent one by the evolution of $\mathbf{A}$ to:

$$\mathbf{A2} = \begin{bmatrix} 0 & 1 \\ 1 & \sqrt{3} \end{bmatrix}$$

which corresponds to final holdings of:

$$\mathbf{H2} = \begin{bmatrix} 0 & 1 \\ 1 & 3 \end{bmatrix}$$

Note that the operation on amplitude space is a length preserving rotation on both the rows and the columns. The lengths of the row zero and column zero in $\mathbf{A2}$ are 1 the lengths of row and column one is 2 just as it was for $\mathbf{A}$. This operation can be effected by the application of an appropriate rotation matrix so that $\mathbf{A2} = \mathbf{M}.\mathbf{A}$. A matrix which produces this particular set of rotations is:

$$\mathbf{M} = \begin{bmatrix} 0 & \frac{1}{2} \\ 1 & \frac{\sqrt{3}}{2} \end{bmatrix}$$

4.2   Price Changes

Price movements are equivalent to the application of scaling operations which can be modeled by the application of diagonal matrices. Thus a 50% fall in the price of corn in our model would be represented by the application of the matrix $\begin{smallmatrix} 1 & 0 \\ 0 & \frac{1}{\sqrt{2}} \end{smallmatrix}$ to the current commodity amplitude matrix. Scaling operations are not length preserving.

The reason for this is that if there is a change in prices an agent holding a vector of physical commodities $\mathbf{b}$ will find that for most arbitrarily chosen commodity vectors $\mathbf{c}$, the quantity of $\mathbf{c}$ that they can exchange $\mathbf{b}$ for will have changed. The length preserving rotations that we have assumed up to now have amounted to assuming that there is no possibility of changing ones net worth by commodity exchanges at a given set of relative prices. If the prices change over time this is no longer the case, hence the introduction of non-conservative operations.

4.3   Modeling Debt

We specified in section 4 that the amplitude matrix must be complex valued. This is required to model debt. Suppose that starting from holdings **H** agent zero buys 2 of corn from agent one. Since agent zero only has 1 of gold to pay for it, the transaction leaves the following holdings:

$$\begin{array}{ccc} \text{Agent} & \text{gold} & \text{corn} \\ 0 & -1 & 2 \\ 1 & 2 & 2 \end{array}$$

The corresponding amplitude matrix is

$$\mathbf{A3} = \begin{bmatrix} i & \sqrt{2} \\ \sqrt{2} & \sqrt{2} \end{bmatrix}$$

It it interesting that this too is the result of applying a unitary rotation operator to the original amplitude vector since the length of row zero $|\mathbf{A3_0}| = i^2 + (\sqrt{2})^2 = 1$, likewise the lengths of all other rows and columns are preserved. The linear operator required to create debts has itself to be complex valued, thus if $\mathbf{A3} = \mathbf{NA}$ we have

$$\mathbf{N} = \begin{bmatrix} i & \frac{1}{\sqrt{2}} \\ \sqrt{2} & \frac{1}{\sqrt{2}} \end{bmatrix}$$

Note that the operators here are not Hermitian. This would appear to preclude the interesting possibility of simulating commodity exchanges on a future quantum computer[5], though there may be renormalization techniques that could be applied to get over this problem.

5   Implications for Similarity Measures

Steedman [20] has proposed that a suitable criterion for assessing similarity of values to market prices is the angle between market price and value vectors, with small angles indicating closeness. If **m, v** are market price and value vectors respectively, the angle between them is given by:

$$ArcCos(\vec{\mathbf{m}}.\vec{\mathbf{v}})$$

where **v** denotes the normalized value vector given by $\mathbf{v} = \frac{\mathbf{v}}{|\mathbf{v}|}$.

If the argument in section 2 is accepted, we should consider using angles between price and value amplitude vectors instead. If we denote the normalized vectors in amplitude space by $\mathbf{m}_a$ and $\mathbf{v}_a$, then the amplitude space angles are given by:

$$ArcCos(\vec{\mathbf{m}_a}.\vec{\mathbf{v}_a^*})$$

where $\mathbf{x}^*$ is the conjugate of $\mathbf{x}$.

What will be the properties of this measure?

In general it will show smaller angles between vectors. For example suppose we have 3 commodities iron, corn, cotton as follows:

			|amplitudes|	
	value	price	value	price	
corn	1	1	1	1	
iron	3	2	$\sqrt{3}$	$\sqrt{2}$	
cotton	1	2	1	$\sqrt{2}$	
angle		30.2°		13.4°	

The fact that smaller angles are shown would be or little significance if the relative sizes of angles in the two spaces was the same. But this need not be the case. Consider the following example:

	value	price	PP	amplitudes		
				value	price	PP
corn	1	1	1	1	1	1
iron	$\frac{1}{2}$	-1	2	$\frac{1}{\sqrt{2}}$	i	$\sqrt{2}$
cotton	0.02	1	1	$\frac{\sqrt{2}}{10}$	1	1
θrelative to price in	74°	0	90°	54°	0	45°

Here we are comparing three hypothetical vectors of values, prices and prices of production (PP). If we treated commodity value space as a vector space, then prices of production would be orthogonal to market prices, whereas in amplitude space they are at 45° to market prices. In this example, when commodity space is treated as a vector space, values appear closer to market prices than do prices of production. When the assumption that commodity space is a vector space is dropped, then prices of production are closer to market prices.

6   Conclusion

We have argued that commodity space can not be directly modeled by a vector space, because of the metric it observes, but that it can be treated as the real valued representation or an underlying vector space. This complex vector space we have, following physics terminology, termed commodity amplitude space. Observed holdings of commodities and money by agents are the squares of corresponding commodity amplitudes. Commodity exchange relations, including the formation of commercial debt can be modeled by unitary rotation matrices operating on this amplitude space. The conceptual model presented borrows extensively from quantum formalism.

It is thus at least arguable that the empirical relation between market prices and labour values should be measured by the angles between their corresponding vectors in commodity amplitude space. The latter space, unlike commodity space, is a linear vector space within which angles of rotation have a clear meaning.

References

1. Cockshott, P., Cottrell, A.: Value's Law, Value's Metric, research report University of Strathclyde, Dept. of Computer Science (1994); Reprinted in: Freeman, A., Kliman, A., Wells, J. (eds.): The New Value Controversy and the Foundations of Economics, Edward Elgar, Cheltenham (2004)

2. Cockshott, W.P., Cottrell, A.: Labour-Time Versus Alternative Value Bases: A Research Note. Cambridge Journal of Economics 27, 749–754 (1997)
3. Cockshott, W.P., Cottrell, A.: A note on the organic composition of capital and profit rates. Cambridge Journal of Economics 21, 545–549 (2003)
4. Cockshott, W.P., Cottrell, A., Michaelson, G.: Testing Marx: Some New Results from UK Data. Capital and Class 55, 103–129 (Spring 1995)
5. DiVincenzo, D.: Quantum Computation. Science 270, 255–261
6. Farjoun, E., Machover, M.: The laws of Chaos. Verso (1983)
7. Iverson, K.: A programming language. Wiley, New York (1966)
8. Iverson, K.: Notation as a tool of thought. In: ACM Turing award lectures, p. 1979. ACM, New York (2007)
9. Kantorovich, L.V.: Mathematical Methods of Organizing and Planning Production. Management Science 6(4), 366–422 (1960)
10. Kliman, A.: The Law of Value and Laws of Statistics: Sectoral Values and Prices in the US Economy, 1977-1997. Cambridge Journal of Economics 26, 299–311 (2002)
11. Kurz, H.D., Salvadori, N.: Von Neumann's Growth Model and the 'Classical' Tradition. Understanding "classical" Economics: Studies in Long-Period Theory (1998)
12. Ochoa, E.M.: Values, prices, and wage–profit curves in the US economy. Cambridge Journal of Economics 13(3), 413–429 (1989)
13. Petrovic, P.: The deviation of production prices from labour values: some methodology and empirical evidence. Cambridge Journal of Economics 11(3), 197–210 (1987)
14. Remak, R.: Kann die Volkswirtschaftslehre eine exakte Wissenschaft werden. Jahrbücher für Nationalökonomie und Statistik 131, 703–735 (1929)
15. Ryabev, L.D., Smirnov, Y.N.: The Atomic Project, Science, and the Atomic Industry. Atomic Energy 99(2), 519–527 (2005)
16. Shaik, A.: The Empirical Strength of the Labour Theory of Value. In: Bellafiore, R. (ed.) Marxian Economics. A reapraisal Essays on Volume III of Capital. Profits, Prices and Dynamics, pp. 225–251. St. Martins Press, New York (1998)
17. Shaikh, A.: The transformation from Marx to Sraffa. In: Freeman, A., Mandel, E. (eds.) Ricardo, Marx, Sraffa, pp. 43–84. Verso, London (1984)
18. Sraffa, P.: Production of commodities by means of commodities. Cambridge University Press, Cambridge (1960)
19. Valle Baeza, A.: Correspondence between labor values and prices: a new approach. Review of Radical Political Economics 26(2), 57–66 (1994)
20. Steedman, I., Tomkins, J.: On measuring the deviation of prices from values. Cambridge Journal of Economics 22, 359–369 (1998)
21. von Neumann, J.: A Model of General Economic Equilibrium. Review of Economic Studies 13(33), 1–9 (1945)
22. von Neumann, J.: Mathematical Foundations of Quantum Mechanics, Engl. transl. of the 1931 German edition by Beyer, R.T. (1955)

Quantum Calculus (q-Calculus) and Option Pricing: A Brief Introduction

Emmanuel Haven

School of Management, University of Leicester, University Road,
LE1 7RH Leicester, United Kingdom
e.haven@leicester.ac.uk

Abstract. q-calculus, also known under the name of h-calculus, has found wide applications in many areas of mathematics. In this paper we provide for a basic financial option pricing application where we try to rationalize the use of a q-derivative. We provide for a brief discussion on how the value of q can be an indicator of either the use (or not the use) of the risk free rate of interest in the option pricing partial differential equation.

Keywords: q-calculus, h-calculus, stochastic differential equation, option pricing.

1 Introduction

q-calculus has had many applications in mathematics. Excellent sources which develop this calculus (also known as quantum calculus) are Kac and Cheung [1] and Andrews [2][3]. We are concerned in this paper with showing how the q-derivative could have a financial interpretation in an option pricing framework.

We define the q-derivative and Hilger delta time derivative in the next section. In the section following, we briefly expand on the link between q-derivatives and delta time derivatives. In section four we briefly argue how q-derivatives can enter Itô's Lemma. In section five we consider how q-derivatives can enter standard financial option pricing methodology. We give financial meaning to the q derivative. In the last section of the paper we provide for a discussion on how quantum physics (via q-derivatives) is linked to the application we propose in this paper.

2 The q-Derivative and the Hilger Delta Time Derivative

The q-derivative and h-derivative have a very simple definition. In definitions 1, 2, 3 and 4, below, we follow Kac and Cheung [1] and we use their notation also.

Definition 1. *For an arbitrary function f(x), the q-differential is:*
$d_q f(x) = f(qx) - f(x)$.

Definition 2. *For an arbitrary function f(x), the h-differential is:*
$d_h f(x) = f(x+h) - f(x)$.

P. Bruza et al. (Eds.): QI 2009, LNAI 5494, pp. 308–314, 2009.

q-calculus and h-calculus are related. The relation $q = e^h$ or $q = e^{ih}$, where i is a complex number, is often used. We will not use any h-calculus in this paper.

Definition 3. *The q-derivative of a function* $f(x)$ *is:*

$$D_q f(x) = \frac{d_q f(x)}{d_q(x)} = \frac{f(qx) - f(x)}{qx - x}.$$

Definition 4. *The h-derivative of a function* $f(x)$ *is*

$$D_h f(x) = \frac{d_h f(x)}{d_h(x)} = \frac{f(x+h) - f(x)}{(x+h) - x}.$$

Note that $\lim_{q \to 1} D_q f(x) = \lim_{h \to 0} D_h f(x) = \dfrac{df(x)}{dx}$.

Let us now define the Hilger delta time derivative. The delta time derivative was developed out of the calculus of Hilger [4][5]. In definition 5 below, we follow Bohner and Peterson [6].

Definition 5. *A time scale T is an arbitrary non-empty closed subset of the real numbers.*

As an example, the usual time derivative, $\lim_{t \to s} \dfrac{f(t) - f(s)}{t - s}$ is obtained when T is the set of real numbers. When T is the set of integers, one uses the forward difference operator: $f(t+1) - f(t)$. We can informally define the Hilger delta time derivative as:

$$f^{\Delta}(t) = \frac{f(\eta(t)) - f(t)}{\eta(t) - t}, \text{ where for } t \in \mathrm{T}, \ \eta(t) \equiv \inf\{s \in \mathrm{T} : s > t\}.$$

3 Link between $D_q f(x)$ and $f^{\Delta}(t)$

The link between the q-derivative and the Hilger delta time derivative is straightforward. Setting $\eta(t) = qt$ in $f^{\Delta}(t)$, we obtain a q-derivative in t. For T being the set of integers, one requires that $qt - t = 1$ wherefrom $q = \dfrac{1}{t} + 1$. For the case where T is the set of real numbers, one requires $qt - t = 0$. Therefore, $q=1$ in that case. We note that henceforth we only use q-derivatives.

4 Itô's Lemma and q-Derivatives

Financial derivative pricing uses the so called Itô Lemma [7]. This Lemma is essential in the derivation of the so called Black-Scholes partial differential equation (PDE) [8], from which one can calculate the value of so called option contracts. Such contracts, in their simplest expression, allow the buyer of such contract to either have the right to buy or sell an underlying asset (such as a stock) for a certain price at a certain time in the future. The Black-Scholes PDE gives (under quite restrictive conditions) the price the seller of such contract should charge.

We can formulate in heuristic terms Itô's Lemma as follows. Assume that X_t is a stochastic integral:

$$dX_t = udt + vdB_t ,\qquad(1)$$

where B_t is a one dimensional Brownian motion, u and v are respectively drift and diffusion factors and t is time.

Consider now the discrete version: $\Delta X_j = u_j \Delta t_j + v_j \Delta B_j$. We can write the Taylor expansion:

$$\Delta g = \frac{\partial g}{\partial X}\Delta X_j + \frac{\partial g}{\partial t}\Delta t_j + \frac{1}{2}\frac{\partial^2 g}{\partial X^2}\left(\Delta X_j\right)^2 + \frac{1}{2}\frac{\partial^2 g}{\partial t^2}\left(\Delta t_j\right)^2 + \frac{\partial^2 g}{\partial X \partial t}\left(\Delta t_j\right)\left(\Delta X_j\right) + R_j .\qquad(2)$$

We can replace $\Delta X_j = u_j \Delta t_j + v_j \Delta B_j$ into Δg.

The Itô Lemma can then be written as:

$$\Delta g = \frac{\partial g}{\partial X}\left(u_j \Delta t_j + v_j \Delta B_j\right) + \frac{\partial g}{\partial t}\Delta t_j + \frac{1}{2}\frac{\partial^2 g}{\partial X^2}v_j^2\left(\Delta t_j\right).\qquad(3)$$

For an excellent in-depth discussion, see Øksendal [9].

Now consider the introduction of q-derivatives (using definition 3) in this Taylor expansion. We could naively write:

$$\Delta_q g = \frac{\partial_q g}{\partial_q X}\left(u_j \Delta t_j + v_j \Delta B_j\right) + \frac{\partial_q g}{\partial_q t}\Delta t_j +$$
$$\frac{1}{2}\frac{\partial_q^2 g}{\partial_q X^2}\left(u_j \Delta t_j + v_j \Delta B_j\right)^2 + \frac{1}{2}\frac{\partial_q^2 g}{\partial_q t^2}\left(\Delta t_j\right)^2 + \frac{\partial_q^2 g}{\partial_q X \partial_q t}\left(\Delta t_j\right)\left(u_j \Delta t_j + v_j \Delta B_j\right) + R_j\qquad(4)$$

It can be shown that this expression is fraught with difficulties when considering q derivatives on time. We can not use such derivatives on time and therefore it is impossible to write out an Itô Lemma with q-derivatives on time. See Haven [10].

There only exists two ways out of this conundrum. Either: i) a q-derivative exists only on position X but not on time or ii) assume a q-derivative exists on both position and time but only first order terms in the Taylor expansion can be used.

Therefore, we can notice two possibilities:

- we assume a q-derivative exists only on position X but not on time
- we assume a q-derivative exists on both position and time but only first order terms in the Taylor expansion can be used.

If we take the first option then we can write, as:

$$\Delta_q g = \frac{\partial_q g}{\partial_q X}\left(u_j \Delta t_j + v_j \Delta B_j\right) + \frac{\partial g}{\partial t}\Delta t_j + \frac{1}{2}\frac{\partial_q^2 g}{\partial_q X^2}v_j^2\left(\Delta t_j\right). \tag{5}$$

Note the mixed use of derivatives in this expansion.

5 q and Its Financial Meaning....

Consider the definition of a financial option.

Definition 6. *A financial option is a contract entitling the buyer of that contract to either having the right to buy or sell an asset (from the seller of that contract) at a certain price at a certain date in the future.*

A 'European' option can only be exercised on the maturity date of the option contract. An 'American' option can be exercised any time before or on the maturity date of the option contract. The intrinsic value (which is the value of the option at maturity) of a so called call option is: *max{S-K, 0}*, where K indicates the price at which the holder of a call option has the right to buy the underlying asset (a stock with price S in this case). The put option has as intrinsic value: *max{K-S, 0}*. K indicates the price at which the holder of a put option has the right to sell the underlying asset (a stock with price S in this case).

For an excellent treatment on the subject of option pricing, please see Wilmott [11] or Baaquie [12].

Let us consider an option with intrinsic value, $\max\{S^2 - K, 0\}$ for the call and $\max\{K - S^2, 0\}$ for the put. McDonald [13] provides for a discussion, albeit without reference to q-derivatives, on claims other than S.

The stochastic differential equation, with q-derivatives, for dS^2 is:

$$dS^2 = \left((1+q)\mu S^2 + \frac{1}{2}(1+q)\sigma^2 S^2\right)dt + (1+q)\sigma S^2 dB. \tag{6}$$

where S^2 is the squared stock price; μ is the drift rate of the stock price (this is the return of the stock) and σ is the standard deviation of the price (also called the volatility of the price).

We set up the usual Black-Scholes portfolio, whose value is, Π. Let $f(S^2, t)$ be the option price. For the particular intrinsic value, we have here, we can write:

$$d\Pi = -df + \frac{\partial_q f}{\partial_q S^2}dS^2, \tag{7}$$

where *df* is defined as:

$$df(S^2,t) = \frac{d_q f}{d_q S^2} dS^2 + \frac{\partial f}{\partial t} dt + \frac{1}{2} \frac{\partial_q^2}{\partial_q S^{2^2}} \left(dS^2\right)^2 . \tag{8}$$

Substituting (6) in (8), one obtains:

$$df(S^2,t) = \frac{\partial_q f}{\partial_q S^2} \left(\left([2]\mu S^2 + \frac{1}{2}[2]\sigma^2 S^2 \right) dt + [2]\sigma S^2 dB \right) + \frac{\partial f}{\partial t} dt +$$

$$\frac{1}{2} \frac{\partial_q^2}{\partial_q S^{2^2}} [2]^2 \sigma^2 S^4 dt \tag{9}$$

Note that we made use of $(dB)^2 = dt$, and $[2]=1+q$.
Substituting (9) and (6) into (7), we obtain:

$$d\Pi = -\frac{\partial f}{\partial t} dt - \frac{1}{2} \frac{\partial_q^2}{\partial_q S^{2^2}} (1+q)^2 \sigma^2 S^4 dt . \tag{10}$$

Equation (10) can be extended into equation:

$$\frac{d\Pi}{\Pi} \frac{1}{dt} = \frac{1}{\Pi} \left(-\frac{\partial f}{\partial t} - \frac{1}{2} \frac{\partial_q f}{\partial_q S^{2^2}} [1+q]^2 \sigma^2 S^4 \right) = r . \tag{11}$$

From a financial perspective, the beauty of equation (11) resides in the fact that the risk free rate of interest, *r* can be used. The extreme left hand side of (11) indicates that the return on the portfolio per unit of time is equal to the risk free rate. Individual decision makers will require differing rates of return on risky investments depending on their attitude towards risk. Hence, no unique rate of return exists on a risky asset. However, with the use of a risk free rate of return, the situation is drastically different. All decision makers can agree upon the rate of return on a risk free investment, such as a government bond (which normally is guaranteed by tax returns).

Note that from a financial perspective, equation (11) will be valid if two necessary conditions are met:

- there exists no arbitrage (this means there do not exist riskless profits)
- $q=1$ (i.e. when we use ordinary derivatives)

All two conditions are necessary so as to be able to write (11). Each condition, taken separately, is not sufficient to ensure the equalities in (11). As an example, it would be insufficient to assume that under no-arbitrage, (11) will hold. Since, when $q \neq 1$ the extreme left hand side of (11) would be an ordinary derivative, while the right hand side would contain a *q*-derivative. Therefore, the value of *q* has financial meaning.

6 How Does Quantum Physics Enter into This Set Up?

Accardi and Boukas [14] have indicated that "Segal and Segal [15] introduced quantum effects into the Black-Scholes model in order to incorporate market features such as the impossibility of simultaneous measurement of prices and their instantaneous derivatives." This clearly introduces the notion of some macroscopic equivalent of a Heisenberg uncertainty principle. Baaquie [12] also argues for the existence of such macroscopic principle. A discussion of possible macroscopic equivalences of Planck constants has already appeared in other papers. See Khrennikov [16] and Choustova [17]. For more of a general overview of how quantum mechanical principles have been used in non-quantum environments, please see for instance: D'Hooghe, Aerts and Haven [18]. We do not expand it on here.

Let us assume, for the sake of argument, we were to know such an equivalent constant. As was remarked in John Baez [19], when using a q-derivative in the momentum operator, the Heisenberg uncertainty relation changes into:

$$PQ - qQP = -i\hbar. \tag{12}$$

where P is the momentum operator and Q is the position operator. The above uncertainty relation (12), can be easily obtained as follows. For a generic function $f(x)$, we write the Heisenberg uncertainty principle without q-derivatives as: $PQf(x) - QPf(x) = -i\hbar \frac{d}{dx}(xf(x)) - x(-i\hbar \frac{d}{dx} f(x)) = -i\hbar f(x)$. The q-derivative of a product of two functions $f(x)$ and $g(x)$ (see Kac and Cheung [1]): $\frac{d_q}{d_q x}(f(x)g(x)) = f(qx)\frac{d_q g(x)}{dx} + g(x)\frac{d_q f(x)}{dx}$. When we write the uncertainty principle with q-derivatives, we can then obtain: $-i\hbar \frac{d_q}{d_q x}(xf(x)) - x(-i\hbar \frac{d_q}{d_q x} f(x)) = -i\hbar(qx\frac{d_q f(x)}{d_q(x)} + f(x)) + xi\hbar \frac{d_q f(x)}{d_q(x)}$. In order to have this expression to be equal to $-i\hbar f(x)$, we need to multiply the term ' $xi\hbar \frac{d_q f(x)}{d_q(x)}$ ' with q. This leads to the Heisenberg Uncertainty principle in (12).

Assuming that we could think of h in $q = e^h$ (please see above definition 3 – section 2) as the Planck constant then $q=1$ indeed occurs when $h=0$. I.e. we move from quantum mechanics to classical mechanics with $q=1$. As we have remarked above, the (non) unity of q has financial meaning. When $q=1$ the option price can be found and will be unique. The non-uniqueness of the option price occurs when q is not equal to one. By analogy this non-uniqueness of the option price (when q is not equal to one) would possibly then also call into existence a so called macroscopic uncertainty principle (we assume we can think of h (in $q = e^h$) as some macro-scopic equivalent of the Planck constant) which has a similar form to (12).

7 Conclusion

A central observation in this paper has been that with the use of q-derivatives the working of the Itô Lemma is impaired. The introduction of q-derivatives on position (but not on time) shows that we need both the no-arbitrage condition and the unitariness of q so as to be able to use the risk free rate of interest. The wider link the existence of q-derivatives have with a (macroscopic) Heisenberg uncertainty principle can be used here.

References

1. Kac, V., Cheung, P.: Quantum calculus. Springer, Heidelberg (2002)
2. Andrews, G.E., Askey, R., Roy, R.: Special functions. Cambridge University Press, Cambridge (1999)
3. Andrews, G.E.: q-series: their development and application in analysis, number theory, combinatorics, physics and computer algebra. In: CBMS Regional Conference Lecture Series in Mathematics, vol. 66. Am. Math. Soc. (1986)
4. Hilger, S.: Ein Maßkettenkalkül mit Anwendung auf Zentrumsmannigfaltigkeiten. Ph.D. thesis, Universität Würzburg, Germany (1988)
5. Hilger, S.: Analysis on measure chains - a unified approach to continuous and discrete calculus. Res. Math. 18, 18–56 (1990)
6. Bohner, M., Peterson, A.: Dynamic equations on time scales: an introduction with applications. Birkhäuser, Basel (2001)
7. Itô, K.: On stochastic differential equations, Memoirs. Am. Math. Soc. 4, 1–51 (1951)
8. Black, F., Scholes, M.: The pricing of options and corporate liabilities. J. of Pol. Econ. 81, 637–654 (1973)
9. Øksendal, B.: Stochastic differential equations. Springer, Heidelberg (1992)
10. Haven, E.: A note on the use of Itô's Lemma with q-derivatives (submitted, 2009)
11. Wilmott, P.: Derivatives: the theory and practice of financial engineering. J. Wiley, Chichester (1999)
12. Baaquie, B.: Quantum finance. Cambridge University Press, Cambridge (2004)
13. McDonald, R.L.: Derivatives Markets. Addison-Wesley, Reading (2003)
14. Accardi, L., Boukas, A.: The quantum Black-Scholes equation. Glob. J. Pure Appl. Math. 2, 155–170 (2007)
15. Segal, W., Segal, I.E.: The Black-Scholes pricing formula in the quantum context. Proc. Natl. Acad. Sci. USA 95, 4072–4075 (1998)
16. Khrennikov, A.: Interpretations of Probability. VSP International Publishers (1999)
17. Choustova, O.: Quantum model for the price dynamics: the problem of smoothness of trajectories. J. Math. Anal. Appl. 346, 296–304 (2008)
18. D'Hooghe, B., Aerts, D., Haven, E.: Quantum formalisms in non-quantum physics situations: historical developments and directions for future research. In: VUB, CLEA (2008), http://www.vub.ac.be/CLEA/workshop/qs08/abstracts08/Haven.pdf
19. Baez, J.: This week's finds in Mathematical Physics (Week 183), http://math.ucr.edu/home/baez/week183.html

Author Index